PROGRAMMING CONTROLLOGIX® PROGRAMMABLE AUTOMATION CONTROLLERS

by
Jon Stenerson
Fox Valley Technical College – Appleton, WI

Australia • Brazil • Japan • Korea • Mexico • Singapore • Spain • United Kingdom • United States

Programming ControlLogix® Programmable Automation Controllers
Jon Stenerson

Vice President, Career and Professional Editorial: Dave Garza

Director of Learning Solutions: Sandy Clark

Acquisitions Editor: Stacy Masucci

Managing Editor: Larry Main

Senior Product Manager: John Fisher

Senior Editorial Assistant: Dawn Daugherty

Vice President, Career and Professional Marketing: Jennifer McAvey

Marketing Director: Deborah Yarnell

Marketing Manager: Erin Coffin

Marketing Coordinator: Shanna Gibbs

Production Director: Wendy Troeger

Production Manager: Mark Bernard

Art Director: Bethany Casey

Technology Project Manager: Joe Pliss

Production Technology Analyst: Tom Stover

© 2009 Delmar, Cengage Learning

ALL RIGHTS RESERVED. No part of this work covered by the copyright herein may be reproduced, transmitted, stored or used in any form or by any means graphic, electronic, or mechanical, including but not limited to photocopying, recording, scanning, digitizing, taping, Web distribution, information networks, or information storage and retrieval systems, except as permitted under Section 107 or 108 of the 1976 United States Copyright Act, without the prior written permission of the publisher.

> For product information and technology assistance, contact us at
> **Cengage Learning Customer & Sales Support, 1-800-354-9706**
> For permission to use material from this text or product,
> submit all requests online at **www.cengage.com/permissions**.
> Further permissions questions can be emailed to
> **permissionrequest@cengage.com**

Library of Congress Control Number: 2009903376

ISBN-13: 978-1-4354-1947-6

ISBN-10: 1-4354-1947-2

Delmar
5 Maxwell Drive
Clifton Park, NY 12065-2919
USA

Cengage Learning is a leading provider of customized learning solutions with office locations around the globe, including Singapore, the United Kingdom, Australia, Mexico, Brazil, and Japan. Locate your local office at: **international.cengage.com/region**

Cengage Learning products are represented in Canada by Nelson Education, Ltd.

To learn more about Delmar, visit **www.cengage.com/delmar**

Purchase any of our products at your local college store or at our preferred online store **www.ichapters.com**

Notice to the Reader
Publisher does not warrant or guarantee any of the products described herein or perform any independent analysis in connection with any of the product information contained herein. Publisher does not assume, and expressly disclaims, any obligation to obtain and include information other than that provided to it by the manufacturer. The reader is expressly warned to consider and adopt all safety precautions that might be indicated by the activities described herein and to avoid all potential hazards. By following the instructions contained herein, the reader willingly assumes all risks in connection with such instructions. The publisher makes no representations or warranties of any kind, including but not limited to, the warranties of fitness for particular purpose or merchantability, nor are any such representations implied with respect to the material set forth herein, and the publisher takes no responsibility with respect to such material. The publisher shall not be liable for any special, consequential, or exemplary damages resulting, in whole or part, from the readers' use of, or reliance upon, this material.

Printed in the United States of America
2 3 4 5 6   16 15 14 13 12

ControlLogix® is a registered trademark of Rockwell Automation Inc.

TABLE OF CONTENTS

Preface v
Acknowledgments vi

CHAPTER 1	Introduction to Control Technology	1
CHAPTER 2	Memory and Project Organization	19
CHAPTER 3	Ladder Logic Programming	45
CHAPTER 4	Timers and Counters	69
CHAPTER 5	Input/Output Modules and Wiring	83
CHAPTER 6	Industrial Sensors	113
CHAPTER 7	Math Instructions	151
CHAPTER 8	Special Instructions	171
CHAPTER 9	Structured Text Programming	199
CHAPTER 10	Sequential Function Chart (SFC) Programming	219
CHAPTER 11	Function Block Diagram Programming	259
CHAPTER 12	Industrial Communications	291
CHAPTER 13	Motion and Velocity Control	315
CHAPTER 14	Risk Assessment and Safety	343
CHAPTER 15	Safety Devices for Risk Reduction	359
CHAPTER 16	Installation and Troubleshooting	387
CHAPTER 17	Lockout/Tagout	411
APPENDIX A	Starting a New Project in ControlLogix	423
APPENDIX B	Configuring I/O Modules in a Remote Chassis	431
APPENDIX C	The Use of Producer/Consumer Tags	439

APPENDIX D	ControlLogix Messaging	451
APPENDIX E	Configuring ControlLogix for Motion	467

Glossary   493

Index   503

PREFACE

ControlLogix technology has had a large impact on automation. ControlLogix controllers can control and integrate complete applications. ControlLogix can be used to act as a communications gateway to other systems. ControlLogix can eliminate the need for separate motion controllers in an application. Multiple languages are available for programming applications. Having it all in one compatible platform makes it much easier to integrate and control complex systems. This book is intended to be a practical and understandable examination of ControlLogix.

Chapters 1 to 4 concentrate on the fundamentals of ControlLogix hardware, project organization, tags, and ladder logic programming. The programming chapters have many examples, questions, and exercises to cement the reader's understanding of the concepts. The tutorials on the DVD are helpful on many of the topics covered in the chapters. Some topics such as the effect of real-world switch states on logic contacts examine if open [XIO] and examine if closed [XIC]) can be very confusing when learning to program. A tutorial is included on the DVD to make this topic clear. There are also tutorials on types of tags and how to create them and on how to start a project, how to program counters and timers, and so on.

Chapter 5 examines ControlLogix I/O modules and wiring. Digital and analog modules are covered. Particular attention is devoted to the concept of sinking and sourcing. Resolution is also examined for analog modules.

Chapter 6 covers a variety of industrial sensors and their wiring. Sensor types and their uses are examined. Digital and analog sensors are covered.

Chapter 7 covers math instructions. The common arithmetic instructions are examined.

Chapter 8 covers special instructions. Instructions such as copy, move, and messaging and diagnostic instructions are covered. In addition proportional, integral, derivative (PID) control and the PID instruction are included. The chapter also covers sequencers.

Chapters 9 to 11 cover the new programming languages. These new languages are rapidly gaining popularity. Each language has its strengths and appropriate uses and can simplify application development and programming. Chapter 9 covers structured text programming. Chapter 10 covers sequential function chart programming. Chapter 11 covers function block programming.

Chapter 12 examines industrial communications. ControlLogix can act as a communication gateway to connect many individual controllers and systems into an integrated system. The chapter covers the fundamentals of communications at all levels in a typical industrial system. It also covers the most common communications protocols.

Chapter 13 examines the fundamentals of motion control and the use of ControlLogix to control motion. The chapter first examines the fundamentals of a typical motion system and then focuses on controlling single and multiaxis systems using ControlLogix.

Safety is becoming more and more important for machines and systems. Chapters 14 and 15 examine safety. Chapter 14 covers risk assessment and risk reduction. Machines and systems should be assessed for safety risks. When unacceptable risks are identified, risk reduction strategies are implemented. Chapter 15 covers safety hardware. The proper design and inclusion of safety hardware is becoming more important as machines and systems become more automated. There is a wide variety of safety hardware available to reduce the risk of injury.

Chapters 16 and 17 focus on installation and maintenance. Chapter 16 covers installation and troubleshooting of automated systems. Chapter 17 examines the use of lockout/tagout in industry.

The appendices are designed to show the reader how to do some of the interesting and unique tasks that ControlLogix is capable of. There should be many aha moments as this material is learned. The power and simplicity of ControlLogix is amazing. Appendix A is a step-by-step look at starting new projects using RSLogix 5000. Appendix B is a step-by-step look at configuring I/O modules in a remote chassis. Appendix C is a step-by-step look at using producer/consumer tags. Appendix D is a step-by-step look at messaging in ControlLogix. Appendix E is a step-by-step look at configuring ControlLogix for motion.

SUPPLEMENTS

An Instructor's Resource CD is available. It contains an Instructor Guide providing answers to the end-of-chapter questions, chapter presentations done in PowerPoint, and test banks (ISBN 1-4354-1948-0).

ACKNOWLEDGEMENTS

We would like to thank Lawrence Ortner for performing a technical edit on the manuscript and providing us with detailed feedback, suggestions, and recommendations.

We would like to express appreciation to the following people for their input as reviewers of this edition:

David Barth, Edison Community College, Piqua, OH
Charles Knox, University of Wisconsin, Platteville, WI
Wade Wittmus, Lakeshore Technical College, Cleveland, WI

CHAPTER 1

Introduction to Control Technology

OBJECTIVES

On completion of this chapter, the reader will be able to:

- Describe the basic history of the development of industrial programmable control.
- Describe the components of a typical PLC system.
- Describe how a PLC system's components are specified.
- Define terminology such as *chassis, module, backplane, CPU, RTB, discrete, analog, gateway,* and so on.
- Describe how a power supply is sized.
- Describe the IEC 61131 standard.

OVERVIEW

PLCs are the backbone of industrial automation. PLCs were designed to be easy for electricians and maintenance technicians to work with. They have been widely used in industry since their introduction in the early 1980s. Their capabilities have expanded tremendously. They have also become easier to program and integrate.

This book will concentrate on ControlLogix™ technology. ControlLogix is exciting and amazing technology. It is easy to use and yet incredibly powerful and versatile. The abbreviations CLX and CL will be used as abbreviations for ControlLogix.

HISTORY OF PLCS

In the old days (only a few decades ago), automation was done with hardwired relays. Relays, being mechanical, also were prone to failure, which would shut down the line. Reliability was low and troubleshooting was cumbersome and time consuming. Control cabinets could contain hundreds of electromechanical relays, timers, and counters. To change a process, production was halted, the manufacturing line was shut down, and wiring and relays were modified. Change was avoided whenever possible.

General Motors Corporation (GM) was one of the first to see the need for a programmable computer that could be used to replace hardwired relay logic and complex control cabinets. GM thought that the program in a programmable control device could be modified to change the way a system operated. GM also needed a programmable device that electricians could program, wire devices to, troubleshoot, and use without learning a computer language.

The first PLCs were simple devices that were designed to replace hardwired relays. The language that was developed for PLCs was called ladder logic. It was a very graphical language and looked a lot like the electrical diagrams that electricians were already very familiar with. PLC input and output modules were designed so that electricians could easily connect inputs and outputs. Imagine an automated automobile line before PLCs. There were many huge electrical control cabinets each filled with miles of wire and relays, motor controls, and so on. The logic to run the system was dependent on how all of the hardware relays and devices were wired. If the assembly process had to be changed, the whole system would be shut down and wiring would have to be changed. This was tremendously costly and made changes expensive and undesirable. If something went wrong, the system would have to be turned off and technicians would have to troubleshoot all of the hardwired logic and devices. Again, very costly and time consuming. PLCs made it possible to wire devices one time and just modify the program to change how the system operated. It also eliminated the need for hardware such as logic relays, hardware timing, counting relays, and so on. The PLC program is used to create the logic, timing, counting, and so on.

Figure 1-1 shows a ControlLogix PLC in a control cabinet. To accomplish the same application with hardwired relays, this cabinet would have been huge before PLCs were invented. Remember also that with the PLC, no wiring has to be changed to change the operation of the system. The programmer simply changes the program.

PLCs enable companies to automate processes rapidly and at low cost. Automating a process improves productivity and quality and dramatically reduces scrap and rework. PLCs can enable a manufacturing process to be flexible so processes can produce products to the individual specifications that each customer has ordered. PLCs also enable very rapid, even instantaneous, product changeover through program logic.

Dramatic advances have been made in PLC capabilities. As PLCs gained more capability and began offering more alternative languages for programming, the term *Programmable Automation Controllers* (*PACs*) has begun to be used to describe the newer, more

CHAPTER 1—INTRODUCTION TO CONTROL TECHNOLOGY

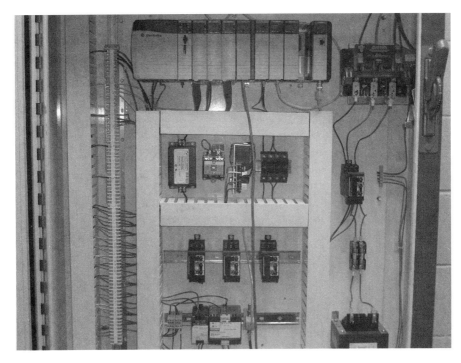

Figure 1-1 CLX control cabinet. The CLX PLC is at the top of the cabinet.

powerful PLCs. ControlLogix (CLX or CL) PLCs fit this new term. While the acronym *PAC* certainly applies to ControlLogix, out of convention the acronym *PLC* will be used in this text. Rockwell Automation has a whole family of Logix products available. Figure 1-2 shows some of the Logix family of controllers.

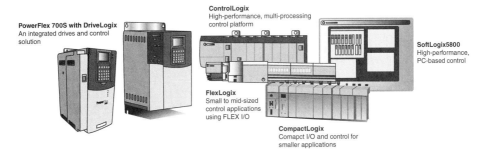

Figure 1-2 Examples of Logix products.

4 PROGRAMMING CONTROLLOGIX® PROGRAMMABLE AUTOMATION CONTROLLERS

What is a PLC?

There are many similarities between a personal computer and a PLC. Note that both have inputs. A computer has a keyboard and a mouse for inputs. The computer can also access the hard drive, CD, DVD, and Internet to get input. A PLC can get input from sensors and other devices such as robots, other controllers, and so on. Computers and PLCs have a central processing unit (CPU) and memory. The CPU runs the user program, evaluates the inputs, and generates outputs. Computers and PLCs both have outputs. A computer can output to a printer, send email out over the Internet, or store a file on a drive. A PLC has outputs such as motors, drives, lights, and so on.

Figure 1-3 shows a simplified overview of a PLC. Switches S1 and S2 provide on or off input signals to the input module of the PLC. The CPU evaluates the logic on the basis of the states of the input/output (I/O) and then changes the states of outputs. The output module turns on outputs on the basis of what the CPU has written to output memory.

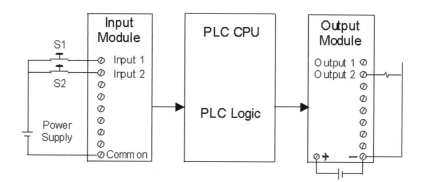

Figure 1-3 Simplified view of a PLC.

Figure 1-4 shows a ControlLogix PLC. The main components have been identified. The power supply is chosen to be large enough to supply current required by the CPU and the modules. The power supply is attached to the left side of the chassis. The chassis size is chosen to hold the number of required modules and the CPU. This system has two CPUs. The first slot in a chassis is slot 0. The CPU is normally installed in slot 0. ControlLogix PLCs can have more than one CPU in a chassis. The second CPU in this example is in slot 1. Slots 3 and 4 have digital I/O modules in this system. Slot 4 has a combination analog I/O module. Slot 5 has a slot filler installed. Slot fillers are available to fill slots that do not have a module installed. Slot 6 has a SERCOS motion control module installed. Slot 7 has a Ethernet bridge module. Slot 9 has a DeviceNet communications module and slot 9 has a Data Highway Plus (DH+) module installed.

PLCs are available with various sizes of memory. CLX CPUs are available with between 750 Kbytes and 16,384 Kbytes of memory.

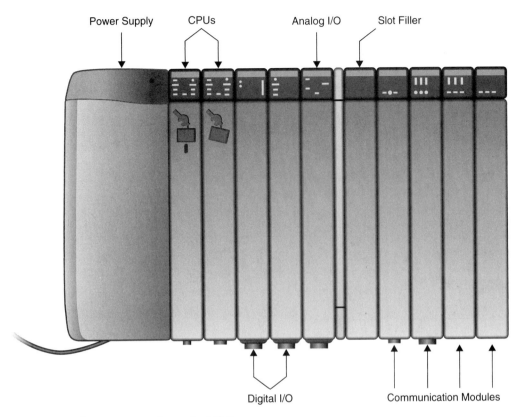

Figure 1-4 Components of a typical PLC.

Some PLC CPU modules allow the user to add additional nonvolatile memory to the CPU module.

Modular PLCs

Almost all PLCs are modular. They allow the user to purchase and install modules to accomplish a task. This enables a user to choose the input and output modules needed for the particular application. A modular PLC begins with a chassis (see Figure 1-5).

Chassis
Chassis, also called racks, are available in various sizes. The user decides which modules will be required for inputs and outputs, communications, and other special purposes such as motion control. The user chooses the size chassis that is needed to hold the number of modules required for the application. Racks have slots, which locate and power the modules. The slots connect the module to the backplane. The backplane passes power to

operate modules and also enables the modules to communicate with the CPU and other modules.

An I/O rack, or chassis, is a housing in which modules are installed (see Figure 1-5). Some applications require more modules than will fit in one rack. Some applications may have some I/O that is located a long distance from the PLC. Remote chassis can be used in these cases.

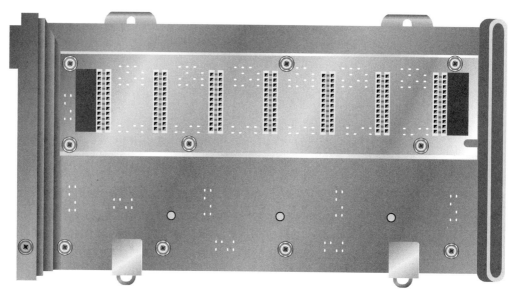

Figure 1-5 A chassis for a CLX system. Note that the power supply would be attached to the left side of the chassis and modules plug into slots. This chassis has ten slots. Chassis are available in different sizes.

Power Supply
When users specify a PLC for a project, they choose the modules that will be needed. They then choose a chassis that will be large enough to hold all of the modules. Then they add up the power requirements for the CPU and all of the modules. This enables them to choose a power supply that is large enough to power all of the modules. The power supply attaches to the left of the chassis in a CLX system (see Figure 1-6).

Central Processing Unit
The central processing unit (CPU) is the brain of the PLC. It can also be called the controller. The CLX controller takes input information, examines the logic in the program in the CPU and then controls the states of outputs.

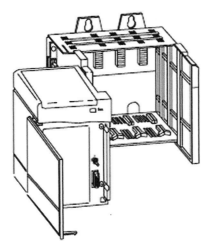

Figure 1-6 A ControlLogix power supply. (Courtesy of Rockwell Automation, Inc.)

The CPU is really just a microcomputer. It has a microprocessor just like a personal computer. The main difference between the PLC CPU and a personal computer is the program. Until recently the language that PLCs were programmed in was a graphical language called ladder logic. Ladder logic was designed to look like a normal industrial electrical print and be easy for electricians and technicians to understand and troubleshoot. There are additional languages that can be used in many PLCs today.

Figure 1-7 shows a ControlLogix CPU. There are three positions that the key switch can be in: RUN, REM, and PROG. If the key is in the RUN position, the CPU will run the program in memory. If the switch is in the REM position, it is in the remote mode. Remote mode means that the computer that is attached to the CPU can control which mode the CPU is in. When writing and testing programs, the REM position is the most convenient. The PROG position stands for program mode. The computer cannot switch modes if the switch is in RUN or PROG modes. The CLX CPU has an RS232 serial programming port, although, since modern programs are quite large, Ethernet is more commonly used.

Figure 1-8 shows the status LEDs on the CPU. These LEDs are either green or red and may be flashing or not flashing. The RUN status LED indicates when the CPU is in the RUN mode. The FORCE LED indicates when forces are in effect that override logic. The FLT LED indicates when there is a fault. The I/O LED indicates the status of communications to local and remote I/O modules. The BAT is the battery indicator. The RS232 status LED indicates the status of the RS232 communication port. Figure 1-9 is a table that explains the LED status indicators. This information is very useful when something goes wrong.

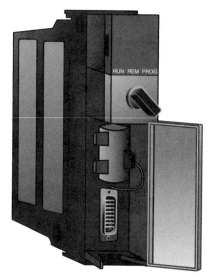

Figure 1-7 A ControlLogix CPU.

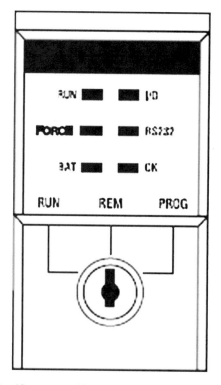

Figure 1-8 CPU LEDs. (Courtesy of Rockwell Automation, Inc.)

LED	Color	Description
Run	Off	The controller is in the program or test mode.
	Solid green	The controller is in the run mode.
I/O	Off	There are no devices in the I/O configuration. Or the controller does not contain a project (memory is empty).
	Solid green	The controller is communicating with all the I/O devices in the configuration.
	Flashing green	One or more of the devices in the I/O configuration are not responding.
	Flashing red	The chassis is defective.
Force	Off	No tags contain I/O force values. I/O forces are disabled (inactive).
	Solid amber	I/O forces are enabled (active). I/O force values may or may not exist.
	Flashing amber	One or more input or output addresses have been forced to an on or off state, but the forces have not been enabled.
RS232	Off	There is no activity.
	Solid green	Data is being transmitted or received.

Figure 1-9 LED indicators for a CLX processor.

Battery Backup
The CPU has a battery that keeps the memory refreshed when the power to the PLC is off. Figure 1-10 shows a CPU and the battery location. The battery provides backup power for the CMOS RAM. Note also the serial and model number information on the module. When a battery is replaced, the date should be noted on the tag so that it is replaced before the battery fails. A larger, longer-lasting external battery pack is also available for 5555 ControlLogix processors. The 556x processors do not need the larger battery since their battery is used to copy the contents of RAM to internal flash memory only on power loss.

Figure 1-11 shows that there are really two CPUs: the Logix CPU and the backplane CPU. The ControlLogix CPU executes application code and messages. The backplane CPU communicates with I/O and sends and receives data from the backplane. The backplane CPU operates independently from the ControlLogix CPU, so it sends and receives I/O information asynchronous to program execution. Note that this really adds to the power and versatility of CLX controllers. A Logix chassis can be used to control I/O and also act as a communications bridge and gateway.

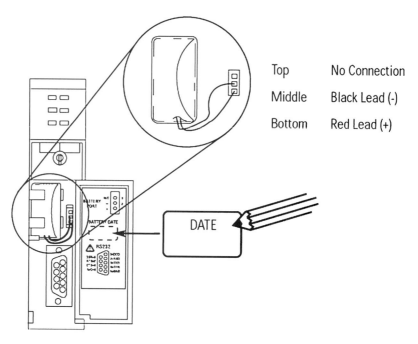

Figure 1-10 Battery location and replacement for a CLX processor. (Courtesy of Rockwell Automation, Inc.)

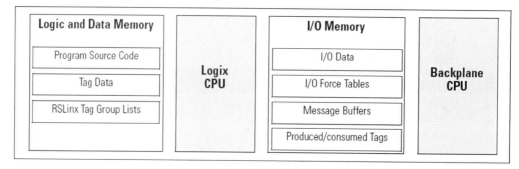

Figure 1-11 This figure illustrates the two CPUs: Logix and backplane. (Courtesy of Rockwell Automation, Inc.)

Memory Cards

Some CLX CPUs have memory boards. Memory is available for some CPUs up to 7.5 megabytes (MB). Some controllers support a removable CompactFlash card for nonvolatile memory. Figure 1-12 shows the location of the specific information about the controller and the memory board on a CLX processor.

CHAPTER 1—INTRODUCTION TO CONTROL TECHNOLOGY 11

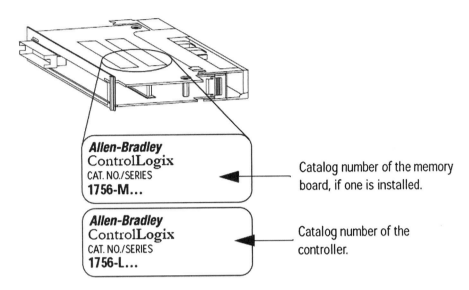

Figure 1-12 Location of catalog numbers for memory and processor. (Courtesy of Rockwell Automation, Inc.)

Input to a PLC

Input modules provide the link between the outside world and the PLC's CPU. The main function of PLC input modules is to take information from the real world and convert it to signals that the PLC CPU can work with. Input modules also protect the CPU from the outside world.

Figure 1-13 shows an example of a sensor connected to an input on an input module. Note the power supply in the circuit. If the sensor is true (on), 24 volts would be seen at input 1. The input module will convert the 24 volts to a 5-volt signal that the CPU will see as a 1 (true). Many input modules run diagnostics to detect broken input wiring. Note also that the input module does not supply the power for the devices. The sensor is powered by an external power supply.

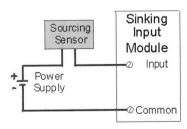

Figure 1-13 Sensor connected to an input.

Figure 1-14 shows how the electric signal is converted to its binary equivalent in memory. The input module converts the electric signal to a binary 1 or 0 and stores it in memory that represents that input's state in memory.

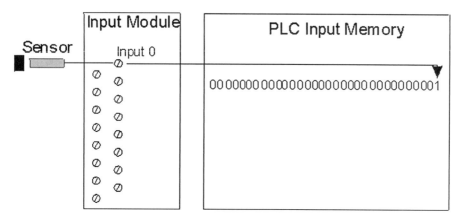

Figure 1-14 How a real-world input's state is stored in memory. The sensor is sensing an object so this sensor's output is true.

A generic CL I/O module is shown in Figure 1-15. The module is installed in the chassis. The backplane connector is used to power the module and also for communications with

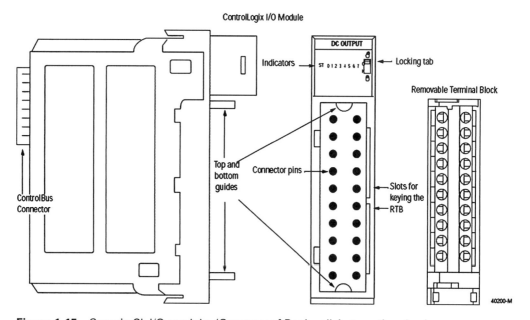

Figure 1-15 Generic CL I/O module. (Courtesy of Rockwell Automation, Inc.)

the backplane and CPU. There is a Removable Terminal Block (RTB) that is used to connect the wiring. If a module needs to be replaced, the RTB is removed without any wiring being removed. The new module is inserted into the chassis and the RTB is reconnected.

Discrete Input Modules

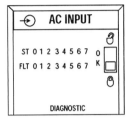

Figure 1-16 Input status LEDs on a module. The Status LEDs are labeled ST and the Fault LEDs are labeled FLT. (Courtesy of Rockwell Automation, Inc.)

Discrete input modules are available for various ranges of AC and DC voltages. Discrete modules are available with various numbers of inputs also. Modules are commonly available with 8, 16, and 32 inputs.

The ControlLogix architecture uses producer/consumer technology, which allows input information and output status to be shared among multiple ControlLogix controllers (see Figure 1-17). This is one of the capabilities that makes CLX so powerful. Input modules are covered in detail in Chapter 5.

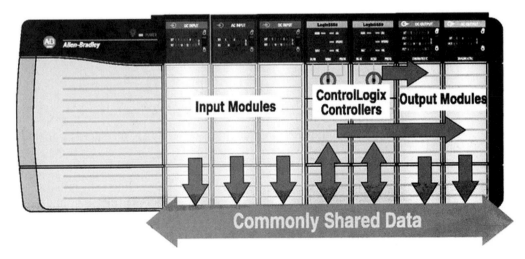

Figure 1-17 Producer/consumer model for data sharing among multiple CLX controllers. (Courtesy of Rockwell Automation, Inc.)

Analog Input Modules

Modules are also available that can input an analog signal. Modules are commonly available to take input signals of 0–10 VDC, −10–+10 VDC, and 4–20 milliamperes (mA). These are useful for taking the input from analog input devices. A temperature-measuring device such as a thermocouple with a converter is one example. For example, low voltage would represent a low temperature; 10 VDC might represent the maximum temperature. Many industrial measurement devices produce a current signal such as 4–20 mA, a flow measurement device for example. For this example 4 mA might be no flow and 20 mA would be maximum flow. Any value between 4 and 20 would represent a different flow rate. Analog modules are covered in detail in Chapter 5.

Outputs from PLCs

Discrete Output Modules

Discrete means that the output module only outputs on- or off-type signals. This is also sometimes called a digital signal. A digital output could be used to turn a valve or a motor on or off. It could be used to turn lights on or off or send an on/off signal to a robot or other equipment.

The output for the module is represented by a 1 or a 0 in memory. Figure 1-18 shows an output connected to output 0 of an output module. The output module converts a binary 1 or 0 that represents the output's state in memory to an electric signal that controls the actual state of the real-world output.

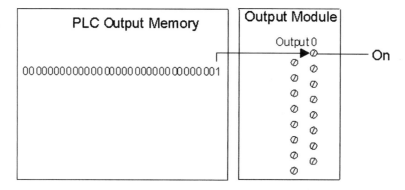

Figure 1-18 How an output state in memory is converted to an electric signal to control an output.

Figure 1-19 shows a generic output module. The module is installed in the chassis. The backplane connector is used to power the module and also for communications with the backplane and CPU. There is a RTB that is used to connect the wiring. If a module needs to be replaced, the RTB is removed without any wiring being removed. The new module is inserted into the chassis and the RTB is reconnected.

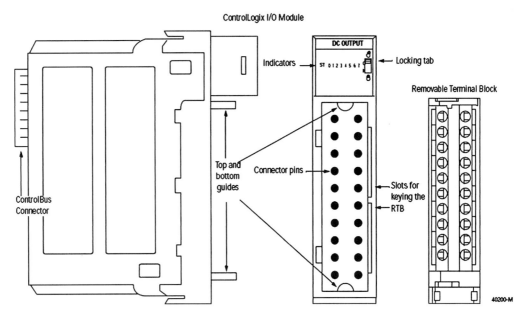

Figure 1-19 Output module. (Courtesy of Rockwell Automation, Inc.)

Output modules are available for AC and DC and various ranges of voltages. Modules are commonly available with 8, 16, and 32 inputs. Output modules are available with relay or solid-state outputs. Solid-state outputs are the most common.

Status LEDs
Figure 1-20 shows several different output module status LEDs. Status LEDs are very useful for troubleshooting. Note that the diagnostic modules provide I/O status LEDs for each I/O point and also fault information for each I/O point. Note also that the electronically fused module (lower left) has status information for the state of each I/O point and a fuse status indicator for each I/O point.

Analog Output Modules
Modules are also available that can output an analog signal. Modules are commonly available to output 0–10 VDC, −10–+10 VDC, and 4–20 mA. These are useful for controlling analog output devices. A motor drive is one example. A drive that is capable of clockwise and counterclockwise rotation at various velocities might require a −10–+10-VDC signal from a PLC to control direction and velocity. If the output is negative, the drive may move in a counterclockwise direction. If the output is positive, the drive may move in a clockwise direction. The magnitude of the signal would control the velocity. For example, low voltage would be slow speed; 10 VDC would represent the maximum speed.

16 PROGRAMMING CONTROLLOGIX® PROGRAMMABLE AUTOMATION CONTROLLERS

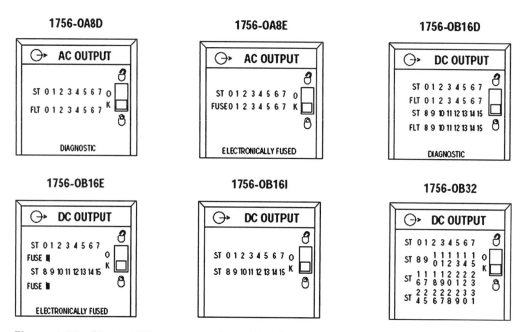

Figure 1-20 Status LEDs on several models of output modules. (Courtesy of Rockwell Automation, Inc.)

Many process devices require a current signal such as 4–20 mA. A valve to control the rate of flow would be an example. For this example, 4 mA would be no flow (valve completely shut) and 20 mA would make the valve wide open. Any value between 4 and 20 would set a different flow rate. Analog modules are covered in detail in Chapter 5.

Programming

The most common programming language for PLCs is ladder logic. Figure 1-1 shows a simple example of ladder logic. It is not important to understand the logic at this point. There is a vertical line on the left and one on the right of the ladder logic. These are sometimes called power rails. The horizontal lines represent rungs of logic. The symbols on the left of the rungs (contacts) represent input states or conditions. The symbol on the right of the rung (coil) represents an output. If the conditions on a rung are true, the output is turned on.

Ladder logic is still the most widely used PLC language. Recently other languages have been rapidly increasing in use. Programs can be written online or offline. Programming software also has error checking to make sure that the program addressing matches the available I/O as well as syntax errors. Programming software also provides many troubleshooting tools to help find and correct errors in program logic.

Figure 1-21 shows an example of ladder logic. The elements on the left are called contacts. Contacts represent input conditions. The element on the right named Out_1 is called a coil. Coils represent outputs. Ladder logic will be covered in Chapter 3.

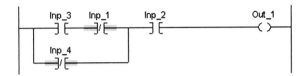

Figure 1-21 Simple example of ladder logic.

International Standard IEC 61131
IEC 61131-3 is an international standard for PLCs. This standard is actually a collection of standards for PLCs and their associated peripherals. The standard consists of eight parts: Part 1: General information, Part 2: Equipment requirements and tests, Part 3: Programming languages, Part 4: User guidelines, Part 5: Communications, Part 6: Reserved for future use, Part 7: Fuzzy control programming, and Part 8: Guidelines for the application and implementation of programming languages.

Part 3 (IEC 61131-3) is the most important to the PLC programmer. It specifies the following languages: ladder diagram, instruction list, function block diagram, structured text, and sequential function chart. CLX has four of the five languages implemented. The instruction list was not implemented. The instruction list is very similar to the language that is used to program the microprocessor. It is very detailed in nature and not friendly to people who have not studied microprocessor programming. The standard is intended to make the languages from different manufacturers more standard. It will never mean that programming software from one manufacturer can be used to program a PLC from another manufacturer. But the logic should be very similar. The standard essentially establishes base languages and elements for each language. Manufacturers must be compatible on these items to be IEC 61131-3 compliant. Manufacturers are free to add additional elements to the languages. Ladder logic, structured text, sequential function chart, and function block languages will be covered in later chapters.

ControlLogix is incredibly powerful and versatile. A CL system can be used for control. It can also be used as just a communications gateway. CL can be used to develop complex networked systems. CL can also handle complex integrated multiaxes motion control. Figure 1-22 shows an overview of a more complex CL system including motion (SERCOS) and communications possibilities. SERCOS is a communications standard commonly used for controlling servos. Note the various communications protocols that can be used.

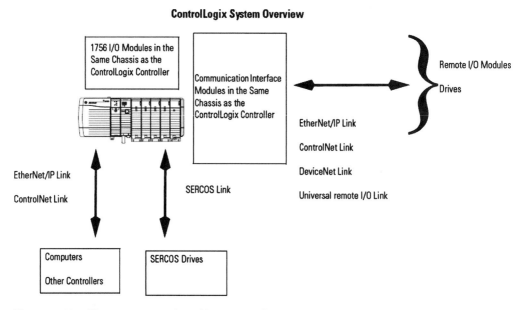

Figure 1-22 CL system overview. (Courtesy of Rockwell Automation, Inc.)

SUMMARY

PLCs are used in any imaginable application. They are used in manufacturing systems of course but also in water treatment facilities, sewer treatment plants, bridge control, power generation and transmission, food production, prison control systems, building environmental control and security, and many others. They are used by inventors and entrepreneurs to start new companies producing newer and faster machines to do almost any task.

1. PLCs were originally intended to replace _____.
2. Name the main components that are found in a basic CLX system.
3. What happens to the memory of a CLX controller when the power is shut off?
4. How is a power supply chosen for a system?
5. What does discrete I/O mean?
6. What is an RTB?
7. What is the difference between discrete and analog?
8. What does producer/consumer technology enable CLX to do?
9. What does a contact typically represent?
10. What does a coil typically represent?
11. What is the most common PLC programming language?
12. What is IEC 61131-3?
13. What four languages can be used to program CLX controllers?
14. What is SERCOS, and what is it typically used for?

CHAPTER 2

Memory and Project Organization

OBJECTIVES

On completion of this chapter, the reader will be able to:

- Describe project organization in CLX.
- Explain the relationship between tasks, programs, and routines.
- List the types of task execution that are possible.
- Describe the base types of tags.
- Create base-, alias-, array-, and User-Defined-type tags.
- Choose the appropriate type of task execution and configure tasks.

INTRODUCTION

ControlLogix was designed to give the programmer a great deal of flexibility in how an application is organized. CLX allows the programmer to keep things simple and program everything as one task or divide it into multiple tasks for efficient operation, clarity, and ease of understanding. There is tremendous flexibility and capability for the programmer. It is very important to have a good understanding of CLX project organization and terminology.

CONTROLLOGIX PROJECTS

Typically in most PLCs we would have a program and maybe some subprograms to control an application. ControlLogix has a different organizational model. The overall application you develop in CLX is called a project. A project contains all of an application's elements and is broken into tasks, programs, and routines.

Tasks

A CLX project can have one or more tasks. Tasks can be used to divide an application (project) into logical parts. Tasks have a couple of important functions. A task is used to schedule the execution of programs in the task. A CLX project can have up to 32 tasks. A task's execution can be configured to be executed continuously, periodically, or on the basis of an event (see Figure 2-1). When the programmer creates a new project, a main task, which is continuous, is created. Continuous tasks are sometimes called the background task since they only execute in leftover time. The name main task is somewhat misleading. It is actually the lowest priority task. It can be renamed.

Task Execution	Function
Continuous	Operates continuously (except while other tasks are executing)
Periodic	Executes at specific intervals. The rate of execution can be set between 1 ms to 2000 seconds. The default execution time is 10 ms.
Event-based	Executes on the basis of an event

Figure 2-1 Task execution types.

A continuous task can be thought of as executing continuously. As only one task can execute at a time, a continuous task executes anytime a periodic or event-based task is not executing. Periodic tasks are set up to operate one time through at specified intervals. Periodic tasks interrupt the operation of the continuous task to operate. When the periodic task is done, the task is executed one more time. The rate for a periodic task can be set between 0.1 ms and 2000 seconds. The default rate for a periodic task is 10 ms.

Figure 2-2 shows a timing diagram for three tasks. The main task is continuous. It is shown in gray. It is always operating if the other two are not. Task 2 (white) is a periodic task. It executes at specific time intervals. The main task (continuous) stops executing and the second task (periodic) executes. The third task (black) is event based. It executes when the specified event occurs. Remember that only one task can execute at a time.

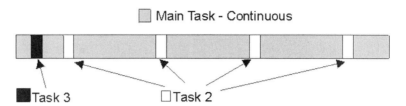

Figure 2-2 Continuous, periodic, and event-based task execution.

Application Example
Imagine a machine that produces packaging material. The machine requires several servosystems for motion, velocity, flow control, temperature control, and many quality control checks as the packaging material is made. This machine application might be broken into several tasks for a CLX project. The main task might be used for overall machine control functions. The company also collects machine production data and displays it for operators on human-machine interface (HMI) monitors. The main task is a continuous task. In this example the servo motion and process control needs to be monitored for safety and for adequate control. This needs to be done in a periodic task. Another operation on this machine is making a perforation. This must occur on the basis of a registration mark on the packaging material. This task would require event-based execution. In this example the project developer might decide to divide the overall application (CLX project) into three tasks, as each has different requirements. Figure 2-3 shows what the project organization might look like for this application. This application (project) was broken into three tasks. One task is continuous, one needs to execute about every 5 ms (periodic), and one is based on the registration mark on the packaging material (event based). Tasks will be covered in greater detail later in the chapter.

Programs

As Figure 2-3 shows, a project consists of all of the things required to control an application. A separate task can be developed to control each logical portion of an application. ControlLogix also enables the programmer to break each task into one or more programs. Each task can have up to 100 programs. In Figure 2-3 the task named Main Task has one program named Control: The Task named Servo and PID has two programs: Servo and Temp. The third Task named Registration has one program. If there is more than one program, the programs will execute in the order they are shown in the controller organizer.

Routines

Each program can also have one or more routines. The application's logic is created in the routines. These are normally organized into a main routine and additional subroutines. In most PLCs the logic is written in programs and subprograms. In CLX they are called routines.

22 PROGRAMMING CONTROLLOGIX® PROGRAMMABLE AUTOMATION CONTROLLERS

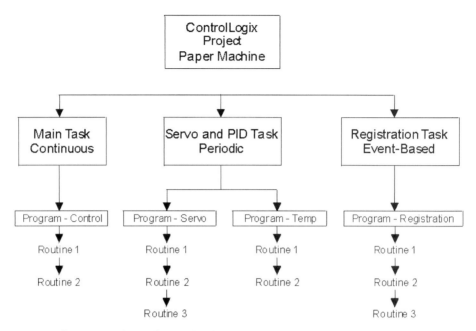

Figure 2-3 ControlLogix project organization with the application organized into tasks, programs, and routines.

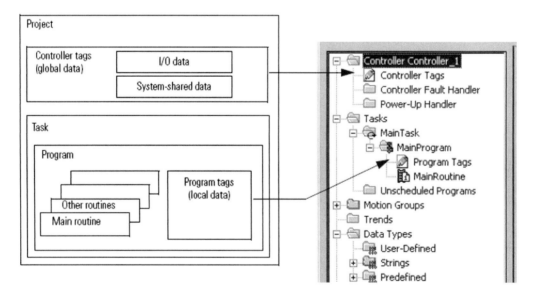

Figure 2-4 Project organization. (Courtesy of Rockwell Automation, Inc.)

Figure 2-4 shows a different representation of project organization. A ControlLogix project does not have to be complex. As shown in Figure 2-4, a simple project might just have one task, one program, and one routine. There is one task named MainTask, one program named MainProgram, and one routine named MainRoutine.

LET'S REVIEW
Project

A project is the overall complete application. It is the file that stores the logic, configuration, data, and documentation for a controller.

Task

A task is a scheduling mechanism for executing programs. An application can be broken into multiple logical tasks. A task enables the programmer to schedule and prioritize one or more programs that execute on the basis of the application requirements.

When a new project is started, a continuous task is created by default. It is preconfigured as a continuous task. The programmer can add additional periodic or event tasks, as needed. Once a task is triggered, every program assigned (scheduled) to the task will execute in the order in which it is displayed in the controller organizer.

Program

In CLX a program has one or more routines. The routines contain the logic in a CLX project. A program could be defined as a set of related routines and tags. Each program contains program tags, a main executable routine, other routines, and an optional fault routine. Programs are contained in a task: When the program's task is triggered, the scheduled programs within the task will execute from the first to the last one.

Routine

A routine is a set of logic instructions written in one programming language, such as ladder logic. A routine in CLX is similar to a program or subprogram in most PLCs. Routines are where the programmer writes the executable code for the project.

The main routine is the first routine to execute when its program is triggered to run. Jump-to-subroutine (JSR) instructions are used to execute other routines. A program fault routine can also be developed. If any of the routines in the associated program produce a major fault, the controller executes the program fault routine, if one was developed.

TAG ADDRESSING IN CONTROLLOGIX

In most PLCs the programmer has to use very specific addressing to specify I/O addresses, bits, variables, timers, counters, and so on. Most PLCs use a physical address for every tag. For example, a SLC 500 PLC would use an address like N7:5 to reference an integer in memory. Addresses typically follow a fixed, numeric format that depends on the type of data, such as B3:6/0, N7:2, and F8:5.

In these PLCs the programmer can use symbolic names to represent the actual address. For example, the programmer might use a symbolic name like Alarm_Light to represent an actual output (O:5/3) in a ladder diagram. The PLC actually uses the O:5/3 address. The symbolic name for it is not even located in PLC memory. It only appears in the program on the computer. It is only for the programmer's use.

In CLX, tags are used to address I/O, bits, variables, timers, counters, and so on. A tag is a user-friendly name for a memory location. For example, we might store a temperature integer value in memory. Temp would be a good name for the tag to hold this data. The processor uses the tag name to address the data.

ControlLogix uses the tag name and doesn't need to cross-reference a physical address. The tag name identifies the data. This enables a programmer to document a program with tag names that clearly represent the application. In CLX the maximum length for a tag name is 40 characters.

Tag names may use alphabetic characters (A–Z or a–z), numeric characters (0–9), and underscores (_). Tag names must start with either an alphabetic character or an underscore. Tags are not case sensitive (A is the same as a). It is wise to use mixed case tag names (upper- and lowercase characters) and underscores because mixed-case tag names are easier to read. Look at the examples in Figure 2-5.

Preferred Tag Name	More Difficult to Read
Temp_1	TEMP_1
Temp_1	TEMP_1
	temp_1

Figure 2-5 The use of upper- and lowercase letters and underscores can make tag names easier to read.

Organizing Tags

RSLogix 5000 organizes tags in alphabetical order. Tag names can be chosen so they keep similar data together. For example, if we are interested in tags related to Machine_1 or tags relating to temperature, we can name them so that they are listed together.

The first column in the table in Figure 2-6 shows an example of naming tags so that similar tags are grouped together. All the tags related to Machine 1 appear together as do the Temp tags. In the second column similar tags are not grouped together because of their names. They are separated from each other. One would have to go through the list to find each tag that related to Machine 1 or Temp.

Logical Organization	No Name Organization
Tag Name	Tag Name
Machine_1_Cyc	...
Machine_1_Hours	Coil_1_Temp
Machine_1_On	Tag names that are between C and E
Machine_1_Stat	...
...	...
...	...
Temp_Coil_1	...
Temp_Extruder	Extruder_Temp
Temp_Heater	
Temp_Machine_Ldr	

Figure 2-6 Tag names. The first column shows an example of careful tag naming so that similar tags are grouped together. In the second column, similar tags are not grouped together by the first word of their name.

Tag Data Types

The data type could be defined as the type of data that a tag stores, such as a bit, integer (whole number), real (floating-point) number, string, and so on. The minimum memory allocation for a tag is 4 bytes (32 bits) plus 40 bytes for the tag name itself. If a tag type that uses fewer than 4 bytes of memory is used, the controller allocates 4 bytes for it anyway. A BOOL-type tag for example only requires 1 bit, but the controller allocates 4 bytes to store it, 1 bit for the actual tag value and 31 unused bits.

Figure 2-7 shows the basic type of tags in a ControlLogix project: base, alias, produced, and consumed.

Tag Type	Use of This Type of Tag
Base	Stores various types of values for use by logic in the project
Alias	Represents another tag
Produced	Sends data to a different controller
Consumed	Receives data from a different controller

Figure 2-7 Basic tag types.

Base-Type Tags

A base-type tag would usually be chosen to create tags that would hold data for logic. For example, we would choose base type for tags to hold temperature, quantities, bits, integer numbers, floating-point (real) numbers, and so on. Figure 2-8 shows some of the numerical types for base-type tags. Figure 2-9 shows the size number each base-type tag can hold.

Type of Tag	Use
BOOL	Bit
BOOL	Digital I/O points
CONTROL	Sequencers
COUNTER	Counter
DINT	Integer (whole number, 32 bit)
INT	Analog device in integer mode (very fast sample rates)
REAL	Floating-point (decimal) number
TIMER	Timer

Figure 2-8 Number types for base-type tags.

Type	Bit Use and Size of Numbers for Each Type					
	31	16	15	8	7	1 0
BOOL						0 or 1
SINT					−128 to +127	
INT			−32,768 to +32,767			
DINT	−2,147,483,648 to +2,147,483,647					
REAL	−3.40282347E^{38} to −1.17549435E^{-38} (negative values) 0 1.17549435E^{-38} to 3.40282347E^{38} (positive values)					

Figure 2-9 Size of numbers that each base type can hold.

The Boolean(BOOL)-type tag is one of the more commonly used tag types. The BOOL is a bit tag that can have a value of 1 or 0. It is 1 bit in length (see Figure 2-2), although it takes up 32 bits in memory.

A single-integer(SINT)-type tag is 8 bits in length, although it uses 32 bits in memory. This type of tag can hold a value between −128 and +127. A SINT tag is used for whole (nondecimal) numbers.

An integer(INT)-type tag can be used to hold a value between −32,768 and +32,767. An INT tag uses 16 bits to hold a value, but uses 32 bits in memory. An INT tag is used for whole numbers. One use of an INT-type tag is when we communicate between a CLX controller and a SLC. The length of an integer in a SLC is the same as an INT in a CLX controller.

A double-integer(DINT)-type tag is used for whole numbers. A DINT tag uses 32 bits to hold a value. A DINT can hold a value between −2,147,483,648 and +2,147,483,647.

A REAL-type tag is used to hold decimal (floating-point) values. A REAL tag is also 32 bits.

Figure 2-9 shows types of tags and the size and range of numbers they can store.

Alias-Type Tags

An alias-type tag is used to create an alternative name (alias) for a tag. The alias tag is often used to create a tag name to represent a real-world input or output. An alias is indeed a tag unto itself, not just another name for the base tag. It is linked to the base tag so that any action to the base also happens to the alias and vice versa. Figure 2-10 shows an example of the use of alias tags. Note that the alias tag name and the actual address (the base tag) of the input and output are shown in the rung.

Figure 2-10 Rung showing a contact and coil. Note the alias name (Fan_Motor) and the base tag (<Local:2:0.Data.5>) of the output coil. The alias name is easier to understand and easier to relate to the application, although the base tag contains the physical location of the output point in the ControlLogix chassis.

Figure 2-11 shows how a tag is configured in CLX. The name is entered first. The tag type is chosen: Base, Alias, Produced, or Consumed. Next the Data Type is chosen. The Tag Properties screen also allows the programmer to choose the Scope of and the Style in which to display the tag.

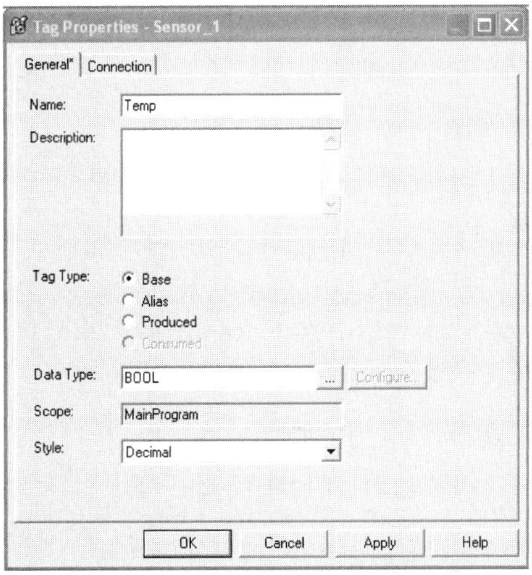

Figure 2-11 How a tag is configured in CLX. Note the choices: Base, Alias, Produced, and Consumed. Note that the programmer also has a choice of the Scope and Style of the tag in this screen.

Scope of Tags

Scope refers to which programs have access to a tag. There are two scopes for tags in CLX: controller scope and program scope (see Figure 2-12). A controller scope tag is available to every program in the project. The controller scope tag data is also available to the outside world, such as SCADA systems. Program scope tags are only available within the program they are created in. Programs cannot access or use a different program's tags if they are program scope. Figure 2-12 shows two programs within a project (Program A and Program B). Note that each program has tags named Tag_1, Tag_2, and Tag_3. Note that the names of tags are the same in both programs. They are not, however, the same tags. There is no relationship between them, even though they have the same name. They are program scope tags. They are only available to routines within that program. Note, however, that there are some controller scope tags: Sensor_1, Temp_1, and CNT. These are available to all programs because they are controller scope. This means, for example, that Temp_1 is available to both programs and is the same tag for both programs.

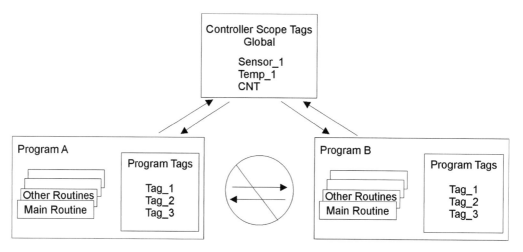

Figure 2-12 Scope of tags. (Courtesy of Rockwell Automation, Inc.)

Routines within a program can only access program tags of the program they belong to and controller tags. Routines are not able to access the program tags of other programs.

Creating a Tag

There is more than one way to create tags. Tags can be created one at a time as you program or tags can be created in the tag editor. The tag editor enables you to create and edit tags using a spreadsheet-like view of the tags. Figure 2-13 shows the tag editor screen with three tags that have been created. Note that these three tags are program scope. The name of the program is MainProgram, and program scope is the scope that was chosen.

Note the Edit Tags tab was chosen on the bottom of the screen. You must be in the edit mode to add or edit tags. The monitor mode is used to monitor tag values.

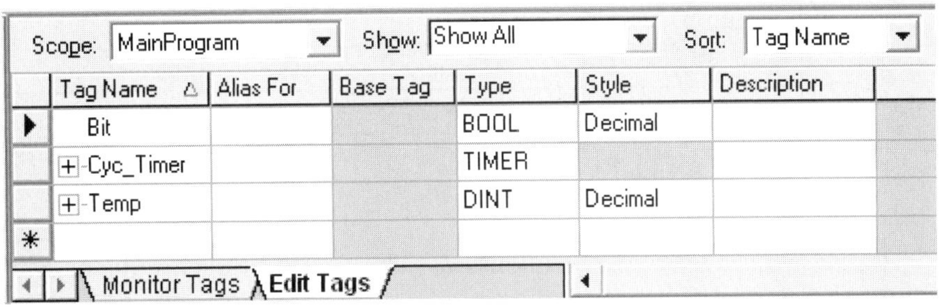

Tag Name	Alias For	Base Tag	Type	Style	Description
Bit			BOOL	Decimal	
+-Cyc_Timer			TIMER		
+-Temp			DINT	Decimal	

Scope: MainProgram   Show: Show All   Sort: Tag Name

Monitor Tags / Edit Tags

Figure 2-13 Tag editor screen. Note that three tags have been defined. Bit is a BOOL-type tag. Cyc_Timer is a TIMER-type tag. Temp is a DINT-type tag. Note the scope of these tags is program scope (named MainProgram). Note also that the two selection tabs at the bottom are Monitor Tags and Edit Tags and that the Edit Tags tab is active.

Arrays

Logix5000 controllers also allow you to use arrays to organize data. Arrays are very important in CLX programming. An array is a tag type that contains a block of multiple pieces of data. An array is similar to a table of values. Within an array of data values, each individual piece of data is called an element. Each element of an array must be of the same data type. An array tag occupies a contiguous block of memory in the controller; each element is in order. Arrays are useful if you want to index (move) through the elements of an array. Arrays can be created with 1, 2, or 3 dimensions.

An array is like a table of tags (see Figure 2-14). It can hold the values of multiple tags. For example, an application might require five different temperatures, one for each different product that is produced. Figure 2-14 shows an example of a 1-dimensional array created to hold five temperatures. The tag name is Temp. A subscript identifies each individual element within the array.

Temp[0]	210
Temp[1]	200
Temp[2]	190
Temp[3]	180
Temp[4]	170

Figure 2-14 This is a 1-dimensional (one column of values) array. There are five members of this array, Temp[0] to Temp[4]. Each member of the array has a different value in this example.

Figure 2-15 shows a 2-dimensional array. Note that there are 3 columns and 5 rows in this example. This would be a 2-dimensional 5 by 3 array and would have 15 members.

Temp[0,0]	225	Temp[0,1]	200	Temp[0,2]	175
Temp[1,0]	220	Temp[1,1]	195	Temp[1,2]	170
Temp[2,0]	215	Temp[2,1]	190	Temp[2,2]	165
Temp[3,0]	210	Temp[3,1]	185	Temp[3,2]	160
Temp[4,0]	205	Temp[4,1]	180	Temp[4,2]	155

Figure 2-15 A 2-dimensional array.

Arrays can also have 3 dimensions. This would be like a cube of values.

Figure 2-16 shows a 1-dimensional array of ten values. The tag name of the array is Temp. There are ten members: Temp[0]–Temp[9]. The values of each member are shown in the second column. The fourth column shows that they are all of type DINT. Remember that arrays can be created for any type of data but an array can only hold one data type.

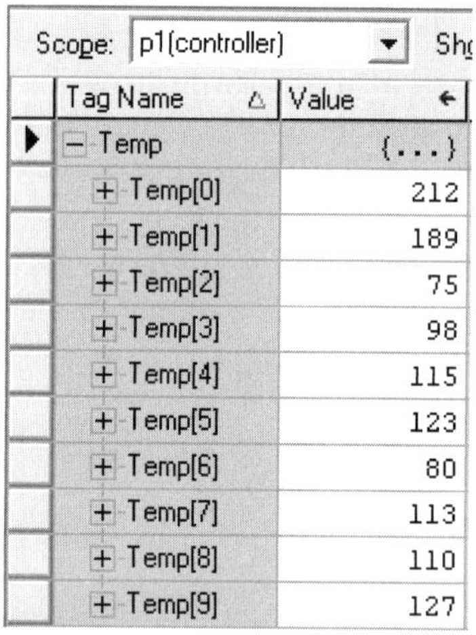

Scope: p1(controller)	Sh
Tag Name	Value
– Temp	{...}
+ Temp[0]	212
+ Temp[1]	189
+ Temp[2]	75
+ Temp[3]	98
+ Temp[4]	115
+ Temp[5]	123
+ Temp[6]	80
+ Temp[7]	113
+ Temp[8]	110
+ Temp[9]	127

Figure 2-16 An array of ten DINT-type tags.

Creating Arrays
To create an array, you create a tag and assign dimensions to the data type. From the Logic menu, select Edit Tags. Type a Name for the tag and select a Scope for the tag (see left side of Figure 2-17). Assign the Array Dimensions (see right side of

Figure 2-17). In this example it will be a 1-dimensional array. There will be five values. Note that in Array Dimensions 5 was entered in Dim 0 and 0 was left in Dim 1.

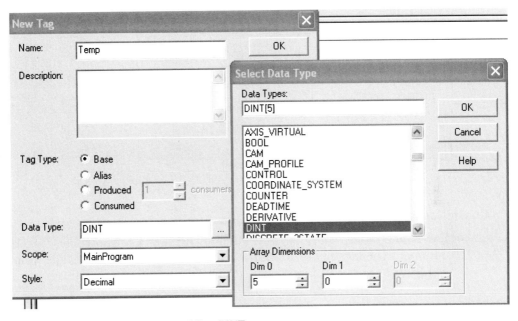

Figure 2-17 Creating an array of five DINTs.

Figure 2-18 shows how the Temp array would appear in the CLX tag editor. Note the + to the left of Temp. If you click on the +, the five members of the array will appear. Figure 2-19 shows the five members. Note that there is a + to the left of each member. You can click on the + to see each bit within the member. There are 32 bits for each array member (DINT).

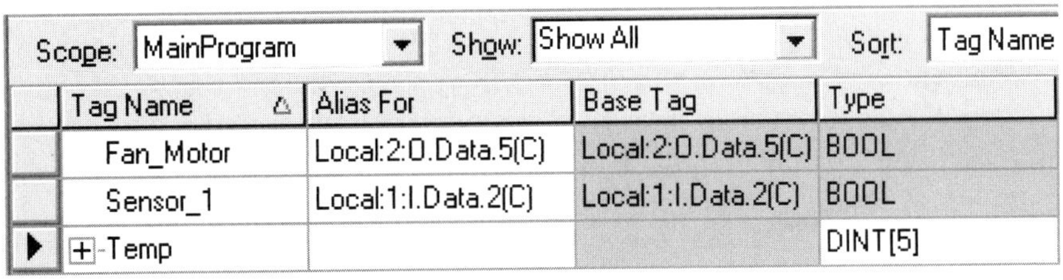

Figure 2-18 Tag editor showing the Temp array.

Scope: MainProgram	Show: Show All		Sort: Tag Name	
Tag Name △	Value	Force Mask	Style	Type
Fan_Motor	0		Decimal	BOOL
Sensor_1	0		Decimal	BOOL
▶ − Temp	{...}	{...}	Decimal	DINT[5]
+ Temp[0]	0		Decimal	DINT
+ Temp[1]	0		Decimal	DINT
+ Temp[2]	0		Decimal	DINT
+ Temp[3]	0		Decimal	DINT
+ Temp[4]	0		Decimal	DINT

Figure 2-19 Tag editor showing an array-type tag that was created. Note there are five members in this array: Temp[0] through Temp[4].

Produced/Consumed Tags

If we would like to share tag information with multiple controllers, produced- and consumed-type tags can be used. If we wanted to make a tag available from PLC 1 to PLC 2, it would be a produced-type tag in PLC 1 and a consumed-type tag in PLC 2.

Structures

Remember that arrays can only hold one data type. CLX offers another type of tag that can hold multiple types of data. This type is a structure. Structures enable the programmer to create a structure-type tag that can hold multiple data types.

A structure can be created to match a specific application's requirements. Each individual data type in a structure is called a member. Members of a structure have a name and data type, just like tags. CLX has several predefined structures (data types) for use with specific instructions such as timers, counters, motion instructions, function block instructions, and so on. Users can create their own structure tags, called a User-Defined data type.

Figure 2-20 shows an example of a structure-type tag. A tag named TMR_1 was created. The type chosen for it was TIMER. Once you have created the tag and tag type for a timer, CLX creates the tag and 9 additional tag members. Note in the figure that the first member shown is TMR_1.PRE. This is the preset value for this timer tag. Note the period between the name of the timer and the tag member (PRE in this example). Note also that PRE is a DINT type. The next tag member is ACC. It is also a DINT type. Next look at the DN member. This would be set to a 1 when the accumulated (ACC) value is equal to the preset (PRE) value. The DN member is a BOOL type. A structure-type tag can hold several pieces of data and each member can be a different type. Structures are also used for counters, motion, and many other purposes in CLX.

				TIMER
–TMR_1	{...}	{...}		TIMER
+TMR_1.PRE	30000		Decimal	DINT
+TMR_1.ACC	0		Decimal	DINT
TMR_1.EN	0		Decimal	BOOL
TMR_1.TT	0		Decimal	BOOL
TMR_1.DN	0		Decimal	BOOL
TMR_1.FS	0		Decimal	BOOL
TMR_1.LS	0		Decimal	BOOL
TMR_1.OV	0		Decimal	BOOL
TMR_1.ER	0		Decimal	BOOL

Figure 2-20 A structure-type tag for a timer named TMR_1. Note there are nine members: PRE, ACC, EN, TT, and so on. Note also that PRE and ACC are DINT types and the rest of the members are BOOLs.

User-Defined Structure Tags

Programmers can create their own structure type for tags. These are called User Defined. An example is shown in Figure 2-21. To create a User-Defined structure tag, the programmer must right click on User Defined in the controller organizer and choose New Data Type and then enter the tag name, members, and their types. This tag will

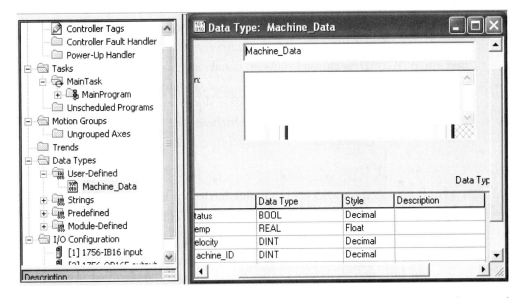

Figure 2-21 A User-Defined stucture-type tag. Note that the new tag type is named Machine_Data and there are four members: Status (BOOL), Temp (REAL), Velocity (DINT), and Machine_ID (DINT).

then be available as a new tag type for programming. Figure 2-21 shows an example of this. The programmer has many machines in his or her company. The machines are all quite similar and have mostly the same type of data available. The programmer created the new User-Defined-type tag to combine the common data for each machine and ease programming. In Figure 2-21 the name of the new User-Defined type is Machine_Data. Note that four tag names were entered as well as the type for each tag. These will be the tag members in the new User-Defined structure tag named Machine_Data.

This new tag type is now available and can be used when new tags are created.

Once the User-Defined-type tag has been created, the programmer can use the new tag type. In Figure 2-22 the programmer created a tag named Weld_Machine in the tag editor. The programmer then chose Machine_Data as the type. Remember that Machine_Data was just created as a User-Defined type. When the programmer created the new User-Defined type, it was automatically added to the Data Types choice list. Now when the programmer creates a tag named Weld_Machine, the software automatically creates the four tag members.

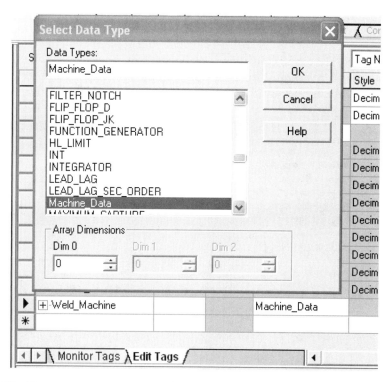

Figure 2-22 The programmer entered a new tag name (Weld_Machine) and chose Machine_Data as the type.

Figure 2-23 shows the new tag that was created (Weld_Machine) and its members. Note that the four members created were of the types specified in the Machine_Data User-Defined tag.

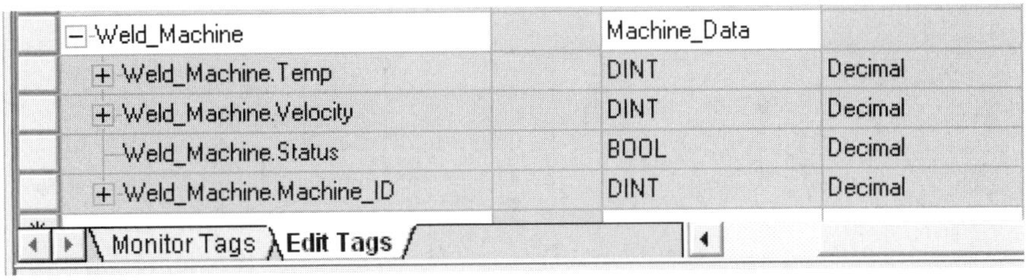

Weld_Machine	Machine_Data	
+ Weld_Machine.Temp	DINT	Decimal
+ Weld_Machine.Velocity	DINT	Decimal
Weld_Machine.Status	BOOL	Decimal
+ Weld_Machine.Machine_ID	DINT	Decimal

Figure 2-23 Tag members for the tag named Weld_Machine.

Guidelines for User-Defined Structure Data Types
When you create a User-Defined data type, keep the following in mind:

> If members that represent I/O devices are used, logic must be used to copy the data between the members in the structure and the corresponding I/O tags.
> If you include an array as a member, it can only be a 1-dimensional array. Multidimensional arrays cannot be used in a User-Defined data type.
> When you use the BOOL, SINT, or INT data types, put members that use the same data type in order. User-Defined data types can be used inside other User-Defined data types.

Figure 2-24 reviews the definition of *data type* and *structure*.

Term	Definition
Data type	The type of data a tag can store (BOOL, SINT, REAL, etc.)
Structure	A tag type that holds more than one tag and more than one type of tag
	Each individual data type in a structure is called a member.
	Members each have a name and a data type.
	CLX has some standard structures available for counters, timers, etc.
	Users can create their own structures for specific uses. These are called User-Defined data types.

Figure 2-24 Definitions of *data type* and *structure*.

REAL-WORLD I/O ADDRESSING

Addressing of real-world I/O is different in CLX than other PLCs. Study Figure 2-25. The first part of the address is the Location. This can be Local, meaning in the same chassis as the controller, or the name of a remote communications adapter or a communications

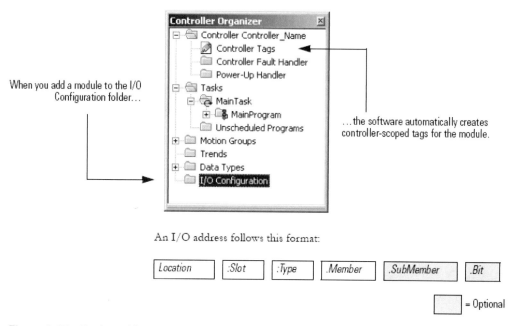

Figure 2-25 Real-world tag addressing. (Courtesy of Rockwell Automation, Inc.)

bridge module. A colon follows the Location. The next part of the address is the Slot number of the I/O module in the chassis. After another colon the Type of data follows. This can be an I (input), O (output), C (configuration), or S (status). A period delimiter is used next, followed by the Member. The Member specifies the specific data from the I/O module. If it is a digital input or output module, it stores the bit values for the module. If it is an analog module, it stores the data for a channel. A period delimiter is next, followed by the Submember. The Submember is specific data related to a member. Another period follows and then the Bit. The bit is a specific point on a digital module, one bit of an output module, for example. Figure 2-26 is a table that explains each part of an address. Occasionally the Member or Submember does not exist.

The good news is that the programmer does not have to type the address for an alias tag. When an alias tag is created, the address can be chosen from the available controller I/O.

CHAPTER 2—MEMORY AND PROJECT ORGANIZATION

Address	Content
Location	Network location Local = same chassis as controller Adapter_Name = name of a remote communications adapter or a bridge module
Slot	Slot number of I/O module in the chassis
Type	Type of data I = input O = output C = configuration S = status
Member	Specific data from the I/O module For a digital module, stores input or output bit values for the module For an analog module, stores the data for a channel
SubMember	Specific data related to a member
Bit	Specific point on a digital I/O module

Figure 2-26 CLX real-world tag addressing.

I/O Module Tags

When you add modules to a project, tags are automatically created for the modules. Figure 2-27 shows the Controller Organizer after an input module was added (slot 3) and an output module was added (slot 4). Note that they were diagnostic modules. Diagnostic modules have more tags than many of the other I/O modules. RSLogix5000

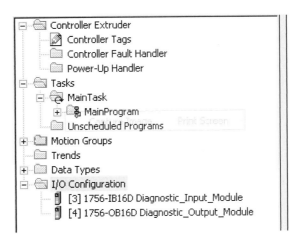

Figure 2-27 Two modules added to a project.

automatically creates the correct controller scope tags for the modules that are installed.

Figure 2-28 shows that tags that were created for the input module that was added. Note that the top portion of the tag editor (Local:3:C) is configuration tags. These are set when the module is configured. Configuration tags determine the characteristics and operation of a module.

Tag Name	Value
− Local:3:C	{...}
Local:3:C.DiagCOSDisable	0
+ Local:3:C.FilterOffOn_0_7	1
+ Local:3:C.FilterOnOff_0_7	1
+ Local:3:C.FilterOffOn_8_15	1
+ Local:3:C.FilterOnOff_8_15	1
+ Local:3:C.COSOnOffEn	2#0000_0000_0000_0000_1111_1111_1111_1111
+ Local:3:C.COSOffOnEn	2#0000_0000_0000_0000_1111_1111_1111_1111
+ Local:3:C.FaultLatchEn	2#0000_0000_0000_0000_1111_1111_1111_1111
+ Local:3:C.OpenWireEn	2#0000_0000_0000_0000_1111_1111_1111_1111
− Local:3:I	{...}
+ Local:3:I.Fault	2#0000_0000_0000_0000_0000_0000_0000_0000
+ Local:3:I.Data	2#0000_0000_0000_0000_0000_0000_0000_0000
+ Local:3:I.CSTTimestamp	{...}
+ Local:3:I.OpenWire	2#0000_0000_0000_0000_0000_0000_0000_0000

Figure 2-28 Tag editor showing tags that were automatically added after the input module in slot 3 was added.

The bottom portion of the tag editor (Local:3:I) shows the input tags that are available. The first ones are Fault bits that can be used in logic or for troubleshooting. The second set, labeled Data, contains the actual input bits from the module. The third set of inputs is the CST Timestamp (CST stands for coordinated system time) information. The last set of inputs are the OpenWire inputs.

Figure 2-29 shows that tags that were automatically created for the output module that was added. Note that the top portion of the tags (Local:4:I) are input-type tags. These include Fault, Data, CST Timestamp, FuseBlown, NoLoad, and OutputVerifyFault.

Next are the actual outputs (Local:4:0.DATA). This module has 16 outputs, given in the first 16 bits.

The last set of tags are configuration tags. These are set when the module is configured. Configuration tags determine the characteristics and operation of a module.

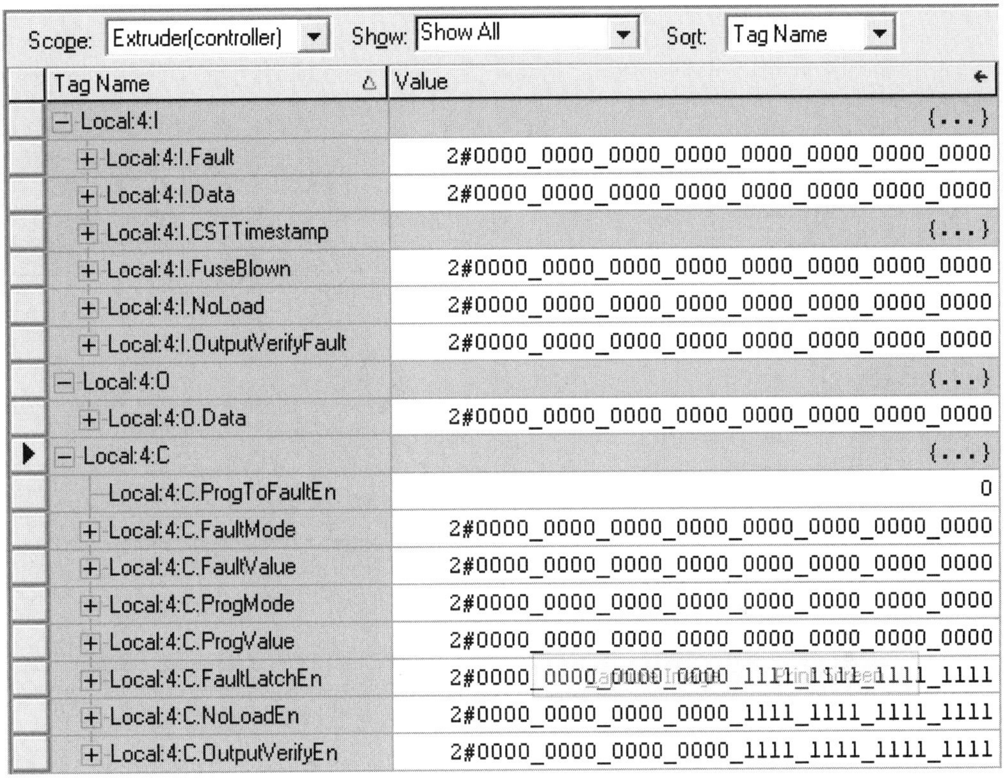

Tag Name	Value
– Local:4:I	{...}
+ Local:4:I.Fault	2#0000_0000_0000_0000_0000_0000_0000_0000
+ Local:4:I.Data	2#0000_0000_0000_0000_0000_0000_0000_0000
+ Local:4:I.CSTTimestamp	{...}
+ Local:4:I.FuseBlown	2#0000_0000_0000_0000_0000_0000_0000_0000
+ Local:4:I.NoLoad	2#0000_0000_0000_0000_0000_0000_0000_0000
+ Local:4:I.OutputVerifyFault	2#0000_0000_0000_0000_0000_0000_0000_0000
– Local:4:O	{...}
+ Local:4:O.Data	2#0000_0000_0000_0000_0000_0000_0000_0000
– Local:4:C	{...}
Local:4:C.ProgToFaultEn	0
+ Local:4:C.FaultMode	2#0000_0000_0000_0000_0000_0000_0000_0000
+ Local:4:C.FaultValue	2#0000_0000_0000_0000_0000_0000_0000_0000
+ Local:4:C.ProgMode	2#0000_0000_0000_0000_0000_0000_0000_0000
+ Local:4:C.ProgValue	2#0000_0000_0000_0000_0000_0000_0000_0000
+ Local:4:C.FaultLatchEn	2#0000_0000_0000_0000_1111_1111_1111_1111
+ Local:4:C.NoLoadEn	2#0000_0000_0000_0000_1111_1111_1111_1111
+ Local:4:C.OutputVerifyEn	2#0000_0000_0000_0000_1111_1111_1111_1111

Figure 2-29 Tag editor showing tags that were automatically added after the input module in slot 3 was added.

MORE ON THE USE OF TASKS

Remember that a CLX project can have multiple tasks and that tasks can be scheduled. The default RSLogix 5000 project provides a single task for all your logic. Although this is sufficient for many applications, some situations may require more than one task.

A Logix5000 controller supports multiple tasks that can be used to schedule and prioritize the execution of programs. This can help balance the processing time of the controller. Remember that

- A controller can only execute one task at one time.
- A task that is executing can be interrupted by another higher priority task.
- Only one program executes at one time in a task.

Figure 2-30 explains the three possible task execution types and the characteristics of each type.

To execute a task	Use a(n)	Description
Continuously	Continuous task	Runs in the background. Any CPU time that is not allocated to other operations is used to execute the programs in the continuous task. Other operations would include CPU time for communications, motion control, and periodic or event-driven tasks.
		Runs all of the time. When a scan is complete a new one begins immediately.
		Projects do not require a continuous task. There can only be one continuous task.
At a periodic rate, multiple times within the scan of the other logic,	Periodic task	Executes at a specific time period. When the time period occurs, a periodic task Interrupts any task with a lower priority and executes once Returns control to where the previous task left off The time period can be configured from 0.1 ms to 2000 seconds. Default time is 10 ms.
Immediately when an event occurs	Event task	Only executes when a specific event (trigger) occurs. When the event occurs, an event task Interrupts any lower-priority task, and executes once Returns control to where the previous task left off The trigger can be a(n) Digital input New sample of analog data Certain motion operation Consumed tag Event instruction

Figure 2-30 Task execution types.

It is important to choose the correct type of execution for each task. The table in Figure 2-31 shows examples of types of applications and which type of task execution they might be best suited to.

Application Example	Type of Task
Fill a tank and control its level (Without PID)	Continuous
Monitor, control, and display application parameters	
Monitor a tag every 0.1 second and calculate a rate of change to be used for control	Periodic
Perform quality measurements every 40 ms	
Control Level with PID	
On a packaging line, seal the package immediately when a registration mark is sensed	Event
If a specific alarm is sensed, shut down the machine immediately	
When a box arrives at the taping position, execute the taping routine immediately	

Figure 2-31 Examples of task execution types.

Number of Tasks

A ControlLogix CPU can support up to 32 tasks. Only one task executes at a time. Only one task can be continuous. It is possible to have too many tasks. Every task takes controller time away from the other tasks when it executes. If there are too many tasks, it is possible that tasks may overlap. If a task is interrupted for too long or too frequently, it may not complete its execution before it is triggered again. It is then possible that the continuous task may take too long to complete.

At the end of a task's execution, the controller performs overhead operations (output processing) for the I/O modules in the system. The output processing may affect the update of the I/O modules in the system. Output processing can be turned off for a specific task; this reduces the elapsed time of that task.

Every task has a watchdog timer that specifies how long a task can execute before it triggers a major fault. It is assumed that something has gone wrong if a task takes too long to execute. The watchdog timer begins to accumulate time when the task is initiated and stops accumulating time when all the programs within the task have executed. A watchdog time can range from 1 ms to 2000 seconds. The default time is 500 ms.

If a task takes longer than the specified watchdog time, a major fault occurs. The time includes interruptions by other tasks. A watchdog time-out fault may also occur if a task is repeatedly triggered while it is still executing. This can occur if a lower-priority task is interrupted by a higher-priority task, and it will delay the completion of the lower-priority task.

It is possible to use the controller fault handler to clear a watchdog fault. However, if the same watchdog fault occurs again during the same logic scan, the controller enters faulted mode, regardless of whether the controller fault handler clears the watchdog fault.

Setting the Watchdog Time for a Task

The watchdog timer is a preset parameter that the programmer can configure. The watchdog timer monitors the scan time of a task. If the watchdog timer reaches the PRE value, a major fault occurs. Depending on the controller fault handler, the controller may shut down. To change the watchdog time of a task, right-click the task and select Properties. Next select the Configuration tab and set the watchdog time-out for the task in milliseconds (see Figure 2-32). The watchdog default time of 500 ms is shown in Figure 2-32.

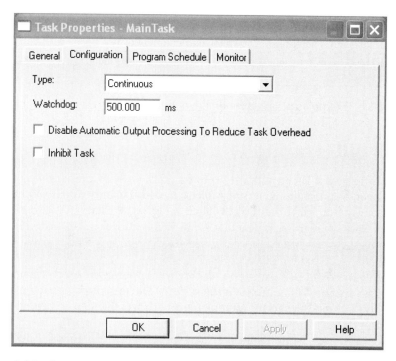

Figure 2-32 Setting the watchdog timer.

QUESTIONS

1. List the three main components that a project is composed of and describe each.
2. Which component in a project contains the logic?
3. What are the three main types of tasks?
4. What is a project?
5. What is the minimum memory allocation for a tag?
6. What is a base-type tag?
7. What is a DINT?

8. What is a SINT?
9. What is a BOOL? What is it typically used for?
10. What is an alias tag, and what is it used for?
11. What is the difference between a produced tag and a consumed tag?
12. What is an array tag? Write down an example of an array tag used to hold seven speeds.
13. What is a structure-type tag?
14. What are the two types of tag scope? What are the differences between them?
15. What is the difference between an array tag and a structure tag?
16. What is a User-Defined tag? What is it used for?
17. True or False: More than one task can execute at one time.
18. True or False: More than one program can execute at one time.
19. What are the three execution types for tasks?
20. What is a watchdog timer?
21. How can the watchdog time be changed?

CHAPTER 3

Ladder Logic Programming

OBJECTIVES

On completion of this chapter, the reader will be able to:

- Define terminology such as rung, contact, coil, scan, examine if open, examine if closed, normally open, and normally closed.
- Explain the difference in operation between normally open and normally closed contacts.
- Write basic ladder logic.
- Describe the relationship between the states of real-world switches and normally open versus normally closed contacts.

LADDER LOGIC

There are several languages that can be used to program industrial controllers. Ladder logic is still the most commonly used language. Ladder logic was designed to be easy for electricians to use and understand. Symbols were chosen that look similar to schematic symbols of electric devices so that a program would look like an electric circuit. An electrician who has no idea how to program a computer can understand a ladder diagram.

The instructions in ladder logic programming can be divided into two broad categories: input and output instructions. The most common input instruction is a contact, and the most common output instruction is a coil. Figure 3-1 shows input and output instructions in logic. Input instructions are on the left and output instructions are to the right of the logic.

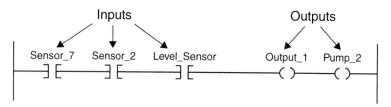

Figure 3-1 A rung of ladder logic.

Contacts

Most inputs to a PLC are discrete. Discrete devices only have two states: on or off. Contacts are used in ladder logic to represent these devices with two states. Symbols for the two types of contacts are shown in Figure 3-2.

Figure 3-2 A normally open and a normally closed contact.

Contacts are like discrete inputs.

Real-World Switches

To understand how a contact will work in ladder logic, we need to take a look at the two main types of mechanical switches: normally open and normally closed. Figure 3-3 shows the two types of switches. A doorbell switch is a normally open switch. A normally open switch will not pass current until it is pressed. A normally closed switch will allow current to flow until it is pressed.

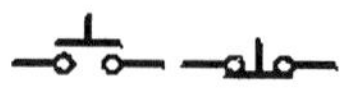

Figure 3-3 Diagram of a normally open switch and a normally closed switch.

Normally Open Contacts
Normally open contacts are also called examine-if-closed (XIC) contacts. Think about a doorbell switch again. If you use a normally open switch, the bell will be off until someone pushes the switch and it allows current to flow. If you use a normally closed switch, the bell will be on until someone pushes the switch to stop the current flow. A normally open switch is true if the real-world input associated with it is true. Think about a home doorbell again. There is a push-button switch (momentary) next to the front door. The real-world switch is a normally open switch. When someone pushes the switch, the bell in the house rings.

The normally open contact in ladder logic is similar to the normally open doorbell switch. Consider a sensor that we would like to control a pump. A normally open contact in a ladder diagram could be used to monitor the real-world level sensor that controls the pump. When the level sensor becomes true, the normally open contact in the rung of logic would allow current flow and the output (pump) would be on while the sensor is true (see Figures 3-4 and 3-5).

Figure 3-4 Normally open contact. Real-world input Sensor_2 is false so the normally open contact is not energized. The rung is false and the output is off.

Figure 3-5 Normally open contact. Real-world input Sensor_2 is true so normally open contact is energized. Thus the rung is true, and the output is on.

Figure 3-6 shows a conceptual diagram of a PLC with an input and output. The logic is shown in the middle in the block that represents the CPU. Note the input and output image table. The sensor attached to input 0 is true, and there is a 1 in the input table with the bit associated with input 0. In the rung of the ladder logic, the normally open contact is true because the bit in memory for input 0 is a 1 (true). This makes the rung conditions true, so the output is energized. Note the 1 in the output table associated with output 11.

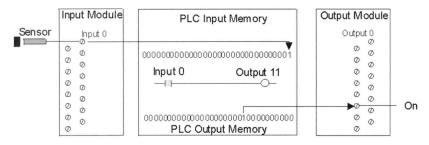

Figure 3-6 Simple conceptual diagram of a PLC.

Normally Closed Contacts
The other type of discrete contact is the normally closed contact. A normally closed contact is also called an examine-if-open (XIO) contact. This type of contact can be confusing at first. A normally closed contact will pass power until it is energized. A normally closed contact in a ladder diagram is only energized if the real-world input associated with it is false. Figure 3-7 shows an example of a normally closed contact in a rung of logic. If

real-world Sensor_2 is off, the normally closed contact in the rung is true and the output (Pump_1) will be on. If Sensor_2 in the real world is on (true), this contact will open and the output (Pump_1) will be off.

Figure 3-7 A normally closed contact is used in the rung. If the input associated with that contact is closed, it forces the normally closed contact open. No current can flow through the rung to the output, so the output is off.

Remember that normally closed contacts are also called XIO contacts. If the bit associated with a normally closed instruction is a 0 (off), the instruction is true and passes power. If the bit associated with the instruction is a 1 (true), the instruction is false and does not allow current flow. Note that the normally closed contact has the opposite effect of the normally open contact.

Figure 3-8 shows two rungs. The first rung has a normally closed (XIO) contact. The second rung has a normally open (XIC) contact. The tag is the same for each: Sensor_2. Sensor_2 is the tag name for a real-world switch that is off. Note that the normally open contact in the second rung is open and will not pass power. In the first rung, the normally closed contact remains closed and does energize the output because the real-world switch is off.

Figure 3-8 Logic example.

Figure 3-8 shows the same logic that was shown in Figure 3-9, but in this example the real-world switch (Sensor_2) is true. Note that the normally open contact in the second rung is true and that the normally closed contact in the first rung is false now because the switch is true.

Coils

A coil is symbol for an output. Rockwell Automation calls the typical output coil an output energize (OTE) instruction. The OTE instruction sets a bit in memory. If the logic in its rung is true, the output bit will be set to a 1. If the logic of its rung is false, the output bit is reset to a 0. Outputs are things such as lights, signals to other devices, motors contactors,

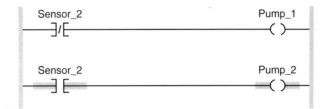

Figure 3-9 Logic example.

pumps, counters, timers, valves, and so on. Coils are only used on the right side of a rung. Contacts are conditions on rungs. If all of the conditions are true, the rung is true and the coil (output) will be true. The symbol for a coil is shown in Figure 3-10. A specific output coil should only be used once in logic. Note that there can be parallel conditions to control the coil on the rung. A specific coil should never be used in more than one rung. The output that the coil represents can be used on the left as contact(s).

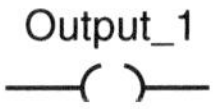

Output_1
—( )—

Figure 3-10 The symbol for a coil.

Figure 3-11 shows the RSLogix 5000 toolbar for ladder logic programming.

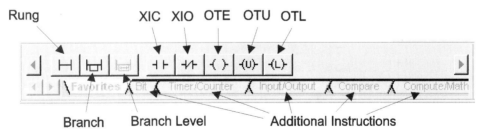

Figure 3-11 Ladder logic toolbar. XIC is the examine-if-closed contact, XIO is the examine-if-open contact, OTE is the output energize instruction, OTU is an output unlatch instruction, and OTL is an output latch instruction.

Ladder Diagrams

The basic structure of a ladder diagram looks similar to a ladder (see Figure 3-12). There are two rails (uprights) and there are rungs. The left and right rails represent power. If the logic on a rung is true, power can flow through the rung from the left upright to the right upright.

Figure 3-12 is a simple application involving a heat sensor and a fan. There is one input and one output for this application. A heat sensor is connected to a PLC input, and a fan is connected to a PLC output. The left and right rails represent voltage that will be used to power the fan if the sensor state is true.

Figure 3-12 Ladder logic example.

When a PLC is in run mode, it monitors the inputs continuously and controls the states of the outputs. This is called scanning. The time it takes for the PLC to evaluate the logic and update the I/O table each occurrence is called the scan time. The more complex the ladder logic, the more time it takes to scan.

Figure 3-13 shows a conceptual view of a PLC system. The real-world inputs are attached to an input module (left side of the figure). Outputs are attached to an output module (right side of the figure). The center of the figure shows that the CPU evaluates the logic. The CPU evaluates user logic by looking at the inputs and then turns on outputs on the basis of the logic.

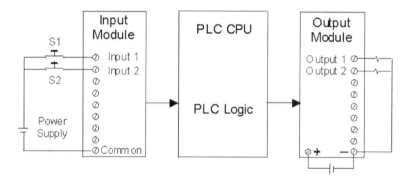

Figure 3-13 Conceptual view of a PLC system.

MULTIPLE CONTACTS

Series Logic

Contacts can be combined to form logic on a rung. A two-hand switch on a punch press is a good example. For safety, the punch press should only operate if both of the operator's hands are on the switches.

Figure 3-14 shows an example of a two-hand safety switch and series logic. The two hand switches are on the left and right. The switch in the middle is a stop switch.

A two-hand switch can help assure that the operator's hands cannot be in a dangerous part of the machine when it punches a part. The switches represent an AND condition. The ladder for the PLC is shown in Figure 3-14. Note that the real-world switches were programmed as normally open contacts in the logic. They are in series in the rung. Both must be true for the machine to punch. This is for illustrative purposes only. Chapter 15 will examine how safety switches are implemented with safety relays.

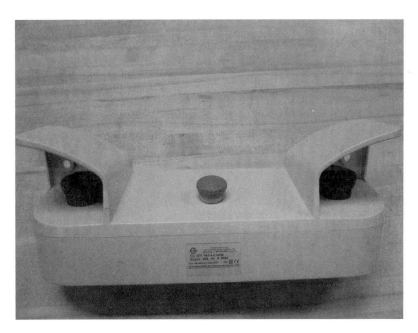

Figure 3-14 A two-hand safety switch. The two hand switches are on the left and the right of the photo.

If the operator removes one hand, the punch press will stop operation. In fact with newer safety relay technology, both switches have to be turned on at almost exactly the same time to make the machine run. Safety relays also prevent an operator from taping one switch closed. Contacts in series are logical AND conditions. In this example, the left-hand switch AND the right-hand switch would have to be true to run the punch press.

Figure 3-15 shows a series circuit. Sensor_1 AND Sensor_2 AND NOT Sensor_3 AND Sensor_4 must be true to turn the output named Fan_Motor on.

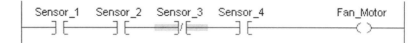

Figure 3-15 This figure shows a series circuit.

Figure 3-16 shows a robotic cell with light curtain protection. The light curtain has two safety outputs. If someone or something blocks any of the light between the light curtain transmitter and receiver, the outputs become false. If everything is normal and no light is blocked, the outputs are both true. There are two outputs to make the system more safe.

This example was an oversimplification of safety technology and logic in order to make contacts more understandable. Light curtains and other safety technology will be covered in Chapter 15.

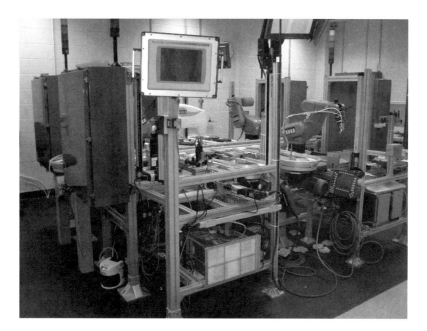

Figure 3-16 Robotic cell with light curtains and other safety devices.

Parallel Logic

Series logic was used to create AND conditions. Parallel logic is used to create OR conditions. These are often called branches. Branching can be thought of as an OR situation. One branch OR another can control the output.

Study Figure 3-17. This logic uses two different inputs to control the doorbell. If either switch is on, we would like the bell to sound. A branch is used to create this logic.

If the front door switch is closed, electricity can flow to the bell. OR if the rear door switch is closed, electricity can flow through the bottom branch to the bell.

Figure 3-17 Ladder logic to control a home doorbell. This figure shows a parallel condition. If the front door switch is closed, the doorbell will sound, OR if the rear door switch is closed, the doorbell will sound. These parallel conditions are also called OR conditions.

ORs allow multiple conditions to control an output. This is very important in the industrial control of systems. Think of a motor that is used to move the table of a machine. There are usually two switches to control table movement: a jog switch and a feed switch. Either switch must be capable of turning the same motor on. This is an OR condition. The jog switch OR the feed switch can turn on the table feed motor.

Series and parallel conditions can be combined in logic. Figure 3-18 shows a simple example. If Inp_1 AND Inp_2 are true, Out_1 will be turned on. OR if Inp_3 AND Inp_2 are true, Out_1 will be turned on.

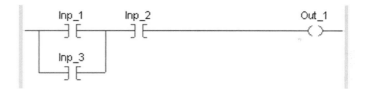

Figure 3-18 Rung using parallel (OR) and series (AND) logic.

Outputs may also be branched. Figure 3-19 shows an example of two outputs being branched. In this example, if the rung is true, both outputs will be turned on.

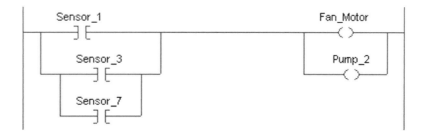

Figure 3-19 Example of parallel outputs.

START/STOP CIRCUITS

Machines typically have a start and a stop switch. The start switch is typically a normally open momentary type. The stop switch is typically a normally closed momentary switch. Knowledge of the ladder logic for a simple start/stop circuit can help us understand several important concepts about ladder logic. It can also help us understand the effect of real-world switch conditions on normally open and normally closed contacts. The normally closed contact is often confusing to many who are beginning to program ladder logic.

Examine Figure 3-20. Notice that a real-world start switch is a normally open momentary push button. When the button is pressed, it closes the switch. When the button is released, the switch opens. A real-world stop switch is a normally closed switch. When pressed, it opens the contacts and stops current flow.

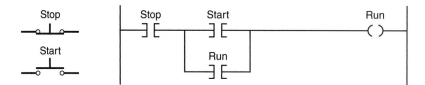

Figure 3-20 Start/stop circuit.

If the start switch is pressed momentarily, the normally open Start contact will become true. The Stop contact is also true, because the real-world stop switch is a normally closed switch. The output coil Run will be true. The Run contact bit is then used in the branch around the Start contact. The Run bit is true so this bit latches around the Start bit.

The output (Run) latches itself on even if the start switch opens. The output (Run) will shut off only if the normally closed stop switch (Stop) is pressed. If the normally closed contact Stop opens, then the rung would be false and Run would be turned off. To restart the system, the Start button must be pushed. Note that the real-world stop switch is a normally closed switch, but that in the ladder, it is programmed as a normally open contact. This is done for safety. It is called a fail-safe. If the stop switch fails, we want the machine to stop. If a wire to the stop switch is cut, we want the machine to stop. By using a normally closed real-world switch with a normally open contact in the logic we fail-safe the logic.

There are many ways to program start/stop circuits and ladder logic. Figures 3-21 and 3-22 show examples of the wiring of start/stop circuits. Safety is always the main consideration in start/stop circuits.

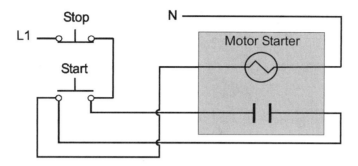

Figure 3-21 Single-phase motor start/stop circuit example.

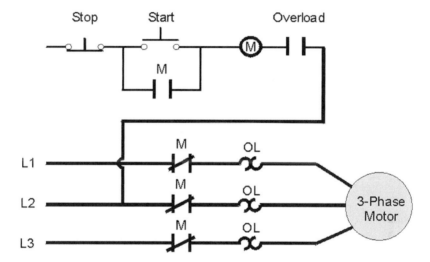

Figure 3-22 Start/stop circuit example for a three-phase motor.

USE OF OUTPUTS IN LOGIC

Output instructions can be placed in series on a rung in a CLX routine (see Figure 3-23). This is equivalent to outputs in parallel.

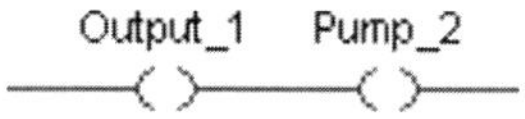

Figure 3-23 OTE coils used in series at the end of a rung.

Output instructions can be placed in branches (parallel) in ControlLogix. Figure 3-24 shows an example of output instructions used in parallel in logic. In ControlLogix there is no limit to the number of branches you may use.

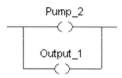

Figure 3-24 Example of OTE instructions used in parallel.

Output branches can also be nested. Figure 3-25 shows an example of a nested branch. Note that if Sensor_3 is true, the Bell will be turned on. The branched rung below that uses Sensor_2 to control the remaining branches. If Sensor_3 is true and Sensor_2 is true, the Bell output and the Pump_2 output will be turned on. The next branch adds another condition (Sensor_7). Branches can only be nested to a maximum of six levels.

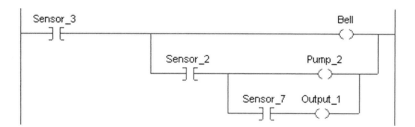

Figure 3-25 Nested output branch.

Output instructions can be placed between input contacts as long as the last instruction on the rung is an output instruction in CLX (see Figure 3-26).

Figure 3-26 Example of OTE coils being used between contacts in logic.

Logic Examples

The states of the real-world inputs are shown in Figure 3-27. Study the logic in Figure 3-27. The output is on. Why?

Real-World Input	State of Real-World Input
Inp_1	False
Inp_2	True

Figure 3-27 Output is true in this example. The output is on because real-world Inp_2 is true, making the normally open contact in the rung true. This is parallel (OR) logic.

The states of the real-world inputs are shown in Figure 3-28. Study the logic in Figure 3-28. The output is energized. Why?

Real-World Input	State of Real-World Input
Inp_1	True
Inp_2	False
Inp_3	True

Figure 3-28 Out_1 is true in this example because real-world inputs Inp_1 and Inp_3 are true making contacts normally open Inp_1 and normally open Inp_3 true.

The state of the real-world intputs are shown in Figure 3-29. Study the logic in Figure 3-29. The output is energized. Why? *Hint:* Remember how normally closed contacts work.

Real-World Input	State of Real-World Input
Inp_1	True
Inp_2	True
Inp_3	False
Inp_4	False

Figure 3-29 Output is energized in this example because Inp_4 contact is true since its real-world input is false and real-world inputs Inp_1 and Inp_2 are true, which make the normally open contacts true in the logic.

The states of the real-world inputs are shown in Figure 3-30. Study the logic in Figure 3-30. The output is off. Why?

Real-World Input	State of Real-World Input
Inp_1	True
Inp_2	True
Inp_3	False

Figure 3-30 Output is off in this example because real-world Inp_2 is true making logic contact Inp_2 false. Contact Inp_3 is also false because the real world Inp_3 is false.

The states of the real-world inputs are shown in Figure 3-31. Study the logic in Figure 3-31. The output is off. Why?

Real-World Input	State of Real-World Input
Inp_1	False
Inp_2	False
Inp_3	True
Inp_4	True

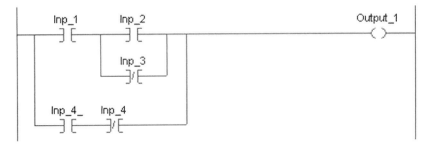

Figure 3-31 Output is off in this example because contacts Inp_2, Inp_3, and Inp_4 are false in logic.

Immediate Outputs

The immediate output (IOT) instruction is used to update output states immediately. In some applications the ladder scan time is longer than the needed update time for certain I/Os. For example, it might cause a safety problem if an output were not turned on or off before an entire scan was complete. In these cases or when performance requires immediate response, IOT instructions are used. When the CPU encounters an IOT instruction, it immediately transfers data to a specified I/O slot.

Figure 3-32 shows an example of an IOT instruction. In this example the outputs for the module in slot 2 would be updated immediately when they were encountered during the scan.

Figure 3-32 Format for an IOT instruction.

One-Shot (ONS) Instructions

An ONS instruction can be used to turn an output on for one scan. Figure 3-33 shows an example. If the Start contact becomes true, the ONS instruction will turn Out_1 on for one scan and then turn it off. One-shot-rising (ONR) and one-shot-falling (ONF) instructions are also available. The ONR instruction requires a low-to-high transition. The ONF requires a high-to-low transition. Figure 3-34 explains the use of some contact and coil instructions.

ONS instructions are usually used to execute things one time. For example, if we need to have an instruction, write parameters once each time we switch products, we would only need to write it once. We would not want to write it every scan while the program runs.

Figure 3-33 An ONS instruction.

To:	Instruction
Enable outputs when bit is set.	XIC
Enable outputs when a bit is cleared.	XIO
Set a bit.	OTE
Set a retentive bit.	OTL
Clear a retentive bit.	OTU
Enable outputs for one scan each time a rung becomes true.	ONS
Set a bit for one scan each time the rung goes true.	OSR
Set a bit for one scan each time the rung goes false.	OSF

Figure 3-34 Instructions.

Latching Instructions

Latches are used to lock in a condition. For example, if an input contact is on for only a short time, the output coil would be on for the same short time. If it were desired to keep the output on even if the input went low, a latch could be used. This could be done by using the output coil to latch itself on (see Figure 3-35).

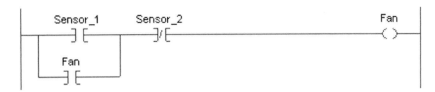

Figure 3-35 Note that even if coil Sensor_1 is only on for a very short time, the output (Fan) will latch around the Sensor_1 contact and keep itself on. It would stay on in this example until the normally closed contact opened. Latching can also be done with a special coil called a latching coil.

Output Latch (OTL) Instruction

The OTL instruction is a retentive instruction. If this input is turned on, it will stay on even if its input conditions become false. A retentive output can only be turned off by an output unlatch (OUT) instruction. Figure 3-36 shows an OTL instruction and an OTU instruction. In this case if the rung conditions for this output coil are true, the output bit will be set to a 1. It will remain a 1 even if the rung becomes false. The output will be latched on. Note that if the OTL is retentive and the processor loses power, the actual output turns off, but when power is restored, the output is retentive and will turn on. This is also true in the case of switching from run to program mode. The actual output turns off, but the bit state of 1 is retained in memory. When the processor is switched to run again, retentive outputs will turn on again regardless of the rung conditions. Retentive instructions can help or hurt the programmer. Be very careful from a safety standpoint when using retentive instructions. You must use an unlatch instruction to turn a retentive output off.

Figure 3-36 OTL and OTU instructions.

When a latching output is used, it will stay on until it is unlatched. Figure 3-37 shows an example of latching an output. If real-world Sensor_1 is true and real-world Sensor_2 is false, the output named Fan will be latched on. The output will remain energized even if the rung conditions become false. Fan will remain energized until an unlatch instruction is used.

Figure 3-37 Use of a latching output.

Output Unlatch (OTU) Instruction

The OTU instruction is used to unlatch (change the state of) retentive output instructions. It is the only way to turn an OTL instruction off. Figure 3-36 shows an OTU instruction. If this instruction is true, it unlatches the retentive output coil of the same name. Figure 3-38 shows an example of an OTU instruction to unlatch the output.

Figure 3-38 Use of an OTU instruction.

Program Flow Instructions

There are many types of flow control instructions available for PLCs. Program flow instructions can be used to control the sequence in which the program is executed. Program flow instructions allow the programmer to control the order in which the CPU scans the ladder diagram. These instructions can be used to minimize scan time or develop more efficient programs. They can be used to troubleshoot ladder logic as well. Program flow instructions can be used to jump around sections of logic for testing. Program flow instructions must be used carefully. Serious consequences can occur if they are improperly used because their use can cause portions of the ladder logic to be skipped.

Subroutine Instructions

Jump-to-Subroutine (JSR) Instructions
Rockwell Automation also has subroutine instructions available like the JSR, subroutine (SBR), and return (RET) instructions. Subroutines can be used to store sections of logic that must be executed in several points in your program. A subroutine saves effort and memory because you only program it once. Subroutines can be nested. This allows the programmer to direct program flow from the main program to a subroutine and then on to another subroutine. Figure 3-39 shows the use of a JSR instruction.

In ControlLogix the JSR instruction specifies the name of the routine to be executed. In this example, a routine named Manual_Mode will execute while the contact named Sensor_2 is true.

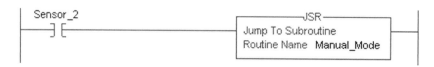

Figure 3-39 Use of a JSR instruction.

Operand	Type	Format	Description
Input parameter	BOOL, SINT, INT, DINT, REAL, Structure	Immediate, Tag, Array Tag	Data from this routine that will be copied to a tag in the subroutine. Input parameters are optional. Multiple input parameters can be used, if needed.
Return parameter	BOOL, SINT, INT, DINT, REAL, Structure	Immediate, Tag, Array Tag	Tag in this routine to which you want to copy a result of the subroutine. Return parameters are optional. Multiple return parameters can be used, if needed.

Figure 3-40 Parameters for subroutine instructions.

If you want to exchange data with a subroutine, the SBR and the RET instructions are used. They are optional instructions that exchange data with the JSR instruction. The SBR instruction is used to input values into a subroutine from the JSR instruction that called the routine. Study Figure 3-41. The SBR instruction must be the first instruction in a ladder logic routine. The RET instruction is used to return parameters with the JRS instruction that called it. Note that there is logic between the two instructions. The usual use would be to send one or more values to a routine, do some manipulation of the values, and return one or more values to be used by other routines.

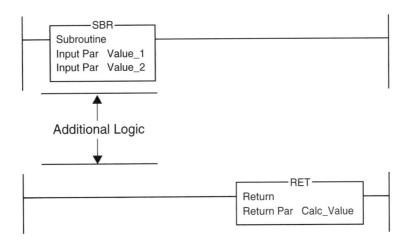

Figure 3-41 Use of an SBR and a RET instruction to exchange parameters with the calling routine.

Jump (JMP) Instruction
CLX has JMP and label (LBL) instructions available (see Figure 3-42). These can be used to reduce program scan time by omitting a section of program until it is needed. It is possible to jump forward and backward in the ladder. The programmer must be careful not to jump backward an excessive amount of times. A counter, timer, logic, or the program scan register should be used to limit the amount of time spent looping inside a JMP/LBL instruction; otherwise the watchdog time may be exceeded and a major fault will occur.

If the rung containing the JMP instruction is true, the CPU skips to the rung containing the specified label and continues execution. You can jump to the same label from one or more JMP instructions.

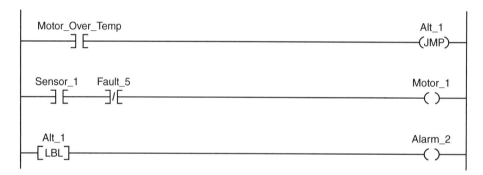

Figure 3-42 Use of JMP and LBL instructions.

QUESTIONS

1. What is a contact? A coil?
2. Explain the term *normally open* (XIC, examine if closed).
3. Explain the term *normally closed* (XIO, examine if open).
4. What are some uses of normally open contacts?
5. Explain the terms *true* and *false* as they apply to contacts in ladder logic.
6. If you were designing a fence with a gate for a robot cell, what kind of real-world switches would you use: normally open or normally closed?
7. Design a ladder that shows series input (AND logic). Use X5, X6, and AND NOT (normally closed contact) X9 for the inputs and use Y10 for the output.
8. Design a ladder that has parallel input (OR logic). Use X2 and X7 for the contacts.
9. Design a ladder that has three inputs and one output. The input logic should be: X1 AND NOT X2, OR X3. Use X1, X2, and X3 for the input numbers and Y1 for the output.

CHAPTER 3—LADDER LOGIC PROGRAMMING

10. Design a three-input ladder that uses AND logic and OR logic. The input logic should be X1 OR X3, AND NOT X2. Use contacts X1, X2, and X3. Use Y12 for the output coil.
11. Design a ladder in which coil Y5 will latch itself in. The input contact should be X1. The unlatch contact should be X2.
12. What is a JMP instruction used for?
13. What is a SBR instruction used for?
14. What is a RET instruction used for?

Examine the rungs below and determine whether the output for each is on or off. The input conditions shown represent the states of real-world inputs.

15.

Real-World Input	State of Real-World Input
Inp_1	True
Inp_2	True
Inp_3	True
Inp_4	False

16.

Real-World Input	State of Real-World Input
Inp_1	True
Inp_2	True
Inp_3	True

17.

Real-World Input	State of Real-World Input
Inp_1	True
Inp_2	True

```
    Inp_1   Inp_2                                Out_1
  ---] [-----]/[--------------------------------( )---
```

18.

Real-World Input	State of Real-World Input
Inp_1	True
Inp_2	False
Inp_3	True

```
    Inp_1   Inp_2   Inp_3                         Out_1
  ---] [-----]/[-----] [------------------------( )---
```

19.

Real-World Input	State of Real-World Input
Inp_1	True
Inp_2	True

20.

Real-World Input	State of Real-World Input
Inp_1	False
Inp_2	True
Inp_3	True

21.

Real-World Input	State of Real-World Input
Inp_1	True
Inp_2	False
Inp_3	True
Inp_4	True

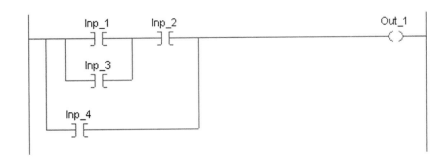

22. Write ladder logic for the application below. Your logic should have a start/stop circuit to start the application and should assure that the tank does not run empty or overflow. Use the I/O names from the table for your logic.

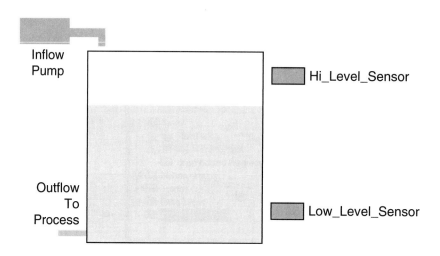

I/O Name	Function
Start	Real-world Switch
Stop	Real-world Switch
Run	BOOL
Pump	Real-world Discrete Output
Hi_Level_Sensor	Real-world Discrete Sensor
Low_Level_Sensor	Real-world Discrete Sensor

CHAPTER 4

Timers and Counters

OBJECTIVES

On completion of this chapter, the reader will be able to:

- Describe the use of counters and timers.
- Understand ControlLogix counter and timer tags and their members.
- Utilize status bits from timers and counters in logic.
- Define terms such as *delay-on, delay-off, preset, accumulated, retentive, cascade,* and so on.
- Correctly use counters and timers.

TIMERS

Timing functions are very important in PLC applications. Timers serve many functions in logic. They can be used to control when events occur. They can be used to delay actions in logic. They can also be used to keep track of elapsed time. Timers can also be misused in logic. Whenever possible, real events should be used to control when things happen in logic. Weak programmers often use timers to make bad logic work. Timers should be used for logic that requires timed events.

Timers have some typical entries. Timers must have a tagname. Timers must typically have a time base and a preset value. Figure 4-1 shows a typical timer. The timer's tagname in Figure 4-1 is Cycle_Time. The Preset is 10000.

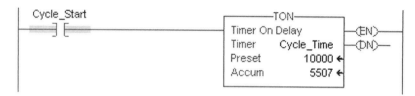

Figure 4-1 A CLX timer.

Preset

Timers have a preset (PRE) value that must be set by the programmer. The PRE value can be thought of as the number of time increments the timer must count before changing the state of the output. ControlLogix timers have a time base in milliseconds.

The actual time delay would equal the PRE value multiplied by the time base. For example, if the PRE value is 10000, this would be a 10-second time (10000 * 1 ms = 10 seconds).

The PRE value is stored in the timers tag member named .PRE. In the example shown in Figure 4-1 the tag member would be Cycle_Time.PRE. This enables the PRE value to be changed in the ladder logic. Timers have one input that enables the timer. When this input is true (high), the timer will accumulate time in the accumulator.

Timers can be retentive or nonretentive. Retentive timers do not lose the accumulated time when the rung conditions go false. A retentive timer retains the accumulated time until the line goes high again. When the rung goes true again, the retentive timer adds time to the accumulator. Retentive timers are also called accumulating timers. They function like a stopwatch. Stopwatches can be started and stopped multiple times and still retain their timed value. There is a reset button on a stopwatch to reset the time to zero. A RES instruction is used to reset a retentive timer.

Nonretentive timers lose the accumulated time every time the rung conditions become false. If the rung conditions become false, the timer accumulated time goes to zero.

Timer-On-Delay Instruction

The TON instruction produces an on-delay timer. An on-delay timer can be used to turn an output on after a delay. An on-delay timer begins accumulating time when the rung conditions become true. If the accumulated time in the timer becomes equal to or greater than the preset time, the timer done (DN) bit is set to a 1. Figure 4-2 shows an example of an on-delay timer.

The DN bit is the most commonly used timer status bit. The timer DN bit is false until the accumulated (ACC) value is equal to the PRE value. If the ACC value is greater than the PRE value, the DN bit is true. The DN bit remains set until the rung goes false or a reset instruction resets the timer. ControlLogix uses milliseconds (ms) for the time base. The preset in this example is 10000. This would be a 10-second time (10000 * 1 ms = 1 second).

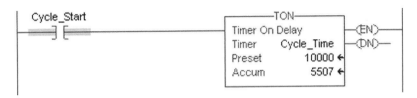

Figure 4-2 Use of an on-delay timer in ControlLogix. Note that the accumulated time has not reached the PRE value so the DN bit is false yet.

Timer Status Bits

Timers have special status bits that can be used in ladder logic. These are also called tag members. Figure 4-3 shows the typical timer bits.

Bit	Use
EN	The timer enable bit indicates that the TON instruction is enabled.
DN	The timer done bit is set when ACC ≥ PRE.
TT	The timer timing bit indicates that a timing operation is in process.

Figure 4-3 Timer bits.

The PRE value can also be used in logic. For example, Cycle_Time.PRE would access the PRE value of the timer named Cycle_Time. The PRE value would be a double integer (DINT). The PRE value can also be modified in logic.

Consider Figure 4-4. The timer is named Cycle_Time. The PRE value is 10000 ms. The timer enable (EN) bit becomes true when the rung conditions are true. The EN bit stays set until the rung goes false or a reset instruction resets the timer. The EN bit indicates that the timer is enabled. The EN bit can be used for logic. For example, Cycle_Time.EN could be used as a contact in a rung.

When the accumulated time in this timer reaches 10000, the DN bit will be true until the rung conditions go false or the timer is reset by a reset instruction. In this example the accumulated time has reached 10000 and the DN bit is true; this turned the output named Mix_Motor on.

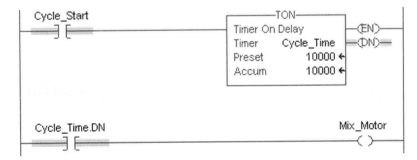

Figure 4-4 Use of a timer DN bit in ControlLogix.

Another useful status bit is the timer timing (TT) bit. The TT bit is set to true when the rung conditions become true. The TT bit remains true until the rung goes false or the DN bit is set (accumulated value equals PRE value). Figures 4-5 and 4-6 show the use of the timer TT bit. In the first figure the ACC value has not reached the PRE value, so Cycle_Time.TT is true. The second figure shows that the TT bit is false when the ACC value reaches the PRE value.

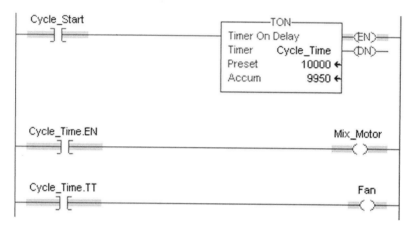

Figure 4-5 Use of timer EN and TT (true) bits in ControlLogix.

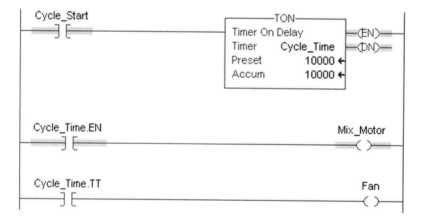

Figure 4-6 Use of timer EN and TT (false) bits in ControlLogix.

Figure 4-7 shows the tag members for the ControlLogix Part_Timer tag. Note that the preset (PRE) member and the accumulated (ACC) member are DINT types. The rest of the members are Booleans (BOOLs).

Tag Name	Value	Type
− Cycle_Time	{...}	TIMER
+ Cycle_Time.PRE	10000	DINT
+ Cycle_Time.ACC	0	DINT
Cycle_Time.EN	0	BOOL
Cycle_Time.TT	0	BOOL
Cycle_Time.DN	0	BOOL
Cycle_Time.FS	0	BOOL
Cycle_Time.LS	0	BOOL
Cycle_Time.OV	0	BOOL
Cycle_Time.ER	0	BOOL

Figure 4-7 Tag members for a timer named Part_Timer in CL.

Figure 4-8 shows an example of using logic to change the PRE value of a timer on the basis of conditions. In this example, if Product_A bit is set, the PRE value is set to 10000. If Product_B is set, the PRE value is set to 20000.

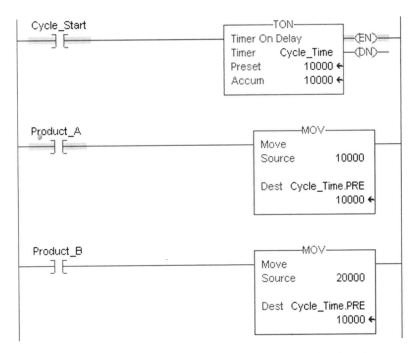

Figure 4-8 Use of move (MOV) instructions to change the PRE value in a timer.

ACC Value Use

The ACC value can also be used by the programmer. Figure 4-9 shows the use of a limit (LIM) instruction to evaluate whether the ACC value of the timer is between the Low Limit of 0 and the High Limit of 1250. The Test value of Part_Timer.ACC would access the ACC value of timer Part_Timer.ACC. In this example the Test value is between the Low and High Limit so the instruction is true and the output is energized.

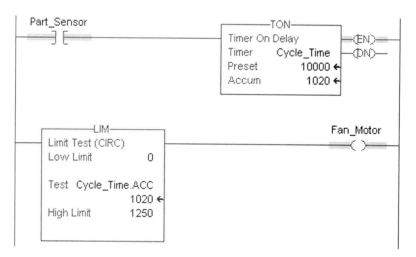

Figure 4-9 Use of the timer ACC value in ControlLogix.

Figure 4-10 shows the use of a reset (RES) instruction to reset a timer's ACC value to zero. If contact Inp_1 becomes true, the value of timer Part_Timer.ACC will be set to zero.

Figure 4-10 Use of a CLX RES instruction to reset a timer's accumulated time.

Figure 4-11 shows an example how the timer DN bit (Part_Timer.DN) can be used to reset the timer accumulator to zero every time the timer accumulated time reaches the PRE value. This would turn Part_Timer.DN on for one scan. This can be used to make things happen at regular intervals. For example, this could be used to execute an instruction or some logic every 10 seconds.

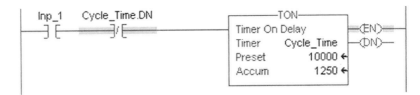

Figure 4-11 Use of a timer DN bit to reset the timer's accumulated time to 0 every time it reaches 10000 (the PRE value). The timer DN bit would be true for one scan every 10 seconds.

Timer-Off-Delay (TOF) Instruction

The TOF instruction can be used to turn an output coil on or off after the rung has been false for a desired time.

Let's use a nonindustrial example to understand the function of an off-delay timer. Think of a bathroom fan. It would be nice if we could just push a momentary switch and have the fan turn on for 2 minutes and then automatically shut off. In the example shown in Figure 4-12, the output (fan) turns on instantly when the input (switch) is momentarily turned on. The timer counts down the time (timed out) and turns the output (fan) off. This is an example of an off-delay timer.

Figure 4-12 shows the use of an off-delay timer in CL. When the Cycle_Start contact becomes true, the timer DN bit becomes true. Note in the second rung in Figure 7-1 that Cycle_Time.DN is on, turning the output (Fan_Motor1) on. The timer's DN bit will stay on forever if the Cycle_Start contact remains true. When Cycle_Start becomes false, it starts the timer timing cycle. When the rung goes false, the timer begins accumulating time. When the accumulator reaches 8000 ms, Cycle_Time.DN will become false. Figure 4-13 shows that when the accumulated time reaches 8000, Cycle_Time.DN is false.

A TOF instruction creates a delay-off timer. The timer turns on instantly, counts down time increments, and then turns off (delay off).

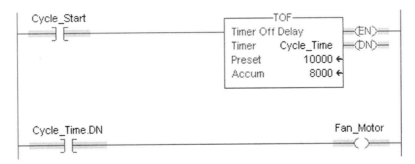

Figure 4-12 Delay-off timing circuit. If the contact named Cycle_Start closes, delay-off timer Cycle_Time.DN immediately turns on; this turns on the output named Fan_Motor1.

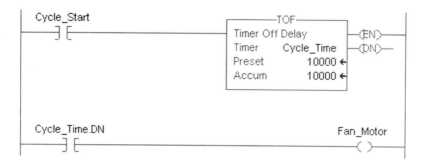

Figure 4-13 Delay-off timing circuit. When the timer accumulated time reaches the preset time, it will turn off; this turns Fan_Motor1 off also.

An off-delay timer starts to accumulate time when the rung becomes false. Off-delay timers accumulate time until the ACC value equals the PRE value or the rung becomes true. The off-delay timer EN bit is set when the rung becomes true. The EN bit is reset when the rung becomes false. If the rung becomes false and the ACC value is less than the PRE value, the TT bit is set to true. If the rung becomes false, the DN bit is reset (ACC = PRE), or a reset (RES) instruction resets the timer, the TT bit is reset to false.

Retentive-Timer-On Instruction

The RTO instruction is used to turn an output on after a preset time has accumulated (see Figure 4-14). The RTO timer is an accumulating timer. It retains the present ACC value when the rung goes false. The RTO timer retains the accumulated time even if power is lost, you switch modes, or the rung becomes false. A RES instruction must be used to zero the ACC value in an RTO timer. The RES instruction is programmed in another rung with the same timer name as the RTO timer you wish to reset.

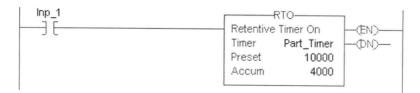

Figure 4-14 Use of an RTO timer.

The table in Figure 4-15 shows the timer instructions and which languages they can be used in.

Instruction	If you need to	Available in
TON	Time how long a timer is enabled	Ladder logic
TOF	Time how long a timer is disabled	Ladder logic
RTO	Accumulate time	Ladder logic
TONR	Time how long a timer is enabled with a built-in reset input	Structured text and function block
TOFR	Time how long a timer is disabled with a built-in reset input	Structured text and function block
RTOR	Accumulate time with built-in reset input	Structured text and function block
RES	Reset a counter or a timer	Ladder logic

Figure 4-15 Timer instructions.

COUNTERS

Counting is a very common function in industrial applications. Actions must often be based on product counts. In case packing, for example, there might be 4 rows of 6 cans making up one case of product. In this simple example, we might need to count to 6 for the 6 products in each row and then 4 for the number of rows in a completed case. Actions would be based on each count. We would also need another counter to count the number of cases that had been produced. We might need another down counter to show how many more need to be produced to complete the order.

Up counters and down counters are available. Up/down counters are also available in some ControlLogix languages. For example, if we were counting the number of filled and capped bottles leaving a bottling line and we were tracking how many parts remain in a storage system, we might use a up/down counter.

Down counters cause a count to decrease by 1 every time there is a pulse. Up/down counters can be used to increase or decrease the count depending on inputs.

Counters have a counter name or address, a PRE value, an ACC value, and several other tag members.

Logic can use counter status bits, PRE values, and ACC values. Bits such as CU, CD, DN, OV, or UN can all be used for logic. The PRE value and the ACC value can also be used in logic.

Count-Up (CTU) Counter

Figure 4-16 shows the use of a CTU counter. Each time contact Part_Counter makes a transition to true, 1 is added to the ACC value of the counter. Note that in this figure the accumulated value is less that the PRE value, so the DN bit of the counter is false in the second rung. The counter's count-up (CU) bit is true, however, because the counter's rung is true.

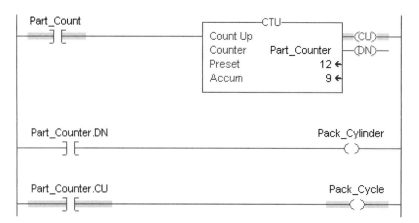

Figure 4-16 Use of the DN and CU bits in CL logic. Note that Part_Counter.CU is true and Part_Counter.DN is false because the ACC value is less than the PRE value.

Figure 4-17 shows the use of a CTU counter in CL. Note that in this figure the ACC value reached the PRE value (12), so the DN bit of the counter is true in the second rung. The counters count-up (CU) bit is now false, because the counter's rung is false.

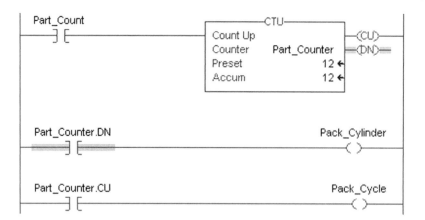

Figure 4-17 Use of the DN and CU bits in CL logic. Note that Part_Counter.DN is true because the ACC value has reached the PRE value. Part_Counter.CU is true anytime the rung condition in front of the counter is true. In this example the rung is false so the CU bit is false.

Figure 4-18 shows the use of a CTU counter's ACC value in a ladder diagram. Each time input Part_Sensor makes a false-to-true transition, the counter ACC value is incremented by 1. The ACC value of the counter is being used in the equal (EQU) math instruction in the second rung to turn on an output when the ACC value reaches 6.

This instruction just compares two values (Source A and Source B). If they are equal the EQU instruction is true making the rung true in this example. Note that math instructions will be covered in detail in Chapter 7. Note that a RES instruction can be used to reset the value of a counter accumulator.

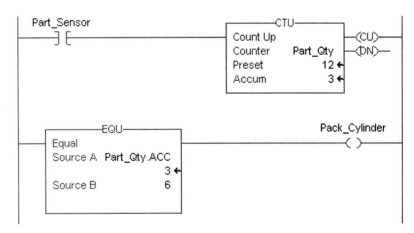

Figure 4-18 Use of a CTU accumulated value.

Count-Down (CTD) Counters

A CTD counter can be used to count down from a preset number. Figure 4-19 shows an example. Although it seems illogical, when the ACC value equals or is greater than the PRE value, the DN bit will be energized, just as with a CTU counter. The CTD counter is almost always used in conjunction with a CTU counter as an up/down counter by assigning the same tag name to them both.

```
   Part_Count                                    -CTD-
─────] [─────────────────────────────────    Count Down           ─(CD)─
                                             Counter    Down_CTR  ─(DN)─
                                             Preset          12
                                             Accum            3
```

Figure 4-19 Use of a CTD counter.

A counter that can count up or down is also available in two CLX programming languages. The count up/down (CTUD) counter is used to count up and count down. It is not available in ladder logic. It can be used in structured text and function block. Figure 4-20 shows the counter instructions that are available and which languages they can be used in.

Instruction	Use	Available in
CTU	Count up	Ladder logic
CTD	Count down	Ladder logic
CTUD	Count up and count down	Structured text and function block
RES	Reset a counter or a timer	Ladder logic

Figure 4-20 Counter instructions.

QUESTIONS

1. What is an on-delay timer?
2. What is an off-delay timer?
3. What is the time base for a CL controller?
4. What is the PRE value used for in a timer?
5. What is the ACC value used for in a timer?
6. Describe two methods of resetting the ACC value of an on-delay timer to 0.
7. What does the term *retentive* mean?
8. Give an example of how the TT bit for an on-delay timer could be used.
9. Give an example of how the EN bit for on- or/and off-delay timer could be used.
10. Give an example of how the DN bit for an off-delay timer could be used.
11. In what way are counters and timers very similar?
12. What is a CTU counter?
13. What is a CTD instruction?
14. What is a CTUD instruction?
15. What languages are CTUD instructions available in?
16. How can the accumulated count be reset in a counter?
17. Complete the descriptions in the tables below.

ControlLogix Timers	Description of Address
T_1.PRE	
T_1.ACC	
T_1.DN	
T_1.TT	
T_1.EN	

ControlLogix Counters	Description of Address
CNT_1.PRE	
CNT_1.ACC	
CNT_1.DN	

18. Write a ladder logic for the following application:

This is a simple heat treat machine application. The operator places a part in a fixture, then pushes the start switch. An inductive heating coil heats the part rapidly to 1500 degrees Fahrenheit. When the temperature reaches 1500, a discrete sensor's output becomes true. The coil turns off, and a valve is opened which sprays water for 5 seconds on the part to complete the heat treatment (quench). The operator then removes the part, and the sequence can begin again. Note there must be a part present or the sequence should not start.

I/O	Type	Description
Part_Present_Sensor	Discrete	Sensor used to sense a part in the fixture
Temp_Sensor	Discrete	Sensor whose output becomes true when the temperature reaches 1500 degrees Fahrenheit
Start_Switch	Discrete	Momentary normally open switch
Heating_Coil	Discrete	Discrete output that turns coil on
Quench_Valve	Discrete	Discrete output that turns quench valve on

CHAPTER

Input/Output Modules and Wiring

OBJECTIVES

On completion of this chapter, the reader will be able to:

- Define terms such as discrete, digital, analog, resolution, producer, consumer, and so on.
- Describe types of digital I/O modules.
- Describe types of analog I/O modules.
- Find the resolution for an analog module.
- Describe how analog modules are calibrated.
- Wire digital and analog modules.

I/O MODULES

There are a wide variety of modules available for CLX systems. Modules are available for digital and analog I/O, communications, motion, and many other purposes. CLX modules have more capability than modules in most PLC systems. Modules are configurable and provide troubleshooting and fault information to the controller.

Every I/O module in a CLX system must be owned by a CLX controller to be used. The controller that owns the module stores configuration data for every module that it owns. Modules can be located in the same chassis as the controller or remotely. The controller that owns the module sends the I/O module configuration data to define the module's behavior. Each individual ControlLogix I/O module must continuously maintain communication with the controller that owns it to operate normally. Output modules are limited to a single owner. Only one controller can own an output module. Input modules can have multiple owners. If multiple owners are connected to the same input module, the controllers must have identical configurations for the module.

Modules are configured when you add them to a ControlLogix project. Local and remote modules are configured by the user. The I/O configuration portion of RSLogix5000 generates the configuration data for each I/O module in the control system on the basis of how the module is configured in the project. A remote chassis contains the I/O module but not the module's owner-controller. Remote chassis can be connected to the controller via scheduled ControlNet or EtherNet/IP networks. Configuration data for modules is transferred to the controller during program downloads and is then transferred to the appropriate I/O modules. When a module is added to a project, tags are created that allow the user to access I/O information, fault information and configuration data.

Any controller in a system can listen to the data from any I/O module even if the controller does not own the module. During module configuration the user specifies one of several Listen-Only modes in the Communication Format field.

Choosing a Listen-Only mode option allows the controller and module to establish communications without the controller owning the module. Remember that only the controller that owns a module sends configuration data to the module.

Controllers using the Listen-Only mode receive data multicast from the I/O module as long as the connection between the owner-controller and the I/O module is maintained. If the connection between the owner-controller and the module fails, the module stops multicasting data. Connections to all listening controllers are also broken.

DIGITAL MODULES

Digital modules are also called discrete modules. Discrete means that each input or output has two states: true (on) or false (off). Most industrial automation devices are discrete.

Digital Input Modules

Digital input modules are used to take input from the real world. Inputs to a discrete module are provided by devices such as switches or sensors that are either on or off.

Input and output modules must be able to protect the CPU from the real world. Assume an input voltage of 110 VAC (volts alternating current). The input module must change the 110 VAC level to a low-level direct current (DC) logic level for the CPU. This is accomplished through opto-isolation. An opto-isolator uses a phototransistor. The LED in the phototransistor is controlled by the input signal. The light from the LED falls on the base of the phototransistor and turns the transistor output on (allows collector/emitter current flow). Figure 5-1 shows how this is done for alternating current (AC) input modules. The light totally separates the real-world electric signals from the PLC internal electrical system. Figure 5-2 shows how opto-isolation is done for a DC input module.

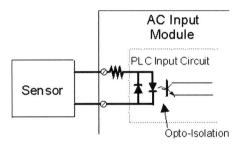

Figure 5-1 Opto-isolation for an AC input module. Note in the figure that the AC input signal is transmitted via an LED to a phototransistor. The output of the photosensor is a low-level DC signal for the CPU.

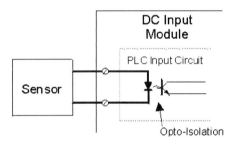

Figure 5-2 Opto-isolation for a DC input module.

All ControlLogix I/O modules may be inserted and removed from the chassis while power is applied. This is often called hot swapping. This feature allows greater availability of the overall control system because, while the module is being removed or inserted, there is not any additional disruption to the rest of the controlled process. Figure 5-3 shows some of the features of a CLX module. Note the Locking tab on the front of the module. The Locking tab locks the RTB or cable onto the module.

LED Status Information

ControlLogix digital I/O modules provide hardware and software indicators when a module fault occurs. Each module's LED fault indicator and RSLogix 5000 software will graphically display the fault. ControlLogix diagnostic digital I/O modules have an

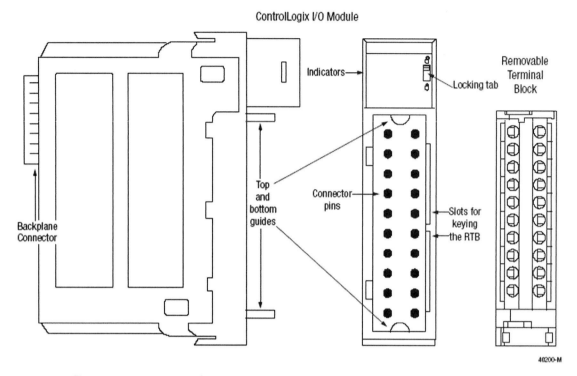

Figure 5-3 A typical I/O module. (Courtesy of Rockwell Automation, Inc.)

LED indicator on the front of the module that enables a technician to check the health and operational status of the module. The LED displays vary for different types of modules. Input modules typically have LEDs for monitoring the inputs. If the input is true, the LED is energized. Some modules also have additional LEDs for troubleshooting. Figure 5-4 shows examples of three different input modules. The middle one and the one on the right have additional diagnostics.

The following statuses can be checked with the LED indicators:

I/O status LEDs- This yellow display indicates the on/off state of each input.
Module status LED- This green display indicates the module's communication status.
Fault status LED- This display is only found on some modules and indicates the presence or absence of faults.
Fuse status LED- This display is only found on electronically fused modules and indicates the state of the module's fuse.

CHAPTER 5—INPUT/OUTPUT MODULES AND WIRING   **87**

  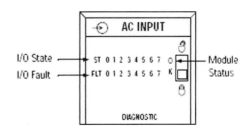

Figure 5-4 A regular input module on the left and two diagnostic input modules. Note that the module on the left only shows the status for each input whereas the diagnostic modules also show the fault status for each input. Note also the location of the module's status LED. All modules have the Module Status LED. (Courtesy of Rockwell Automation, Inc.)

Figure 5-5 has a table that shows the states for the LED status indicator on input modules.

LED Status	Indication	Meaning	Action
Green	OK	Inputs are being multicast and are in a normal operation state.	None.
Flashing green	OK	Module passed internal diagnostics but is either inhibited or is not multicasting.	None.
Flashing red	OK	Previously established communication has timed out.	Check the controller and chassis communication.
Red	OK	Module must be replaced.	Replace the module.
Yellow	I/O state	The input is active.	None.
Red	I/O fault	A fault occurred for this point.	Check this point at the controller.

Figure 5-5 Status indicators for input modules.

Keying Modules

Many PLCs have had the capability to key I/O modules mechanically. The user would attach pegs or plastic keys to modules and the matching configuration to a rack slot. Each type of module would have a specific key pattern. This prevents putting the wrong type of module in a slot. For example, an output module could not be inserted in an input slot. This protects the application from replacing a module with the wrong type.

Electronic Keying

ControlLogix PLCs have electronic keying. Electronic keying enables the ControlLogix controller to control what modules belong in the various slots of a system. Electronic keying is much more powerful and configurable than mechanical keying.

When a module is configured the user can choose Exact Match, Compatible, or Disable keying. Exact Match keying would require an exact match of the module and also the firmware version of the module be the same as the one it was replacing. Compatible would allow any module that would be compatible. With ControlLogix digital I/O modules, the module can emulate older revisions. The module will accept the configuration if the configuration's major/minor revision is less than or equal to the physical module's revision. For example, if the configuration contains a major.minor revision of 2.7, the module inserted into the slot must have minor revision of 2.7 or higher for a connection to be made. Disable keying allows any module to be installed in the slot, whether it is the correct replacement or not.

Removable Terminal Blocks

ControlLogix modules have plug-on wiring terminal strips that can be mechanically keyed. These are called Removable Terminal Blocks (RTBs) for ControlLogix PLCs. All wiring is connected to the RTB. The RTB is plugged onto the actual module. If there is a problem with a module, the entire RTB is removed, a new module inserted, and the RTB is plugged into the new module, without any rewiring. A module can be replaced in a very short time, thus reducing downtime. This is very important when one considers the cost of a system being down. Keying the RTBs also assures that the technician cannot install the wiring harness on the wrong module. Figure 5-6 shows how a wiring harness is keyed.

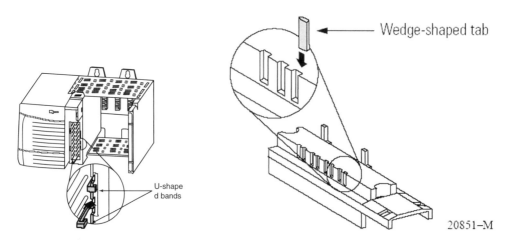

Figure 5-6 Keying an RTB. (Courtesy of Rockwell Automation, Inc.)

Time-stamping Inputs

The system clock can be used to time-stamp inputs and schedule outputs.

ControlLogix controllers generate a 64-bit Coordinated System Time (CST) for their chassis. The CST is a chassis-specific time that is not synchronized with or connected to the time generated over ControlNet to establish a network update time (NUT). Digital input modules can be configured to access the CST and time-stamp input data with the value of the CST when that input data changes state.

Time-stamping for a Sequence of Events
The CST can be used to establish a sequence of events occurring at a particular input module point by time stamping the input data. To determine a sequence of events, you must

- Configure the input module's communications format to CST time-stamped input data
- Enable change of state (COS) for the input point where the sequence will occur
- Use time-stamping on only one input point per module. Disable COS for all other points on the module, because only one CST value is returned to the controller when any input point changes state.

If multiple input points are configured for COS, the module generates a unique CST each time any of those input points changes state, as long as the changes do not occur within 500 µs of each other. If multiple input points, configured for COS, change state within 500 µs of each other, a single CST value is generated for all of them. This makes it appear that they all changed at the same time.

Rolling Time Stamp

Every module maintains a rolling time stamp. This rolling time stamp is not related to the CST. The rolling time stamp is a continuously running timer that counts in milliseconds.

When an input module scans its channels, it also records the value of the rolling time stamp at that time. The user's program can use the last two rolling time stamp values to calculate the interval between the receipts of data or the time when new data has been received. For output modules the rolling time stamp value is only updated when new values are applied to the digital-to-analog converter.

Input Wiring

Figure 5-7 shows an example of the wiring of a CLX DC input module. Note that the module must be connected from the negative side of the power supply to the ground terminals of the module. The positive side of the power supply is not directly connected to the module. Positive is connected to one side of the inputs. The other side of each input connects to the desired input terminal. In this example, when the switch is closed, there is a complete series path from the positive side of the supply through the switch to the input terminal and then to the ground on the module and finally back to the negative side of the supply.

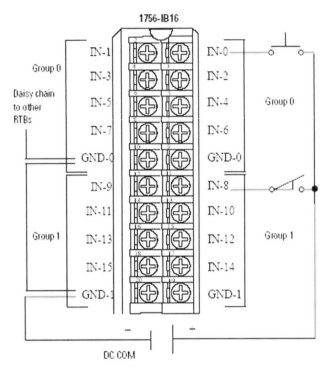

Figure 5-7 A wiring diagram for a ControlLogix sinking digital input module. (Courtesy of Rockwell Automation, Inc.)

Figure 5-8 shows a wiring diagram for a Rockwell input module. It is a good idea to first determine what the module requires for power before we worry about the input wiring. The power supply is not directly connected to the module shown in Figure 5-10. There is a direct connection from the negative side of the power supply to the module. The negative is connected to one of the grounds (common). Note also that the grounds are wired together to make them all common.

Two inputs are shown on the right of the module. The top switch is a normally open switch. The right side of the switch is connected to the + side of the power supply. The left side of the switch is connected directly to the desired input terminal (0 in this example).

The bottom switch symbol means that this switch is a limit switch. It is a normally closed switch that is held in the open position in this diagram. Note that the left sides of the switches are connected to input terminals and the right sides of the switches are connected to positive DC.

Sinking versus Sourcing Modules

Input modules can be purchased as either sinking or sourcing. The easiest way to understand this is to remember that a sinking PLC input module would require a positive input signal. For example, if we have a sensor that has a positive output signal (sourcing), we would need a sinking PLC input module. Figure 5-9 shows an example of a sinking PLC input module.

CHAPTER 5—INPUT/OUTPUT MODULES AND WIRING   91

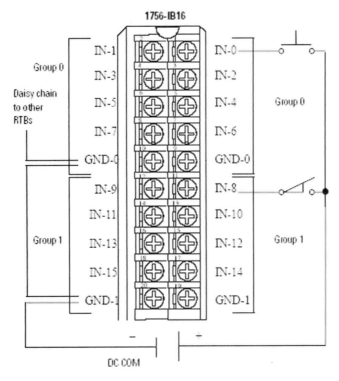

Figure 5-8 CLX DC digital input module wiring. (Courtesy Rockwell Automation Inc.)

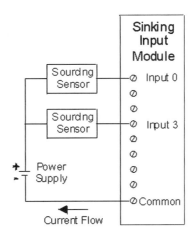

Figure 5-9 Sinking input module. Note that the sensors have sourcing (positive) outputs.

If we have a sensor or device with negative output, we would need a sourcing PLC input module (see Figure 5-10). This often occurs when connecting drives, vision, or robot outputs to a PLC. You must be careful to check the output polarity required and match it to the correct type of PLC module. Remember opposites attract. If you have a sensor with a sourcing (positive) output, you need a sinking input module. If you have a sensor with a sinking (negative) output, you need a sourcing PLC input module.

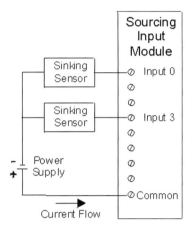

Figure 5-10 Sourcing input module. Note the sensors have sinking (negative) outputs.

When two-wire sensors are used there is always a small leakage current that is required for the sensor's operation. This does not normally cause a problem because the leakage current is very small and not enough to be seen as an input by the input card. In some cases, however, this leakage current is enough to trigger the input of the PLC module. If leakage current is seen as an input, a resistor can be added that will bleed the leakage current to ground (see Figure 5-11). When a bleeder resistor is added, most of the current goes through it to common. This assures that the PLC input is only triggered when the sensor's output is on.

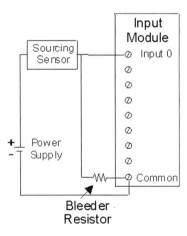

Figure 5-11 Use of a bleeder resistor.

Discrete Output Modules

Discrete output modules are used to turn real-world output devices off or on. Output modules are available for AC and DC devices. They are also available in various voltage ranges and current capabilities. The actual output device used for each output includes transistors, triac output, or relay output. The transistor output would be used for DC outputs. Triac outputs are used for AC devices. Transistor-transistor logic (TTL) output modules are also available.

The current limit specifications for an output module are normally given for each individual output and as an overall module current limit. Figure 5-12 shows the specifications for a 1756-OB16E output module. Note that each output has a current limit of 1 ampere. The overall current limit is 8 amperes. This module has 16 outputs. If each output were at the maximum current, the module would exceed the overall current limit (16 *1 > 8). The total current for the module must not exceed the total that the module can handle. Normally each PLC output will not draw the maximum current, nor will they all be on at the same time. Consider the worst case when choosing an appropriate output module.

Output Voltage Range	10–31.2 VDC
Output current rating Per point Per module	1-A maximum @ 60 degrees C 8-amp maximum @ 60 degrees C
Surge current per point	2 A for 10 ms each
Minimum load current	3 mA per output

Figure 5-12 Specifications for a 1756-OB16E output module.

Figure 5-13 shows the LED troubleshooting panel from an electronically fused output module. Note the status lights—one for each output and the two fuse LED indicators (one for outputs 0–7 and one for outputs 8–15).

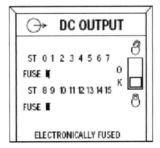

Figure 5-13 Electronically fused output module. Note the LED indicators for the fuses. (Courtesy of Rockwell Automation, Inc.)

Many output modules are fused to protect each output. This fuse is normally intended to provide short circuit protection for wiring only to external loads. If there is a short circuit on an output channel, it is likely that the output transistor, triac, or relay associated with that channel will be damaged. In that case the module must be replaced or the output moved to a spare channel on the output module. The fuses are normally easily replaced. Check the specifications for the particular module. Check the technical manual for the module to find the correct fuse and procedure.

Output Wiring

Figure 5-14 shows the wiring for a DC output module. One of the first questions when you are figuring out the wiring for a module is: Does the module require connections to a power supply? In Figure 5-14 you see that the positive side of the power supply must be connected to two terminals: DC-0 and DC-1. The negative side of the supply must be connected to terminals RTN OUT-0 and RTN OUT-1.

Now that we have the module powered correctly, we can examine the wiring of the outputs. The output devices are connected to an output terminal and then tied to the RTN OUT (negative) terminals.

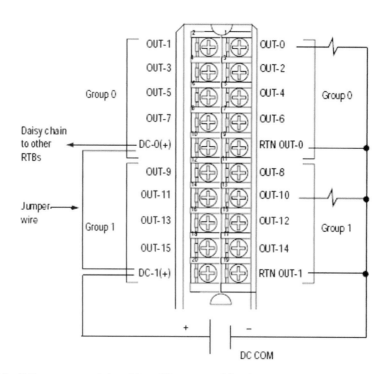

Figure 5-14 DC output module wiring. (Courtesy of Rockwell Automation, Inc.)

Figure 5-15 shows the wiring for an AC output module. Let's first examine the power wiring for the module. L1 from the AC supply is connected to L1-0 and L1-1. L2 from the AC supply is connected to L2-0 and L2-1. Next let's look at the wiring for an output. Each output is connected to an output terminal and tied to terminals L2-0 and L2-1.

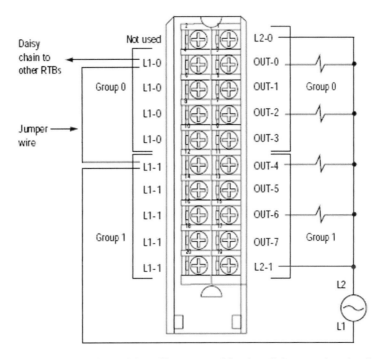

Figure 5-15 AC output module wiring. (Courtesy of Rockwell Automation, Inc.)

Sinking versus Sourcing DC Outputs

Just as with input modules, there are two choices for output: negative output and positive output. This is often an issue when connecting outputs to other devices such as vision systems, robots, and so on. The devices might specifically require a positive or a negative signal for their input. Remember that an output from the PLC becomes an input to another device.

A sourcing output module supplies a positive signal to an output. Figure 5-16 shows that each output for a sourcing output module is connected to the output terminal and the other side of the output is tied to the negative side of a supply.

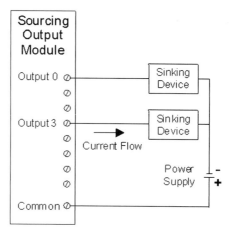

Figure 5-16 AC output module wiring.

Figure 5-17 shows a sinking output module. One side of the output is connected to an output terminal and the other side of the output is tied to the positive side of a supply.

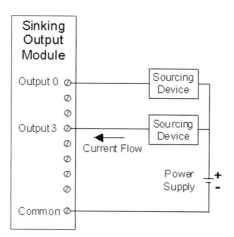

Figure 5-17 Wiring a sinking output module.

No-Load Detection

Diagnostic output modules also have the capability to sense the absence of field wiring for every output. This is called no-load detection. No-load detection can also detect a missing load from each output point in the off state, but only in the off state.

High-Density I/O Modules

The most common I/O modules have 16 inputs or outputs. A high-density module may have up to 32 inputs or outputs. The advantage is that there are a limited number of slots in a PLC rack. Each module uses a slot. With the high-density module, it is possible to install 32 inputs or outputs in one slot. The only disadvantage is that the high-density output modules typically cannot handle as much current per output.

Fusing of Digital Output Modules

Many output modules are internally fused to provide protection. Some modules fuse every output individually. Other output modules may fuse a group of outputs. For example, one fuse for the first eight outputs on the module and one fuse for the second eight outputs on the module. Other modules may only provide one fuse for the whole module.

Some PLC modules have LEDs to indicate a blown fuse. The fusing may be for each output, a group of outputs, or one LED for all of the outputs.

Electronic fuses can be reset with the programming software (see Figure 5-18) or by your logic. Check the appropriate manual to find the correct fusing for your particular module.

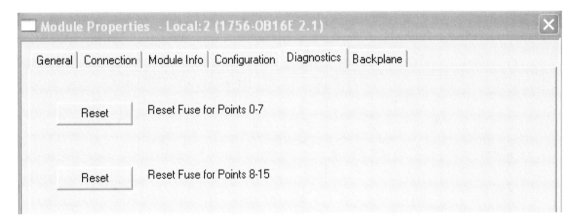

Figure 5-18 Resetting fuses in the programming software.

ANALOG INPUT MODULES

Analog input modules are designed to take analog information from devices and convert the analog signal to digital information. The two most common types are current sensing and voltage sensing. These cards will take an analog current or voltage and change it to digital data for the PLC.

Analog Module Timing Parameters

When a module resides in the same chassis as the owner-controller, the Real Time Sample (RTS) and the Requested Packet Interval (RPI) configuration parameters will affect how and when the input module multicasts data. The RTS is concerned with how often to sample, and the RPI deals with how often to multicast (communicate) the information to other modules.

Real Time Sample (RTS)
This RTS parameter is configured in software by the user. The RTS instructs the module to perform the following operations (see Figure 5-19):

1. Scan all input channels and store the data into module memory.
2. Multicast the updated channel data and status data to the backplane of the local chassis.

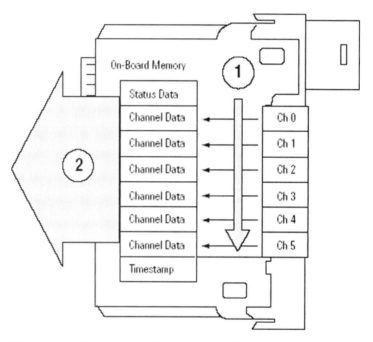

Figure 5-19 RTS sequence for an analog input module. (Courtesy of Rockwell Automation, Inc.)

Requested Packet Interval (RPI)
The RPI is a configurable parameter. The RPI instructs the module to multicast the current contents of its onboard memory when the RPI expires. The module does not update its channels prior to the multicast.

When remote analog I/O modules are connected to the owner-controller on a ControlNet network, the RPI and RTS intervals define when the module multicasts data within its own chassis. The RPI value, however, determines how often the owner-controller will

CHAPTER 5—INPUT/OUTPUT MODULES AND WIRING 99

receive the data over the network. Figure 5-20 shows how the data are transferred over a ControlNet network.

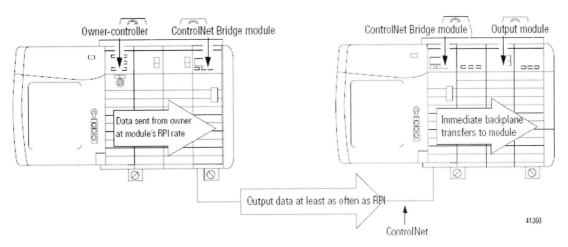

Figure 5-20 Transfer of I/O data over a ControlNet network. (Courtesy of Rockwell Automation, Inc.)

Remote Input Modules Connected via EtherNet/IP

When EtherNet/IP is used to connect remote analog input modules to the owner-controller, data are transferred to the owner-controller in the following manner:

At whichever is faster, the RTS or the RPI, the module multicasts its data within its own chassis. The Ethernet module in the remote chassis then immediately sends the module's data through the network to the owner-controller as long as it has not sent data within a timeframe that is 1/4 the value of the analog input module's RPI.

For example, if the RPI = 100 ms, the Ethernet module will only send module data immediately on receiving it if another data packet was not sent within the last 25 ms.

Analog voltage input modules are available in two types: unipolar and bipolar. Unipolar modules can take only one polarity for input. The bipolar card will take input of positive and negative polarity. Analog input modules are commonly available in 0 to 10 volts (unipolar) and −10 to +10 VDC (bipolar).

Current-sensing analog modules are also available. The most common input range is 4 to 20 mA, although other ranges are available.

Some analog modules will accept voltage or current input. These are called combination modules.

Figure 5-21 shows an example of a module properties screen for a analog input module. Note that the input range can be selected. In this example, -10 to 10 VDC was chosen for the input range. This screen also enables you to scale the input values to your application. In this example the values were scaled −10 VDC input (low signal) should equal -10-volt engineering units, and 10 volts input should equal 10 engineering units. Note also in this example that we are configuring channel 0.

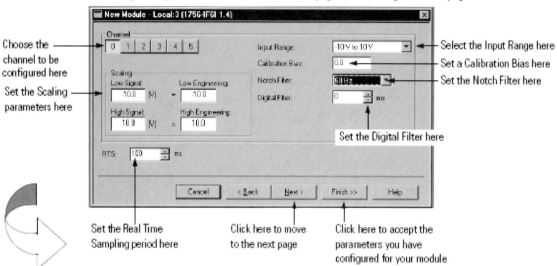

Figure 5-21 Configuring channels in an analog module. (Courtesy of Rockwell Automation, Inc.)

Scaling

Scaling enables you to modify the input or output from a module. Scaling can only be used in the floating-point data format in ControlLogix analog I/O modules. When you scale a channel, you must choose two points along the module's operating range and apply low and high values to those points. For example, imagine an application where you are using an input module that has a 0- to 21-mA range capability, but your sensor is 4–20 mA (see Figure 5-22). You can scale the module so that 4 mA is the low signal and 20 mA is the high signal. Scaling allows you to configure the module to return data to the controller so that 4 mA returns a value of 0 percent in engineering units and 20 mA returns a value of 100 percent in engineering units.

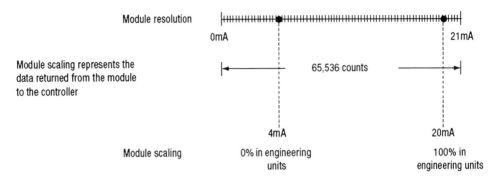

Figure 5-22 Scaling an analog module input. (Courtesy of Rockwell Automation, Inc.)

Analog Resolution

Resolution has to do with how closely something can be measured. Imagine a ruler. If the only graduations on the ruler were inches, the resolution would be 1 inch. If the graduations were every 1/8 inch, the resolution would be 1/8 inch. The closest we could measure any object would be 1/8 inch. The CPU in a PLC only works with digital information. The analog-to-digital (A/D) card changes the analog source into discrete steps. The higher the resolution, the finer the measurement. Another way to think of resolution is in terms of a pie. If you have people over for Thanksgiving the pie will be divided on the basis of the number of people. The pie represents what we are measuring (maybe 0–10 volts); the number of people represents the size of each piece of pie. So the higher the number of people (bits of resolution), the smaller each piece is.

Resolution is the smallest amount of change that a module can detect. Analog modules are available in different resolutions. Output modules are typically available in 13–16-bit resolution. A 16-bit module would have 65,536 counts. This can be calculated by raising 2 to the number of bits the module has. For example a 16-bit module would be 2^{16}, or 65,536.

Figure 5-23 shows an example of resolution. In this example the module's input range is 0–21 mA. The module is 16 bits; this means that there are 65,536 counts. If we divide the 21 mA by 65,536, we get the measurement resolution. In this example the resolution is 0.0003204 mA.

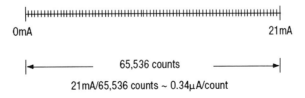

Figure 5-23 Resolution for a specific module. (Courtesy of Rockwell Automation, Inc.)

Field Calibration

ControlLogix analog I/O modules allow you to calibrate each channel individually or module-wide. Modules can be calibrated in RSLogix 5000 software.

Calibration is done to make sure a device is making accurate measurements. Analog input and output modules are calibrated to improve the module's accuracy and repeatability. Calibration procedures are different for input and output modules. Analog modules come calibrated from the factory but they can also be recalibrated. They can also be calibrated so that they meet the requirements of a particular application.

Analog I/O modules can be calibrated individually or with the channels grouped together. Calibration is done to correct any hardware inaccuracies that may be present on an I/O channel.

A calibration procedure is designed to compare a known accurate standard with the actual I/O channel's performance. A linear correction factor between the measured and the ideal is then calculated and applied to the channel. The calibration correction factor is applied on every input or output to obtain maximum accuracy.

You must be online to calibrate your analog I/O modules through RSLogix 5000. When you are online, you can use Program or Run Mode as the state of your program during calibration. Program Mode is preferred. The module should be in Program Mode and not be controlling a process when it is calibrated.

To calibrate input modules, you must provide accurate known current, voltage, or ohm values to the module. When you calibrate output modules, you use a calibrated digital multimeter (DMM) to measure the output from the module. The table in Figure 5-24 shows the recommended accuracies for calibration instruments. If you calibrate your module with an instrument that is less accurate than those recommended in Figure 5-24, the following may occur:

- Calibration appears to be normal but the module gives inaccurate data during operation.
- A calibration fault may occur, which forces you to abort calibration.
- The calibration fault bits are set for the channel you attempted to calibrate. The bits remain set until a valid calibration is completed. In this case, you must recalibrate the module with a more accurate instrument.

Modules	Recommended instrument ranges
1756-IF16 & 1756-IF8	0- to 10.25-V source +/−150 V voltage
1756-IF6CIS	1.00- to 20.00-mA source +/−0.15 A current
1756-IF6I	0- to 10.00-V source +/−150_V voltage 1.00- to 20.00-mA source +/−0.15 A current
1756-IR6I	1.0 and 487.0_ resistors(1) +/− 0.01%
1756-IT6I & 1756-IT6I2	−12- to 78-mV source +/−0.3 V
1756-OF4 1756-OF8	DMM with accuracy better than 0.3 mV or 0.6 A
1756-OF6VI	DMM with resolution better than 0.5 mV
1756-OF6CI	DMM with resolution better than 1.0 A

Figure 5-24 Recommended accuracies for calibration equipment for specific modules.

Analog modules freeze the state of each channel and do not update the controller with new data until after calibration ends. This can be dangerous if active control is attempted during calibration.

Typical Calibration Procedure

1. Connect a voltage calibration instrument to the module.
2. Go to the Calibration page in RSLogix 5000 (see Figure 5-25).
3. Choose the channels to be calibrated.
4. Set the calibrator for the low reference signal and apply it to the module. The screen should display the status of each channel after calibration for the low reference. If all channels are OK, continue with the calibration. If there is an error, repeat step 4.
5. Set the calibrator for the high reference signal and apply it to the module (see Figure 5-26). This screen should display the status of each channel after calibration for the high reference. If all channels are OK, continue. If there is an error for any channel, repeat step 5.

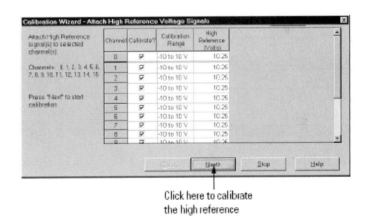

Figure 5-25 Calibration screen. (Courtesy of Rockwell Automation, Inc.)

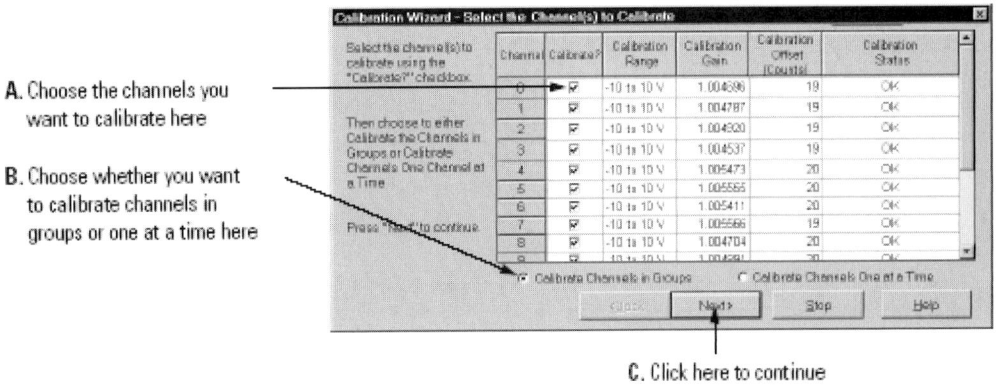

Figure 5-26 Calibration screen. (Courtesy of Rockwell Automation, Inc.)

Sensor Offset

You can add an offset value to inputs or outputs during calibration. An offset enables you to you to compensate for any I/O offset errors that may exist in an application. For example, offset errors are common in thermocouple applications.

Setting Alarms

Figures 5-27 and 5-28 show alarm configuration screens. The configuration screen in Figure 5-28 has been labeled with descriptions. Note that four alarms can be set: Low-Low; Low; High; and High-High. You can also set a deadband or a rate alarm. You may also disable alarms or latch the alarms. Note that each channel can be configured separately.

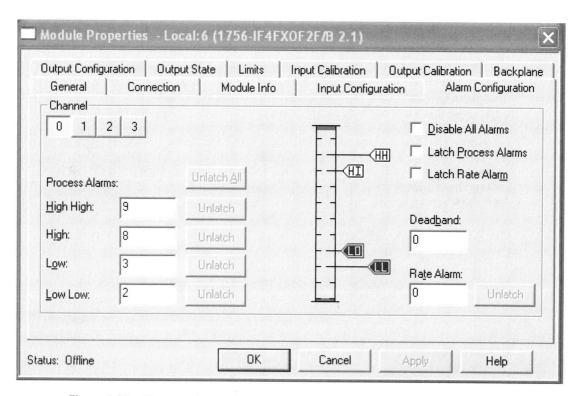

Figure 5-27 Alarm configuration screen. (Courtesy of Rockwell Automation, Inc.)

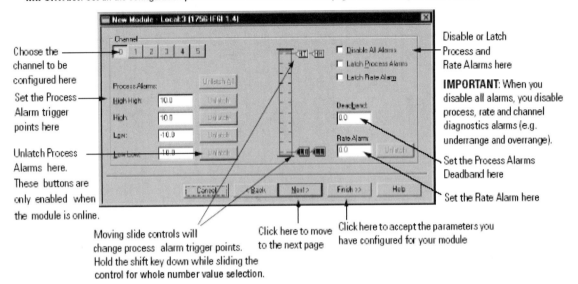

Figure 5-28 Screen to configure alarms with explanations. (Courtesy of Rockwell Automation, Inc.)

Figure 5-29 shows the wiring for an analog input module. Note the shield around the signal wires. Also note that it is only grounded at one end. It is normally grounded to the chassis at the control end.

Single-Ended Inputs

Single-ended analog is an electric connection where one wire carries the signal and another wire or shield is connected to electric ground. All of the analog input commons are tied together.

Single-ended wiring compares one side of the signal input to ground. This difference is used by the module to generate the digital data for the controller. In addition to the common ground, the use of single-ended wiring maximizes the number of usable channels on an analog module.

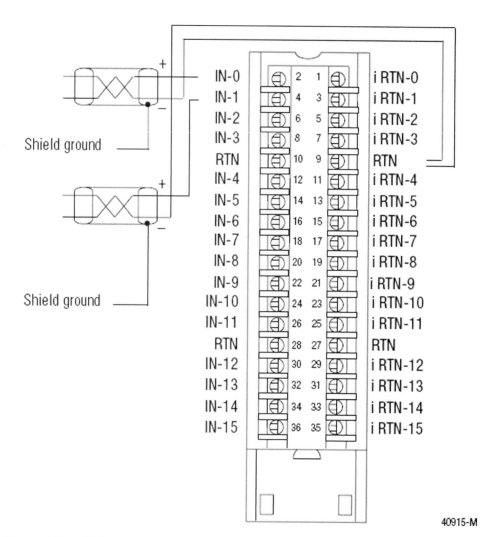

Figure 5-29 Wiring diagram for a Rockwell Automation analog input module. This module has been wired with single-ended inputs. (Courtesy of Rockwell Automation, Inc.)

Differential Inputs

The use of differential wiring is recommended for applications in which it is advantageous or required to have separate signal pairs. Differential wiring improves the noise immunity of a signal. Figure 5-30 shows an example of differential wiring. Note that each analog input has two leads to the module inputs. The first analog input is connected to terminals IN-0 and IN-1. None are tied together to ground. This is called differential wiring.

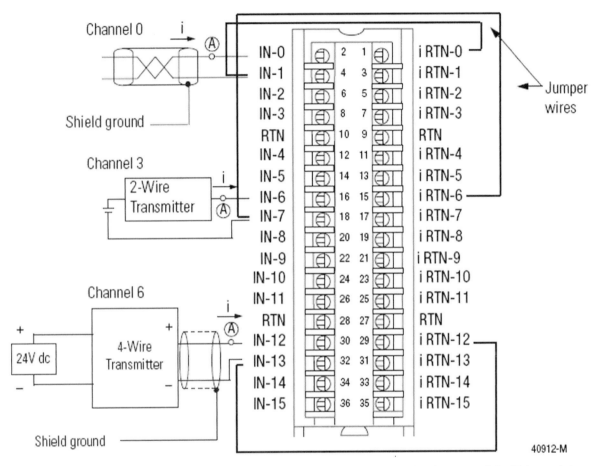

Figure 5-30 Wiring diagram for a Rockwell Automation analog input module. This module has been wired with differential inputs. (Courtesy of Rockwell Automation, Inc.)

If differential wiring is used you can only use of half of a module's channels. For example, you can only use eight channels on a 1756-IF16 module and four channels on a 1756-IF8 module.

High-Speed-Mode Differential Wiring Method

Some analog modules can be configured for a high-speed mode that will provide the fastest data updates. The high-speed mode can only be used with differential wiring. If the high-speed mode is used only one out of every four channels on the module can be used.

Analog Data Format

The user can determine whether the data returned from the module to the owner-controller will be integer or floating point. This is done when you choose a Communications Format for the module during configuration. The integer mode uses a 16-bit signed format and allows faster sampling rates and uses less controller memory. The use of integer mode limits the availability of certain features of an analog module. Check the manual for the specific module for more information.

Analog Output Modules

Analog output modules are used to convert digital values to analog output signals. Analog output modules are available with voltage or current output. Typical outputs are 0 to 10 volts, −10 to +10 volts, and 4 to 20 mA.

The RPI value for an analog output module tells the controller when to broadcast the output data to the module. If the module resides in the same chassis as the owner-controller, the module receives the data almost immediately after the controller sends it (see Figure 5-31).

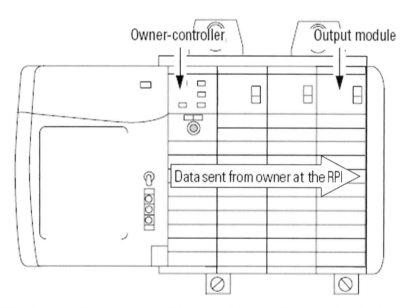

Figure 5-31 Broadcasting output data. (Courtesy of Rockwell Automation, Inc.)

If an output module resides in a remote chassis, the role of the RPI changes slightly with respect to getting data from the owner-controller, depending on which type of network is being used to connect to the modules.

When remote analog output modules are connected to the owner-controller via a scheduled ControlNet network, the controller multicasts the output data within the local

CHAPTER 5—INPUT/OUTPUT MODULES AND WIRING   **109**

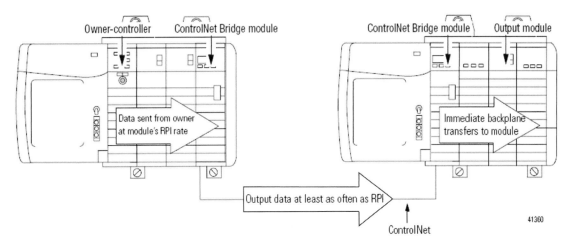

Figure 5-32 Use of a ControlNet network. (Courtesy of Rockwell Automation, Inc.)

chassis, and the RPI also reserves a spot in the stream of data flowing across the ControlNet network (see Figure 5-32).

The timing of this reserved spot in the ControlNet stream may or may not coincide with the exact value of the RPI, but the control system will guarantee that the output module will receive data at least as often as the specified RPI.

Remote Output Modules Connected via EtherNet/IP

When an EtherNet/IP network is used to connect remote analog output modules to the owner-controller, the owner-controller multicasts data within its own chassis at the RPI rate. The Ethernet module in the local chassis then immediately sends the data over the network to the analog output module as long as it has not sent data within a time frame that is 1/4 the value of the analog module's RPI.

Output Resolution

Most modules are capable of 16-bit resolution. The 16 bits represent 65,536 counts.

Scaling

Scaling is used to change a quantity from one notation to another. Scaling is only available with the floating-point data format in CL modules. When a channel is scaled, two points along the module's operating range are chosen and low and high values are applied to the points.

For example, if you are using a module that has a 0- to 21-mA range, but your device requires a 4- to 20-mA signal, you can scale the signal to meet the requirement (see Figure 5-33). The 4 mA would be set as the low signal and 20 mA as the high signal. Scaling enables the module to be configured so that 4 mA has a value of 0 percent in engineering units and 20 mA has a value of 100 percent in engineering units.

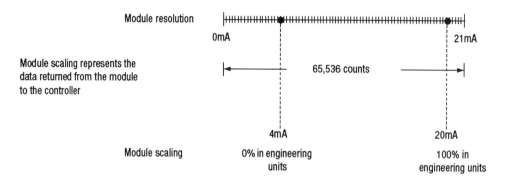

Figure 5-33 Scaling. (Courtesy of Rockwell Automation, Inc.)

When you choose the low- and high-point values for your application, it does not limit the range of the module. The module's range and its resolution remain constant regardless of how the module is scaled for the application. You can choose integer mode or floating-point mode for an analog module.

Integer Mode

Integer mode provides the most basic representation of analog data. Scaling is not available in integer mode. The low signal of your application range equals −32,768 counts while the high signal equals 32,767 counts. Output modules allow you to generate an analog signal at the terminals that correspond to a range from −32,768 to 32,767 counts (see Figure 5-34).

Module	Range	Low Signal and User Counts	High Signal and User Counts
1756-OF4/OF8	0–20 mA	0 mA = 0 counts	21.2916 mA = 32767
	+/− 10 V	−10.4336 V = −32768 counts	10.4336 V = 32767 counts
1756-OF6CI	0–20 mA	0 mA = 0 counts	21.074 mA = 8192 counts
1756-OF6VI	+/− 10 V	−10.517 V = −8192	10.517 V = 8192 counts

Figure 5-34 Selected analog output ranges and counts.

Differences between Integer and Floating-Point Modes

The main difference between choosing integer or floating-point mode is that integer is fixed between −32,768 and 32,767 counts and floating-point mode provides scaling to represent I/O data in specific engineering units to match your application.

Ramping/Rate Limiting

Ramping limits the speed at which an analog output signal can change. This prevents fast transitions in the output that could damage some devices. The maximum rate of change in outputs is expressed in engineering units per second and called the maximum ramp rate. Figure 5-35 shows three types of CLX ramping.

Ramp Type	Occurrence
Run mode ramping	Module is in the run mode and begins operation at the configured maximum ramp rate when it receives a new output level (only available in floating-point mode).
Ramp to program mode	Present output value changes to the program value after a program command is received from the controller.
Ramp to fault mode	The present output value changes to the fault value after a communications fault occurs.

Figure 5-35 Ramping.

Analog Output Module Wiring

Figure 5-36 shows the wiring for a typical analog output module. Note that this example is for a current output. Note the shield around the signal wires. Note also that this module is also capable of outputting voltage.

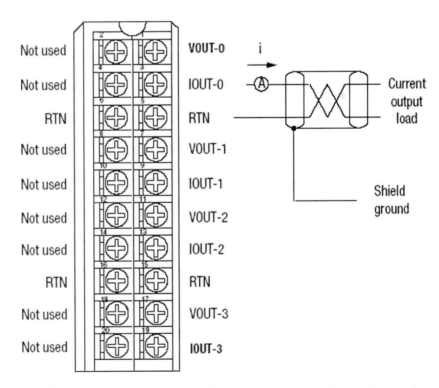

Figure 5-36 Typical analog output wiring. (Courtesy of Rockwell Automation, Inc.)

QUESTIONS

1. Who owns an I/O module?
2. How is the ownership of a module established?
3. Can more than one controller get I/O information about a module?
4. Do modules have to reside in the same chassis of the controller to be owned by that controller?
5. What is opto-isolation?
6. What is an RTB, and what is its purpose?
7. Why would you key a CLX module?
8. Describe how CLX modules are keyed.
9. Which two networks can be used for remote I/O?
10. Explain the term *resolution*.
11. If a 16-bit input module is used to measure the level in a tank and the tank can hold between 0 and 15 feet of fluid, what is the resolution in inches?
12. What is the purpose of calibrating analog modules?
13. Describe a typical procedure for calibrating an analog module.
14. What are the four alarms that can be set for analog inputs?
15. What is a differential input?
16. What is scaling?
17. A 1756-OF4/OF8 module will be used to output a 4–20-mA output to control a valve. Look up the module to find its resolution. Calculate the counts that would be used to output 4 and 20 mA. Calculate the resolution in milliamperes/count.
18. A 1756-OF6VI module will be used to output a 0–10 VDC output to control a valve. Look up the module to find its resolution. Calculate the counts that would be used to output 0 and +10 VDC. Calculate the resolution in volts/count.

CHAPTER 6

Industrial Sensors

OBJECTIVES

On completion of this chapter, the reader will be able to:

- Describe the typical uses of various types of industrial sensors.
- Choose appropriate sensors for various applications.
- Explain terminology such as *sourcing, sinking, range, hysteresis, light on, dark on, normally open, normally closed, load powered, line powered,* and so on.
- Explain the wiring of two- and three-wire sensors.
- Explain how capacitive and inductive sensors function.

INTRODUCTION

World competition is forcing industries to automate. Automation must be fast and flexible to compete. Programmable control devices such as PLCs and programmable automation controllers can be integrated with industrial sensors to create smart, flexible systems.

Automatic operation increases the need for safety devices to ensure the safety of the operator. Many sensors have been developed for safety applications—laser scanners to make sure no one enters a cell, and special interlocks to make sure safety doors cannot be opened during cell operation, for example. Sensors can also be used to check for the presence or absence of parts, to measure size or proper fill, and so on.

Mechanical Switches

A simple limit switch is an example of a mechanical sensor. When the lever on the switch is moved, the switch changes state (see Figure 6-1). The part contacting the switch creates a change in state that the PLC can monitor. Mechanical devices are less reliable than electronic devices. Mechanical devices are prone to failure. Contacts wear and can weld together. Mechanical devices are also slow to actuate. It takes time to open and close the contacts. The contacts can also chatter on opening or on closing, giving many on and off signals instead of one crisp, timely change of state.

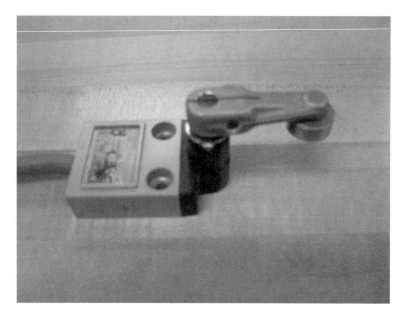

Figure 6-1 A mechanical limit switch.

Electronic Sensing

Electronic devices operate much more quickly than mechanical devices. Electronic sensors can perform at very high production rates. This is very important in many automated processes. Mechanical switches are too slow in many cases. Electronic sensors can detect objects without physically touching the object. Mechanical sensors must contact the object to be sensed. Electronic sensors are also much more dependable. They can operate for a much longer time before failure. This is important to keeping production running. Breakdowns are very expensive. It takes time to find and remedy problems.

Digital sensors are on/off devices. They are sometimes called discrete sensors because there are two discrete states: on or off. If a discrete sensor senses an object, the output changes state. Discrete sensors usually use transistors for the output.

Analog sensors can provide much more information than digital sensors. Analog sensors are also called linear output sensors. Sensors with an analog output have a variable output that is proportional to the input. For example, if we have a sensor that can sense temperatures between 0 and 500 degrees Fahrenheit, a common one might then have an output that would be between 4 and 20 mA depending on the temperature. An analog sensor can provide an output that enables the controller to determine the exact state of the input to be measured.

There is a need for digital and analog sensors in industrial applications. Digital sensors are more widely used because of their cost, simplicity, and ease of use. There are, however, applications that require more information about a process.

Optical Sensors

Optical sensors sense objects with light. They are typically called photosensors. A photosensor has a light source (emitter) and a photodetector to sense the presence or absence of light. LEDs are used for the light source. An LED is a semiconductor diode that emits light. The type of material used for the LED determines the wavelength of the emitted light. LEDs can be turned on and off at extremely high speeds. They are able to keep up with the high speeds required in production applications. LEDs are also reliable, small, and energy efficient. LEDs operate in a narrow wavelength and are not sensitive to shock, temperature, or vibration. They also have a very long life.

Photosensors are very immune to ambient light. The photoemitter and photoreceiver are both tuned to a common frequency. The photodetector essentially ignores ambient light and looks for the correct frequency. The frequencies chosen are typically invisible to the human eye. Manufacturers choose wavelengths so that the sensors are not affected by other lighting in the plant. Color mark sensors use the ability to detect different wavelengths to differentiate between colors. Visible sensors are usually used for this purpose.

Some applications utilize ambient light for sensing. Red-hot materials such as metal or glass emit infrared light. Photoreceivers that are sensitive to infrared light can be used in these applications to sense temperature.

Types of Optical Sensors

There are three general types of photosensors: reflective, retro-reflective, and thru-beam. They all function in the same basic way. The type differences are based on the way in which the light source (emitter) and receiver are housed in the sensor.

Reflective Sensors

One of the common types of optical sensors is the reflective type. They are also called diffuse sensors. The emitter and receiver are housed in the same unit (see Figure 6-2). The emitter sends out light, which reflects off the product to be sensed. The receiver senses the reflected light. Reflective (diffuse) sensors have less sensing distance (range) than other types of optical sensors because they rely on reflected light from the product.

The light emitter and receiver are in the same housing. When the light from the emitter reflects off an object, it is sensed by the receiver and the output of the sensor changes state. The broken-line style of the arrows in Figure 6-2 represents the pulsed mode of lighting, which is used to assure that ambient lighting does not interfere with the application. Because the sensing distance (range) is limited by how well the light reflects off the product, reflective photosensors have the shortest sensing range of the three types.

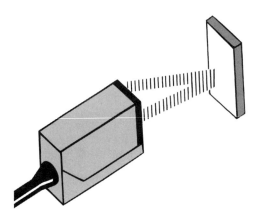

Figure 6-2 Reflective sensor. (Courtesy ifm efector inc.)

Retro-reflective Sensors
The retro-reflective sensor is similar to the reflective (diffuse) sensor except that a reflector is used to reflect the emitted light back to the receiver (see Figure 6-3). The emitter and receiver are both housed in the same package. A retro-reflective sensor bounces the light off a reflector instead of the product. The reflector is similar to the reflectors used on bicycles but is of a higher quality.

If an object obstructs the beam, the output of the sensor changes state. A retro-reflective sensor has a longer sensing range than a reflective sensor because the reflector is more efficient at returning light than an object. The broken line in Figure 6-3 represents the pulsed method of lighting that is used.

Figure 6-3 Retro-reflective sensor. (Courtesy ifm efector inc.)

Thru-Beam Sensors
The thru-beam photosensor is the third type of configuration (see Figure 6-4). In this configuration the emitter and receiver are housed separately. The emitter sends out light through a space, and the light is sensed by the photoreceiver. If an object passes between the emitter and receiver, it prevents the light from arriving at the receiver and the sensor knows there is product present. This is probably the most reliable sensing mode for nontransparent objects.

The emitter and receiver are housed separately. The broken line in the figure symbolizes the pulsed mode of the light that is used in optical sensors. Thru-beam sensors have the longest sensing range.

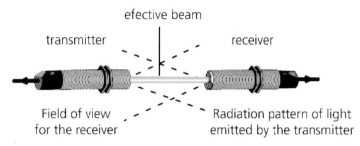

Figure 6-4 Thru-beam sensor. (Courtesy ifm efector inc.)

Fiber-Optic Sensors
Fiber-optic sensors can be purchased in the same configurations as other photosensors: reflective , retro-reflective, and thru-beam. Fiber-optic cables are very small and flexible. They are clear strands of plastic or glass fibers that are used as light pipes. The light from the emitter passes through the fiber and exits from the other end. The light enters the end of the fiber attached to the receiver, passes through the cable, and is sensed at the receiver.

One of the main advantages of fiber-optic sensors is that they can be used in applications where there is very little room. The fiber cable can be extremely small, and the electronics for the sensor can be mounted at a different location where more space is available. Another advantage is that, being very small, the fiber cable can direct the light so that small objects can be sensed. Figure 6-5 shows a fiber-optic sensor being used to check for resistor leads. In this example, the fibers are being used in a thru-beam mode.

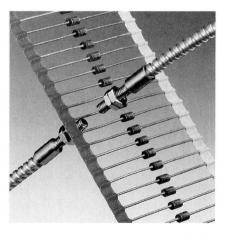

Figure 6-5 Fiber-optic photosensor used to sense very small objects in a tight area. (Courtesy ifm efector inc.)

There are many types of fiber-optic cables available. One type is the bifurcated cable. Figure 6-6 shows the ends of a bifurcated cable. In a bifurcated cable one end has two cables, one to attach to the emitter and the other to attach to the receiver. At the other end of the cable the two are combined into one cable. The emitting fibers and the receiving fibers are run in the same end.

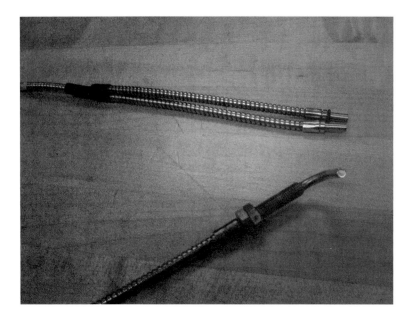

Figure 6-6   Bifurcated fiber-optic cable.

Light/Dark Sensing

The terms *light-on* and *dark-on* are often used to describe the outputs on a photosensor. Dark-on means that the output is on when there is no light at the sensor's receiver. Light-on means that the sensor's output is on when there is light at the receiver. Dark-on is also called dark-operate. Light-on is also called light-operate. Photosensors are available in either light (light-on) or dark (dark-on) sensing. In fact, some sensors can be switched between light and dark modes. Figure 6-7 shows how these terms apply to a thru-beam sensor.

CHAPTER 6—INDUSTRIAL SENSORS

Thru-beam "dark operate" mode
output on when target is present

Thru-beam "light operate" mode
output on when target is not present

Figure 6-7 Dark-on and light-on for thru-beam-style sensor. (Courtesy ifm efector inc.)

Special Photosensors

Polarizing Photosensors

A polarizing photosensor is used for special applications such as sensing shiny objects or sensing clear plastic film. This sensor uses a special polarizing reflector. The reflector has small prisms that polarize the light from the sensor. The sensor emitter emits horizontally polarized light.

The polarizing reflector vertically polarizes the light and reflects it back to the sensor's receiver. For example, if a shiny object moves between the sensor and reflector and reflects light back to the sensor, the light will be ignored because it is not vertically polarized (see Figure 6-8).

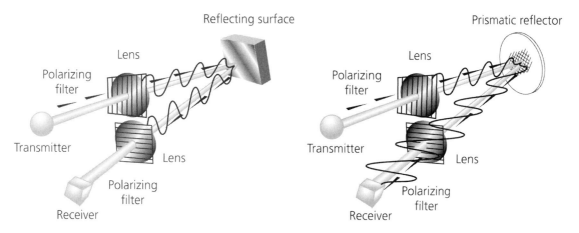

Figure 6-8 Polarizing sensor. (Courtesy ifm efector inc.)

Convergent Photosensors

Convergent photosensors are a special type of reflective sensor. They direct the emitted light to a specific area in space (focal point or field of view). The light must be reflected from that area (focal point) to reach the receiver (see Figure 6-9). If the object to be sensed is not in the field of view or focal point, the light will not be returned to the receiver. Convergent photosensors are good for sensing for objects that must be in a particular area in space.

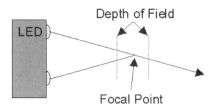

Figure 6-9 A convergent-style photosensor.

Laser Sensors

Laser sensors utilize laser LEDs as their light source. Laser light is very coherent and can provide a very narrow beam of light. Laser sensors are used to sense very small objects because of their coherent beam. Resolution can be as small as a few microns.

Laser sensors are also used to sense distance. Laser sensors can be used to make very accurate measurements. The output from a laser sensor can be analog or digital. Digital outputs can be used to signal pass/fail or other indication. The analog output can be used to make actual measurements. Figure 6-10 shows a laser sensor that is used for measurement. This sensor can make accurate distance measurements from 0.2 to 10 meters.

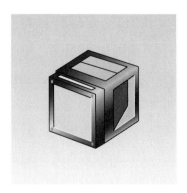

Figure 6-10 Laser measurement sensor.

Inspection Photosensors

Sensors have been developed to provide the capabilities of object recognition and object inspection. One example of an object recognition and inspection sensor is the efector dualis object recognition sensor. Figure 6-11 shows an object recognition and inspection sensor. The efector dualis object sensor provides 100 percent inspection testing. Many vision and inspection tasks are quite simple and do not require a full-blown vision system. Simple, inexpensive, easy to program devices are being developed. This particular inspection sensor has a resolution of 640 × 480 pixels. It can provide 100 percent inspection in an application. Objects can also be detected and evaluated, regardless of orientation, by the object sensor.

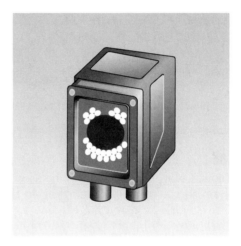

Figure 6-11 A efector dualis object recognition sensor.

The efector dualis sensor is a compact sensor. The image sensor has an integrated infrared LED lighting source. The sensor's lighting provides the correct amount of brightness at close range. For longer distances, external light sources can be used. The evaluation electronics are built into the sensor.

The sensor can be set up using a menu-guided software package on a computer over Ethernet. Programming essentially consists of teaching the sensor what a good part looks like. The sensor detects and compares defined shapes and provides up to five configurable outputs that include counting, sorting, logic functions, and pass/fail. Figures 1-12 and 1-13 show examples of inspections. The object sensor also has an LED display that indicates active outputs. The software enables the sensor to find objects with extreme accuracy and speed.

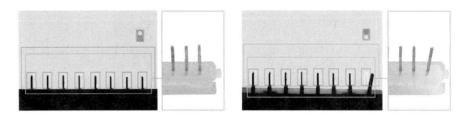

Figure 6-12 Inspection to check for bent or missing pins on a chip. (Courtesy ifm efector inc.)

Figure 6-13 Inspection to check for the correct part profile. (Courtesy ifm efector inc.)

Color Mark Sensors

Color mark sensors are a special type of optical sensor. A color mark sensor can differentiate between colors. Color mark sensors can be used to check labels and so on.

Color mark sensors are chosen according to the color that needs to be sensed. Color mark sensors work by detecting the contrast between two colors. The background color (behind the object) is an important consideration in any color mark application. Sensor catalogs typically have charts available to select the proper color mark sensor for various colors.

Incremental Encoders

Encoders can be used for position feedback and also for velocity feedback. The most common type of encoder is incremental (see Figure 6-14). The resolution of an encoder is determined by the number of lines on the encoder disk. The more lines there are, the higher the resolution. LEDs are used as light sources in encoders. Light shines through the lines on the encoder disk and a mask and is sensed by light receivers (phototransistors). While the disk turns, the receivers sense the pulses of light as they become visible and invisible.

An incremental encoder produces a series of square wave pulses as it turns. The number of square waves in one turn of the shaft determines the resolution of the encoder. Incremental encoders rotate a disk in the path of a light source. The disk acts as a shutter to alternately block or transmit the light to photodetectors.

The resolution of an encoder is equal to the number of lines on the encoder's disk. An encoder that has resolution of 500 will have 500 lines on it, and one turn of the encoder shaft will produce 500 complete square wave cycles. This would be 500 square wave pulses for 360 degrees of rotation. An encoder disk is shown in Figure 6-15.

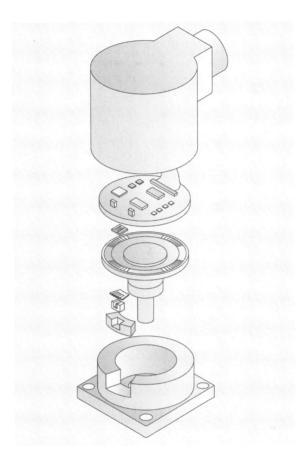

Figure 6-14 An exploded view of an encoder. (Courtesy BEI Industrial Encoders.)

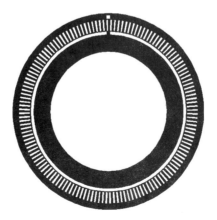

Figure 6-15 An incremental encoder disk. (Courtesy BEI Industrial Encoders)

Quadrature Encoders
Most encoders use two channels for sensing the position. A third channel is used for establishing the home position. The two channels are called the A and the B channel. When light passes through the slits, as the disk rotates, the channel produces a pulse. The pulses generate a square wave.

The transitions from high to low and low to high on the square wave can be sensed by a controller. The LEDs used to sense the A and B signals are offset so that the B signal lags the A signal by 90 degrees (see Figure 6-16).

The index track has only one pulse per revolution. This is usually called a zero or index pulse. This is used for establishing the home position.

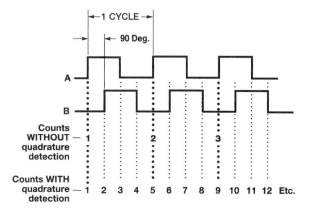

Figure 6-16 Encoder pulses for the A and B signals. (Courtesy BEI Industrial Encoders)

Study Figure 6-16. If just the rising edge of the A pulse were used, there would be one count for every pulse (without quadrature detection). If the rising and falling edge of the A and B pulse were counted, there would be four counts for every cycle. This would be quadrature detection. So, if the encoder had 500 pulses per revolution, the encoder's resolution could be 4 times greater ($4 \times 500 = 2000$).

Direction Sensing
Pulses from the A and B channels can be fed to a PLC input card. Many PLC high-speed input modules have the capability of accepting encoder input.

The A and B channels are used to determine the direction of rotation and position. One channel is used as a reference. The direction of rotation can be determined by whether the A or B channel's output signal leads or lags (see Figure 6-16).

Single-Ended versus Differential Wiring
Encoders are available in two wiring types: single ended and differential. In single-ended encoders, there is one wire for each ring (A, B, and X), a wire for +5 VDC, and a ground wire (0 VDC) that is used as a common for all of the signals. In a single-ended encoder, the pulses have a 5-volt (0–5 VDC) difference.

The differential encoder is more immune to noise because the rings do not share a common ground. The differential encoder has two outputs for each channel.

The A channel, for example has an A+ voltage and an A- voltage. For a 5-VDC encoder, this results in a difference of 10 volts between a high and a low (+5 to −5 volts). Because of its noise immunity, the differential type is more common in industry.

Absolute Encoders

The absolute encoder provides a word of output with a unique pattern for each position. A diagram of an absolute encoder disk is shown in Figure 6-17. The LEDs and receivers are aligned to read the disk pattern (see Figure 6-18). There are several types of coding schemes that can be used for the disk pattern. The most commonly used patterns are gray, natural, binary, and binary-coded decimal (BCD). Gray code is popular because it is a nonambiguous code. Only one track changes at a time. Any indecision that occurs during an edge transition is limited to plus or minus one count. If the output changes while it is being read, a latch option locks the code to prevent ambiguity.

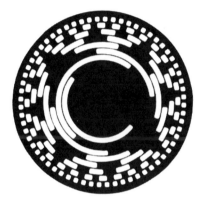

Figure 6-17 A disk from an absolute encoder. (Courtesy BEI Industrial Encoders.)

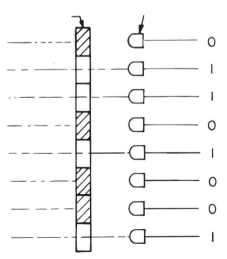

Figure 6-18 How photodetectors read the position of an absolute encoder. (Courtesy BEI Industrial Encoders.)

Incremental encoders can provide more resolution at a lower cost than absolute encoders. Incremental encoders have a simpler interface because they have fewer output lines. A simple incremental encoder would have four lines: two quadrature (A and B) signals and power and ground lines. A 13-bit absolute encoder would have 13 output wires plus 2 power lines. If the absolute encoder had complementary outputs, the encoder would require 28 wires.

Field Sensors

The two types of field sensors are capacitive and inductive. Both sensors are very similar. Field sensors have a coil that is used to generate a field in the front of the sensor. Figure 6-19 shows a block diagram of a field sensor. Capacitive sensors can be used to sense any material. Inductive sensors can only sense metallic objects.

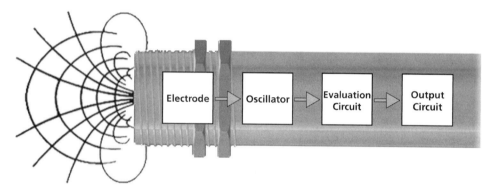

Figure 6-19 Block diagram of field sensor components. (Courtesy ifm efector inc.)

Figure 6-20 shows a special purpose capacitive-type sensor. This sensor can be used to measure the level in a tank.

Hysteresis in Field Sensors
Hysteresis is a term for sensors that means there is an on point and a separate off point. The sensor output will not turn on until the target crosses the on point. It will then stay on until the target moves away and crosses the off point. Figure 6-21 shows a diagram of hysteresis. Note the difference between the turn-on and turn-off points. Imagine an object moving horizontally above the sensor in the sensing range. The object would first enter the turn-off point of the field. The output would not change state. As the object continued to move, it would enter the turn-on point and the output would change state. For this example let's assume the output is now on. The object would continue to move and the trailing edge of the object would leave the turn-on point on the other side of the sensor. The output would not change state. As the object continued to move, the trailing edge of the object would leave the turn-off point and the output would change state (to off).

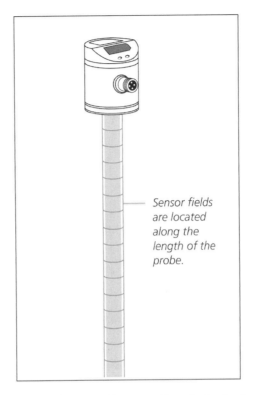

Figure 6-20 Capacitive tank level sensor. (Courtesy ifm efector inc.)

Hysteresis is a benefit in sensing. It prevents an object from teasing the sensor. In other words, imagine a bottle moving down a conveyor line. Without hysteresis it might be vibrating as it passed the sensor and might be sensed multiple times. Hysteresis assures that it is only counted once.

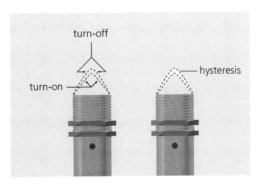

Figure 6-21 Sensor hysteresis. (Courtesy ifm efector inc.)

Sensing Range
Sensing range is primarily dependent on the size of the coil in the sensor. This means that the larger the diameter of the sensor, the larger the sensing range. Figure 6-22 illustrates the sensing range for three different-diameter sensors.

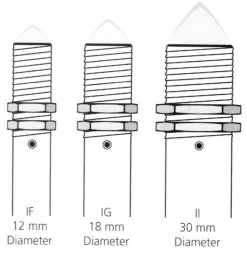

Figure 6-22 Note the difference in sensing range from the smallest diameter of sensor to the largest. (Courtesy ifm efector inc.)

The target and the target material can also affect sensing range. Figure 6-23 shows how range is affected by part size. If the part is the smallest as shown, the range will only be 25 percent of the specified range.

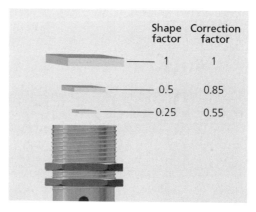

Figure 6-23 How size of the object can affect sensing range. (Courtesy ifm efector inc.)

In inductive sensors the material also affects sensing range. Ferrous (iron-based) metals are sensed the best. Nonferrous metals such as aluminum are not sensed as well so the sensing range must be reduced.

Shielding
A regular field sensor generates a field that is not just on the end of the sensor. The field extends out from the side of the sensor also. This means that it can sense objects on the side of the sensor. This is usually not desirable and can cause problems. The sensor for example might sense the fixture that it is mounted in.

Sensors are available in shielded and nonshielded styles. Shielded sensors are also called flush sensors because they can be mounted flush in their mounting and not sense the mount. Nonshielded sensors are called nonflush (see Figure 6-24). A shielded sensor has a brass or copper ring around the outside of the coil. It prevents the field generated by the sensor from going out to the side. This reduces the sensing range, but it does allow them to be flush mounted.

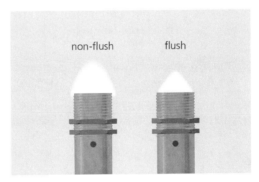

Figure 6-24 Difference in field size for nonflush and flush sensor. (Courtesy ifm efector inc.)

Mounting Field Sensors

Mounting wells are available for field sensors. Figure 6-25 shows a capacitive sensor mounted in a protective well in a tank. The sensor is adjusted to ignore sensing the well and only senses the material in the tank. Mounting wells can protect sensors from corrosive environments.

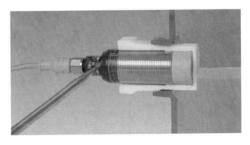

Figure 6-25 A capacitive sensor mounted in a mounting well. (Courtesy ifm efector inc.)

Figure 6-26 shows mounting specifications for a nonshielded field sensor. Note that the sensor must be mounted so that the surrounding material is sufficiently far away from the field. Note also that if there is anything in front of the sensor, it must be sufficiently far away so it does not affect the sensing.

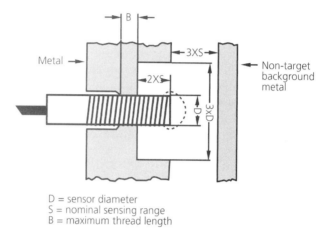

D = sensor diameter
S = nominal sensing range
B = maximum thread length

Figure 6-26 Mounting a nonshielded sensor. (Courtesy ifm efector inc.)

Figure 6-27 shows specifications for mounting shielded sensors. Note that they can be mounted flush with the surface. Note also that less distance is required between shielded sensors.

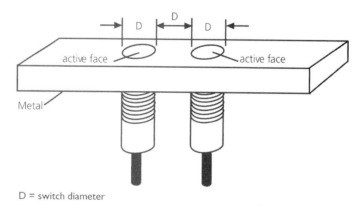

D = switch diameter

Figure 6-27 Mounting a shielded sensor. (Courtesy ifm efector inc.)

Radio Frequency Identification (RFID) Sensors

RFID sensors can help implement closed-loop manufacturing applications including pallet tracking, component identification, intelligent part routing, and assembly verification.

Inductive RFID Technology

An RFID system generates an electromagnetic field for reading and writing data. The electromagnetic field emitted by the antenna induces voltage in the passive identification (ID) tag (transformer principle). This activates the ID tag (transponder), which returns its code. The read/write module processes the code and sends the transmission to the interface networking system or to a ControlLogix process or through a special purpose module. Figure 6-28 shows an example of a RFID reader and a tag. The tag is shown on the lower left of the figure. The reader is on the upper right. Tags can be installed on the product, fixtures, pallets, and so on. Tags can be read from or written to.

Figure 6-28 RFID reader and a tag. (Courtesy ifm efector inc.)

Pressure Sensors

Strain Gauges

Strain gauges are used for pressure/force sensing. They are very versatile and have a wide variety of uses. They are based on the principle that a thin wire has more resistance than a thick wire.

If an elastic wire were stretched, the diameter of the wire would decrease in the middle of the wire. The decrease in wire diameter would increase the resistance of the wire. If we measured the change in resistance, we could relate it to the force applied to the wire. The change in resistance is proportional to the change in the force that is applied. Assume that a constant current is sent though the strain gauge. When a force is applied, the resistance of the strain gauge changes. The constant current and the change in resistance produce a voltage change that can be measured. Thus a change in force can be made proportional to the voltage change.

A transmitter is often used with a strain gauge. The transmitter is designed for the strain gauge. The strain gauge provides the current for the strain gauge and monitors the change in resistance. The transmitter outputs a current or voltage signal that is proportional to the force exerted on the strain gauge.

Strain gauges used in pressure measurement are typically bonded to a membrane. Pressure strain gauges are mounted so that a change in pressure distorts the membrane proportionally with the force. Strain gauges can also used to measure weight or acceleration.

Strain gauges must be mounted in the correct orientation because they are only sensitive to change in one direction. An arrow is typically used on a strain gauge to show which direction it should be mounted in. Adhesives are commonly used to mount strain gauges.

Pressure is another example of something that often requires analog readings. Pressure sensors are available with digital and analog output. They provide a range of output voltage (or current), proportional to the pressure. Figure 6-29 shows pressure sensors on the top. These pressure sensors provide an output and also a digital readout for the operator. The diagram on the top right and at the bottom of Figure 6-29 shows the sensor's components. Note that the sensor is programmable. It is programmed with switches on the front of the sensor. The display provides feedback. The diagram at the bottom shows the pressure-sensing cell.

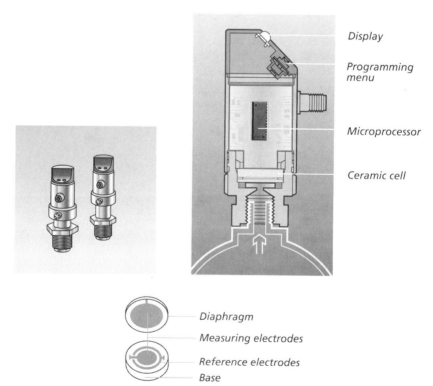

Figure 6-29 On the top left are two pressure sensors. On the top right is a diagram showing the components of a pressure sensor. The diagram at the bottom shows the pressure-sensing cell. (Courtesy ifm efector inc.)

Flow Sensors

There are many types of sensors that can be used to measure flow. Flow sensors are very important in process control. This chapter will only examine two of them.

Magnetic Inductive Flow Meters
The operating principle of a magnetic flow sensor is based on Faraday's law of electromagnetic induction. The fluid to be sensed must be conductive. Faraday's law states that the voltage induced across any conductor as it moves at right angles through a magnetic field is proportional to the velocity of that conductor. The sensor generates a magnetic field around the fluid. The flow of an electronically conducting fluid through a defined pipe diameter is detected by the sensor and used to determine the velocity of the fluid.

The magnetic flow sensor shown in Figure 6-30 has a display that includes a numeric display that indicates the flow rate (gal/min or l/min), total volume (gallons or liters), and temperature (°F or °C) of all conductive media. The sensor can provide a switching output, analog output, and pulsed output. The application parameters are programmed through push buttons on the front of the sensor. Magnetic flow sensors are accurate and repeatable.

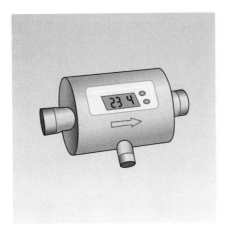

Figure 6-30 Magnetic flow meter.

Calorimetric Principle Flow Sensors
A calorimetric flow sensor detects the cooling effect of a flowing fluid or gas to monitor the flow rate of a fluid. Figure 6-31 shows the principle of operation. The diagram on the left shows that there are two transistors and a heating element in the tip of the sensor.

The tip of the probe is heated. As the fluid begins to flow, heat will be carried away from the sensor tip as shown in Figure 6-31. In the middle figure there is no flow, so both transistors sense the same temperature. In the diagram on the right there is flow, so there is a temperature difference between the transistors. The faster the flow, the more difference in temperature between the transistors. The difference in temperature between the two transistors provides a measurement of the flow.

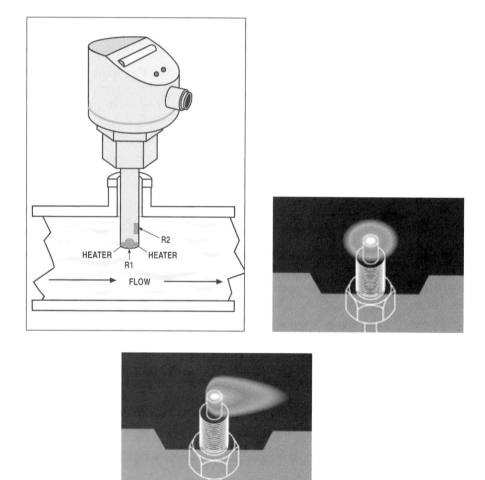

Figure 6-31 Calorimetric flow sensing. (Courtesy ifm efector inc.)

Ultrasonic Flow Sensors
Ultrasonic flow sensors utilize sound waves to measure flow rate. Ultrasonic technology is based on the differential transit time principle. Sound pulses are alternately emitted and detected with and against the direction of flow through the use of sound transducers.

The flow rate is calculated from the difference of the transit time–the time it takes for the sound wave to be transmitted and received. Figure 6-32 shows a diagram of the operation of an ultrasonic flow sensor.

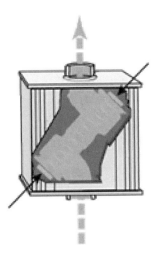

Figure 6-32 Ultrasonic sensing. (Courtesy ifm efector inc.)

The length of time it takes for the sound to return is proportional to the distance. The ultrasonic sensor can then output a signal to represent the distance to the object. Ultrasonic sensors can be very accurate. Note the use of the gate sensor to notify the PLC when a part is present.

Temperature Sensors

Thermocouples

The thermocouple is one of the most common temperature sensors. Thomas J. Seebeck discovered the principle of thermocouples in 1821.

Temperature is typically analog information. A thermocouple is a very simple sensor: two pieces of dissimilar metal wire joined at one end. The other ends of the wire are connected via compensating wire to the analog inputs of a control device such as a PLC. The principle of operation is that when dissimilar metals are joined, a small voltage is produced. The voltage output is proportional to the difference in temperature between the cold and hot junctions. Thermocouples are colorcoded for polarity and also for type. The negative terminal is red, and the positive terminal is a different color that can be used to identify the thermocouple type. Figure 6-33 shows a typical thermocouple.

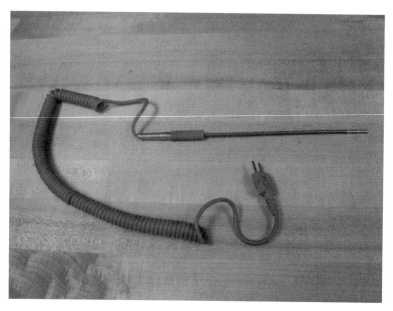

Figure 6-33 Common thermocouple.

Current output is often used for process measurement and control. A current signal has many advantages in industrial measurement and control. A 4–20-mA current loop system can be used when the sensor needs to be mounted a long distance from the control device. A 4–20-mA loop is useful to about 800 meters. Current signals are more noise immune also. Process sensors are often connected to transmitters. A transmitter takes the small signal from the sensor and converts it to a voltage or current signal. Current signals are very common. Figure 6-34 shows a thermocouple transmitter.

Assume this 4–20-mA thermocouple transmitter varies its output between 4 and 20 mA. There must be an adjustment on the sensor transmitter to adjust the zero and span so that the sensor transmitter can be calibrated. Remember the example of the thermocouple transmitter that should output 4–20 mA on the basis of a 0–200-degree-Fahrenheit temperature. The zero adjustment is used to make sure the transmitter outputs 4 mA for a 0-degree signal. The span adjustment is used to make sure that the transmitter outputs 20 mA for 200 degrees. Note that when you make these adjustments, you must always check, adjust, recheck, and readjust to make sure both are accurate after each is adjusted.

For this example, assume the thermocouple transmitter outputs 4–20 mA for temperatures between 0 and 200 degrees Fahrenheit.

Figure 6-34 A thermocouple transmitter.

The output from the analog sensor can be any value in the range from 4 to 20mA based on the 0–200-degree temperature. Thus the PLC can use the signal from the thermocouple transmitter to monitor temperature very accurately and closely control a process.

Industrial thermocouple tables use 75 degrees Fahrenheit for the reference temperature.

Temperatures vary considerably in an industrial environment. If the cold junction varies with the ambient temperature, the readings will be inaccurate. This would be unacceptable in most industrial applications. It is too complicated to try to maintain the cold junction at 75 degrees. Industrial thermocouples must therefore be compensated. This is normally accomplished with the use of resistor networks that are temperature sensitive. The resistors that are used have a negative coefficient of resistance. Resistance decreases as the temperature increases. This adjusts the voltage automatically so that readings remain accurate. PLC thermocouple modules automatically compensate for temperature variation.

The thermocouple is an accurate device. The resolution is determined by the device that takes the output from the thermocouple. The device is normally a PLC analog module. The typical resolution of an industrial analog module is 14 bits; 2 to the 14th power is 32,768. This means that if the range of temperature to be measured were 1200 degrees Fahrenheit, the resolution would be 0.03662109 degrees/bit (1200/32768 = 0.0367); this would mean that our PLC could tell the temperature to about 4 hundredths of 1 degree.

Thermocouples are the most widely used temperature sensors. There are a wide variety of thermocouples available. Figure 6-35 shows a few of the available types. This figure also shows the composition of the thermocouples.

Type	Materials	Temperature Range (degrees Fahrenheit)	Temperature Range (degrees Celsius)
B	Platinum and rhodium	32 to 3272	0 to +1,800
E	Chromel and constantan	−310 to 1832	−190 to +1000
J	Iron and constantan	−310 to 1472	−190 to +800
K	Chromel and alumel	−310 to 2498	−190 to +1370
R	Platinum or platinum and 13 % rhodium	32 to 3092	0 to +1700
S	Platinum or platinum and 10 % rhodium	32 to 3209	0 to +1765
T	Copper and constantan	−310 to 752	−190 to +400

Figure 6-35 Temperature ranges for some common thermocouples.

Resistive Temperature Devices (RTDs)

An RTD is a sensor that changes resistance with a change in temperature. RTDs are more accurate than thermocouples.

An RTD a precision resistor that is temperature sensitive. RTDs are made from a pure metal that has a positive temperature coefficient.

Platinum is the most popular material for RTDs. Platinum has a very linear change in resistance versus temperature. Platinum RTDS also have a wide operating range. Platinum is very stable; this makes the RTD a very stable device.

The most common resistance for an RTD is 100 ohms.

Wiring RTDs

An RTD is basically a two-wire device. The lead wires from the RTD affect the accuracy of the RTD. RTDs can also be purchased in a three- or four-wire configuration (see Figure 6-36). Three- and four-wire RTDs compensate for lead wire resistance and are more accurate.

Figure 6-36 RTD wiring configurations.

Thermistors
A thermistor is a temperature-measuring sensor. Thermistors are semiconductors that are constructed from human-made materials. Thermistors are very precise and stable. They have a negative coefficient of temperature. This means that the resistance decreases as temperature increases. Thermistors produce a large change in resistance for changes in temperature. The output of thermistors is not very linear; it is only linear within a small temperature range. A thermistor is a good choice if the range of temperature to be measured is relatively small. Thermistors are more sensitive than RTDs. They are often used in motor applications to monitor temperature. This can then be used to shut off the motor circuit if the temperature gets too high. Thermistors cannot typically be used above 300 degrees Celsius.

Sensor Wiring

Sensors are available in two- and three-wire types. Two-wire sensors are called load powered and three-wire sensors are called line powered.

Load-Powered Sensors
Load-powered sensors are two-wire sensors. One wire is connected to power; the other wire is connected to one of the load's wires (see Figure 6-37). The load represents whatever device is being used to monitor the output. The load is usually a PLC input. The load must limit the output current to an acceptable level for the sensor, or the sensor output will be destroyed.

Wiring diagrams are usually located on the sensor or its leads.

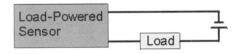

Figure 6-37 Load-powered sensor.

In a load-powered sensor the current required for the sensor to operate must pass through the load. Think of the load as being a PLC input. A small current must be allowed to flow to enable the sensor to operate. This operation current is called a leakage current.

The leakage current to operate the sensor is typically under 2.0 mA.

The leakage current is sufficient to operate the sensor but not enough to turn on the input of the PLC. (The leakage current is usually not enough current to activate a PLC input. If it is enough current to turn on the PLC input, it will be necessary to connect a bleeder resistor as shown in Figure 6-38.) When the sensor turns on, it allows enough current to flow to turn on the PLC input.

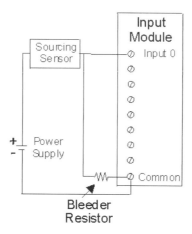

Figure 6-38 Use of a two-wire (load-powered) sensor. In this case the leakage current was enough to cause the input module to sense an input when there was none. A resistor was added to bleed the leakage current to ground so that the input could not sense it.

Line-Powered Sensors

Line-powered sensors typically have three-wires (see Figure 6-39), two wires for power and one wire for the output signal. Figure 6-39 shows the load. The load is connected to the output signal. The load is normally a PLC input. The load must limit the output current to an acceptable level.

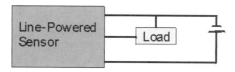

Figure 6-39 Three-wire (line-powered) sensor.

The load current is the output from the sensor. If the sensor is on, there is load current. This load current turns the load (PLC input) on. The maximum load current is typically between 50 and 200 mA for most sensors. Make sure that you limit the load (output) current or the sensor will be ruined. Note that it is possible for the output LED on a sensor to function and the output to still be bad.

Sinking versus Sourcing

The terms *sinking* and *sourcing* confuse even experienced technicians. Sensors are available with outputs that are either positive or negative. A sensor with a positive output is called a sourcing sensor. Sourcing sensors are also called PNP sensors because they utilize a PNP transistor for the output device. So this concept thus far can be relatively straightforward if you remember that: a sourcing sensor is one that has a positive output polarity. (See Figure 6-40.)

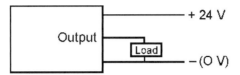

Figure 6-40 A sourcing or PNP sensor. Note that 24 VDC and 0 VDC are connected directly to the sensor to power it. The output wire is connected directly to one side of the load. The other side of the load is connected to the 0-VDC side. Don't forget the importance of the load. The load is usually a PLC input. It must limit the output current to protect the sensor. Assume the output current limit for this sensor is 100 mA. The load must draw less than 100 mA.

The other way to think about this is to remember that if the sensor output is positive, the only way there is any potential voltage is if we connect to the negative side. In other words, if we connected a voltmeter to the output lead and also to 0 VDC, there would be a potential voltage difference. If we connected a meter to the output lead and to 24 VDC, there would be no potential voltage difference. Figure 6-41 shows sinking sensors connected to a sourcing input module.

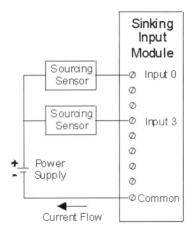

Figure 6-41 A sourcing sensor connected to a PLC input.

NPN (Sinking Type)
When the sensor is off (nonconducting), there is no current flow through the load. When the sensor is conducting, there is a load current flowing from the load to the sensor. The choice of whether to use an NPN or a PNP sensor is dependent on the type of load. In other words, choose a sensor that matches the PLC input module requirements for sinking or sourcing.

A sensor with a negative output is called a sinking sensor. Sinking sensors are also called NPN sensors because they utilize an NPN transistor for the output device. So this can be easily understood if you remember that: a sinking sensor is one that has a negative output polarity. (See Figure 6-42.) Figure 6-43 shows how a sinking sensor is connected to a sourcing input module.

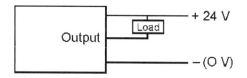

Figure 6-42 A sinking or NPN sensor. Note that 24 VDC and 0 VDC are connected directly to the sensor to power it. The output wire is connected directly to one side of the load. The other side of the load is connected to the 24 VDC. Don't forget the importance of the load. It must limit the output current to protect the sensor. Assume the output current limit for this sensor is 100 mA. The load must draw less than 100 mA.

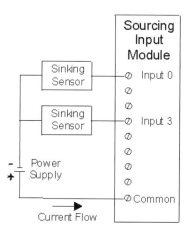

Figure 6-43 A sinking sensor connected to a PLC input.

Output Current Limit
The output current limit for most sensors is quite low. Output current must often be limited to under 100 mA. If you exceed the output current limit for a sensor, the sensor's output will be destroyed. The indicator LED on the sensor may still indicate on or off, but the output may be bad. If the sensor's output is connected to a PLC input, the PLC input will limit the current to a safe amount.

Timing Functions for Outputs
Timing functions are available on some sensors. They are available with on delay and off delay. On-delay delays turning the output on by a user-selectable amount of time after the sensor senses the part. Off delay holds the output on for a user-specified time after the part leaves the sensing area.

Normally Open and Normally Closed Outputs
There are two different types of outputs available for sensors: normally open and normally closed. The output in a normally open sensor is open (off) until the sensor senses

an object. The output in a normally closed sensor is on until it senses an object; then the output turns off.

Switching Frequency
When choosing a sensor, one of the considerations is that the sensor be capable of sensing and switching the output fast enough to meet the demands of the application. In most applications, speed will not be an issue as electronic sensors are very fast. In very high speed applications, however, you must consider the sensor's switching time. Figure 6-44 shows a diagram of switching speed for a sensor. You must consider the time that the target is present and also the time that represents the time between targets.

Response time is the lapsed time between the target being sensed and the output changing state. Response time can be crucial in high-production applications. Sensor specification sheets will give response times.

$$\text{Switching Time } (t_s) = \frac{2 \times \text{Smallest Distance}}{\text{Maximum Target Speed}}$$

$$\text{Switching Frequency } (f_s) = \frac{\text{Maximum Target Speed}}{2 \times \text{Smallest Distance}}$$

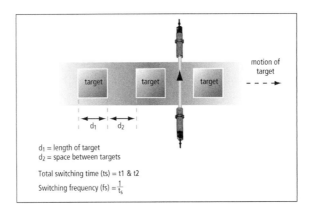

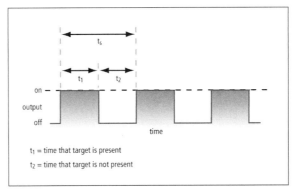

Figure 6-44 Diagram of switching frequency. (Courtesy ifm efector inc.)

Installation Considerations

Electric Mounting Considerations

The most important wiring caution for a sensor is to limit the output load current. The load (output) current must be limited to a very small amount. The output limit is typically 100 mA or less. If the load draws more current than the sensor current limit, the sensor output is blown. You must limit output current to a level less than the sensor's output current limit. PLC input modules draw very little current and will not exceed a sensor's output limit.

Sensor wiring should be run separately from high-voltage wiring and in a metal conduit if the high-voltage wiring is in close proximity to the sensor wiring. This will help prevent false sensing or malfunction. The other main consideration is to specify the proper polarity for the sensors.

Mechanical Mounting Considerations

Sensors should not be mounted vertically. In a vertical position dirt, oil, and chips can accumulate on the sensing surface and cause false reads. If the sensor is mounted horizontally, the debris will have more of a tendency to fall off. Air blasts or oil baths can be used to remove chips and dirt if sensors are prone to accumulating debris.

A sensor must be mounted so that it does not detect its own mount or another sensor. If a sensor is unshielded, it cannot be mounted flush in a fixture. Sensors must not be mounted too close together, as they can interfere with each other.

APPLICATIONS

Figure 6-45 shows a smart level sensor. The sensor's microprocessor and push button are used to teach the empty and full conditions of the container.

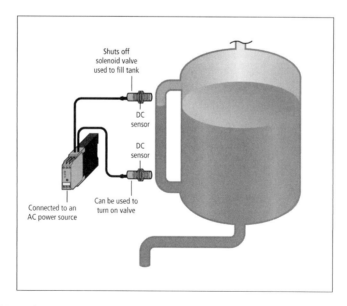

Figure 6-45 Use of two sensors and a sensor controller sensor to sense the high and low level in a container. (Courtesy ifm efector inc.)

Figure 6-46 shows the use of a laser distance sensor to measure the diameter of a paper roll and also to measure the distance of paper coming off the roll to aide the controller with tensioning.

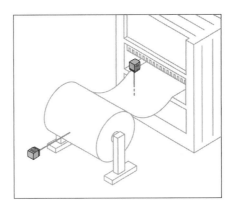

Figure 6-46 Laser distance application. (Courtesy ifm efector inc.)

Figure 6-47 shows the use of sensors to check for the proper number of bottles in a case. As the cases pass by, each of the three sensors should sense a bottle. As the cases move, if one or two sensors sense a bottle and the other one or two do not, bottles are missing.

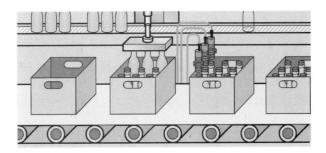

Figure 6-47 Checking for the presence of bottles. (Courtesy ifm efector inc.)

Figure 6-48 shows pneumatic cylinders with sensors to sense the position of the piston in the cylinder.

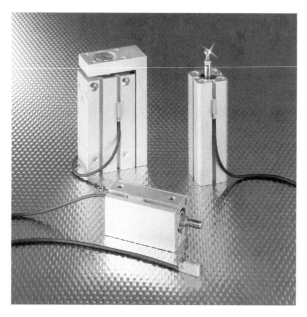

Figure 6-48 Position sensing in a cylinder. (Courtesy ifm efector inc.)

Figure 6-49 shows an application for a polarizing photosensor. The sensor is being used to sense a transparent bottle. A regular photosensor would not work well in this application.

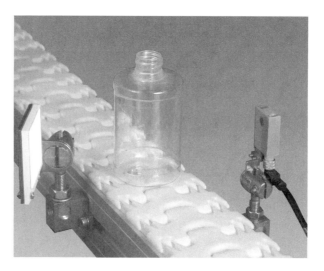

Figure 6-49 A polarizing sensor application. (Courtesy ifm efector inc.)

Figure 6-50 shows a fiber-optic thru-beam sensor being used to check for individual components. The beam from the fiber is obstructed by one of the wires of the electric component.

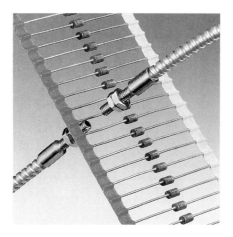

Figure 6-50 Fiber-optic thru-beam sensor application. (Courtesy ifm efector inc.)

In Figure 6-51 a laser sensor in an application is being used to sense very small glass components. Laser sensors have a very small diameter coherent beam so that even very small parts can be sensed.

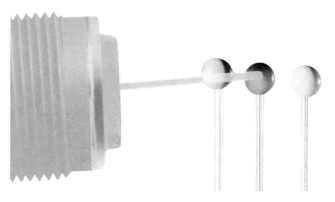

Figure 6-51 Laser sensor application. Note the small size being sensed. (Courtesy ifm efector inc.)

A pressure sensor application is shown in Figure 6-52. A pressure sensor is used to provide feedback to the controller. Pressure sensors are available with digital or analog output.

Figure 6-52 Pressure sensor checking pressure in a line. (Courtesy ifm efector inc.)

In Figure 6-53 an inductive sensor being used to sense the speed of the gear. The sensor senses every tooth as it passes by. The controller divides the number of pulses it receives by the number of teeth on the gear in a period of time to calculate the revolutions per minute (RPM) of the gear.

Figure 6-53 An inductive sensor used to sense teeth on a gear. (Courtesy ifm efector inc.)

Considerations in the Choice of Sensors

There are several important considerations when choosing a sensor.

The characteristics of the object are crucial. Is the material metallic? Is it ferrous (iron-based) metal? Is it nonmetallic? Is it a large object or a very small one? Is the object transparent or reflective? The particular characteristics will exclude many types of sensors and may limit the choice to one particular type.

What type of output is needed: analog or digital? How much accuracy is required? How much range is required from the sensor? Is there excessive electrical noise present, such as in a welding application?

Are there problems with contaminants such as oils or metal chips, sawdust, dirt, and so on? Is the sensing area very small?

Is the speed of response important? Response time is the time between the object being sensed and the output changing state. Some applications require very fast response times.

The answers to these considerations will narrow the possible choices. One sensor can then be chosen on the basis of factors such as the cost of the sensor, the cost of failure, and the reliability of the sensor.

QUESTIONS

1. Describe the output of a digital sensor.
2. Describe the output of an analog sensor.
3. Describe a thru-beam sensor.
4. Describe a retro-reflective sensor.
5. Describe a diffuse sensor.
6. Which type of photosensor has the longest range?
   a. Thru-beam
   b. Diffuse
   c. Retro-reflective
   d. Convergent
7. What does the term *light-on* mean?
8. If a reflective sensor's receiver is not receiving light and the sensor's output is on, is it a light-on or dark-on sensor?
9. If a thru-beam sensor's receiver is receiving light and the sensor's output is on, is it a light-on or dark-on sensor?
10. Explain the principle of a field sensor.
11. Explain hysteresis.
12. Draw and explain the wiring of a load-powered sensor.
13. Draw and explain the wiring of a sourcing line-powered sensor.

14. True or false: A sinking sensor has an output that has positive polarity.
15. What is leakage current?
16. What is the importance of limiting the load current?
17. Explain the color-coding system for thermocouples.
18. How are changes in ambient temperature compensated for with a thermocouple?
19. What determines which thermocouple type should be chosen?
20. Explain at least two electric precautions as they relate to sensor installation.
21. Explain at least two mechanical precautions as they relate to sensor installation.

CHAPTER

Math Instructions

OBJECTIVES

On completion of this chapter, the reader will be able to:

- Utilize math instructions in programs.
- Explain terms such as *comparison instructions, precedence, logical instruction,* and so on.
- Explain how to determine which instructions are available in each programming language.

INTRODUCTION

Arithmetic instructions are very useful in programming industrial applications. Many types of instructions are available. This chapter will cover many of the more commonly used instructions. CLX has a wide variety of instructions that can ease application development. Once you become familiar with some of the more common instructions, it will be easy to learn new ones.

OPERATION INSTRUCTIONS

Math instructions are very useful in developing automated systems. For example, many times a bit count in memory must be changed to a more useful value for display on an operator screen. Bit counts may not make much sense to an operator, but the actual RPM does. Input values must usually be modified before a value is sent to an analog output module. This chapter will examine some of the math instructions available in RSLogix 5000.

The instruction help file in RSLogix is very useful in explaining each instruction and also showing which languages they are available in.

Add (ADD) Instruction

The ADD instruction is used to add two numbers. The instruction adds these values from Source A and Source B. The source can be a constant value or an address (tag in CL). The result of the ADD instruction is put in the destination (Dest) address (tag in CL).

An example of an ADD instruction is shown in Figure 7-1. If contact Inp_1 is true, the ADD instruction will add the number from Source A (Val_1) and the value from Source B (Val_2). The result will be stored in the Dest address (Val_3). In this example, 12 was added to 14 and the result (26) was stored in Val_3.

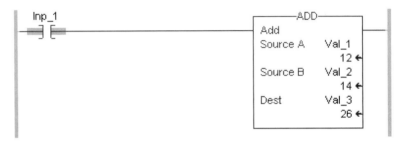

Figure 7-1 Use of an ADD instruction.

Subtract (SUB) Instruction

SUB instructions are used to subtract two numbers. The source of the numbers can be constants or addresses (tags in CL). A SUB instruction subtracts Source B from Source A. The result is stored in the Dest address (tag).

The use of an SUB instruction is shown in Figure 7-2. If the tag named Calculate is true, the SUB instruction is executed. Source B is subtracted from Source A; the result (798) is stored in the Dest tag named Number_Left. Source A in this example is a constant (1250). Source B is a tag named Qty_Parts. Qty_Parts has a value of 452. Note that a constant (1250) was used for Source A. A tag could have been used.

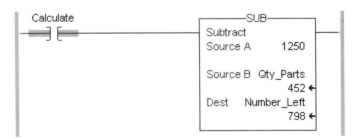

Figure 7-2 Use of a SUB instruction.

Divide (DIV) Instruction

The DIV instruction can be used to divide two numbers. The source of the two numbers can be constants or addresses (tags). When a DIV instruction is executed, Source A is divided by Source B and the result is placed in the Dest tag.

Figure 7-3 shows the use of a DIV instruction. If contact Calculate is true, the DIV instruction will divide the number from source A (569) by the value from source B (12.456). The result is stored in the Dest tag named Answer_Int. Note that the tag is an integer-type tag, so the result was rounded to an integer.

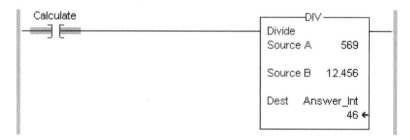

Figure 7-3 Use of a DIV instruction in RSLogix 5000. In this example the result is a whole number because the Dest tag is an Int.

Figure 7-4 shows an example of a division that has a Real-type tag as its destination. In this example, the result is a decimal (Real) number. Note that a constant (569) was used for Source A and a constant (12.456) was used for Source B. Tags could have been used for Source A or Source B.

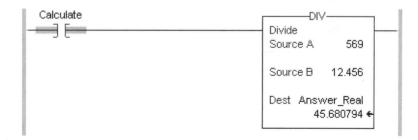

Figure 7-4 Use of a DIV instruction. In this example the result is a decimal number because the Dest tag is a Real.

Multiply (MUL) Instruction

A MUL instruction is used to multiply two numbers. The first number (Source A) is multiplied by the second number (Source B). The result of the multiplication is stored in the Dest address. Source A and Source B can be numbers or tags. Figure 7-5 shows the use of a MUL instruction. If the tag named Calculate is true, Source A (the value in tag

Cases_Prod) is multiplied by Source B (24) and the result is stored in the Dest tag Bottles_Prod. Note that a constant (24) was used for Source B. A tag could have been used.

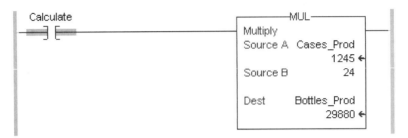

Figure 7-5 Use of a MUL instruction in RSLogix 5000.

Average (AVE) Instruction

The AVE instruction calculates the average of a set of values. The simplest example would be a 1-dimensional array. In the example shown in Figure 7-6, the name of the array is Data_Array. There is only one dimension to this array so 0 is entered for the dimension (Dim) to vary. The result of the AVE instruction will be put into the tag named Ave_Value. The AVE instruction also requires a tag name for a control tag. In this example the tag name is Control_Tag. The Length parameter tells the instruction how many elements to include in the calculation, five in this example. The Position contains the position of the current element that the instruction is accessing.

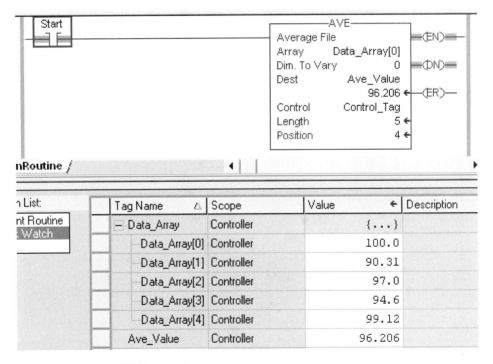

Figure 7-6 Use of an AVE instruction.

An AVE instruction can also be used on a multidimensional array. When used on a multidimensional array, the dimension to vary determines which row or column within the array is used for the calculation. The name of the array specifies the start position, and the Dimension to Vary specifies the direction to use (row or column). DINT in the example is a double integer type tag.

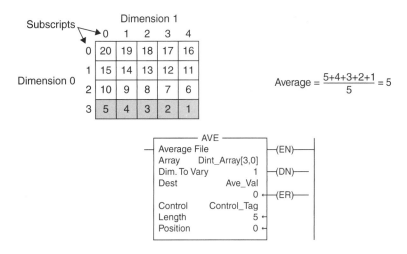

Example of AVE Instruction for Dint_Array Which is a DINT[4,5]

Figure 7-7 Use of an AVE instruction on a 2-dimensional array.

Modulo (MOD) Instruction

The MOD instruction divides Source A by Source B and places the remainder in the Dest. In the example shown in Figure 7-8, two constant values were used for the source values. Tags could have been used. In this example the MOD instruction would divide 8 by 3 for an integer result of 2 with a remainder of 2. The remainder is put in the Dest tag. The Dest tag name is Answer in this example.

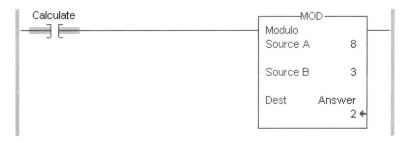

Figure 7-8 Use of a MOD instruction in RSLogix 5000.

Negate (NEG) Instruction

The NEG instruction is used to change the sign of a value. If it is used on a positive number, it makes it a negative number. If it is used on a negative number, it makes it a positive number. Remember that this instruction will execute every time the rung is true. Use transitional contacts if needed. The use of a NEG instruction is shown in Figure 7-9.

If contact Inp_1 is true, the value in source A (Val_1) will be given the opposite sign and stored in the Dest tag (Val_2). Note that in this example Val_1 contained 5 and the NEG instruction changed it to −5 and stored it in Val_2.

Figure 7-9 Use of a NEG instruction.

Square Root (SQR) Instruction

The SQR instruction is used to find the square root of a value. The result is stored in a destination address. The source can be a value or the address of a value. Figure 7-10 shows the use of a SQR instruction. If the tag named Calculate is true, the SQR instruction will find the square root of the Source (64). The result will be stored in the tag at the Dest (tag named Answer in this example). Note that a tag could have been used for the Source.

Figure 7-10 Use of a SQR instruction in RSLogix 5000.

Compute (CPT) Instruction

The CPT instruction can be used for copy, arithmetic, logical, and number conversion operations. The operations to be performed are defined by the user in the Expression and the result is written in the Dest. Operations are performed in a prescribed order. Operations of equal order are performed left to right. Figure 7-11 shows the order in which operations are performed. The programmer can override precedence order by using parentheses.

Operation	Precedence
( )	1
ABS, ACS, ASN, ATN, COS, DEG, FRD, LN, LOG, RAD, SIN, SQR, TAN, TOD, TRN	2
**	3
− (Negate), not	4
*, /, MOD	5
− (Subtract), + (Add)	6
AND	7
XOR	8
OR	9

Figure 7-11 Precedence (order) in which math operations are performed. When precedence is equal, the operations are performed left to right. Parentheses can be used to override the precedence order.

Figure 7-12 shows the use of a CPT instruction. The mathematical operations in the CPT are performed when the tag named Calculate is true. In this example a formula is used to convert a Fahrenheit temperature to a Celsius temperature. The result of the equation is put into the Dest (tag named Answer). The tag Temperature held a value of 212. When 212 is put into the equation, the result is 100, which the CPT instruction put into the tag named Answer. Note that parentheses were used to assure that the subtraction was done before the multiplication and division.

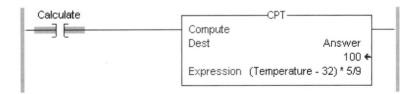

Figure 7-12 Use of a CPT instruction in RSLogix 5000.

It might be noted that the reverse of this formula, Temperature * 9/5 + 32, gives an incorrect answer if Temperature is an integer even though the formula is correct. This is because the processor uses integer arithmetic since all parts are integers. Changing to Temperature * 9.0/5 + 32 gives the correct answer, since 9.0 is a real.

RELATIONAL INSTRUCTIONS

Relational instructions have many uses in programming. In most cases they are used to compare two values. They can be used to see if values are equal, if one is larger, if one is between two other values, and so on. Figure 7-13 shows relational instructions.

Instruction	Use
CMP	Compare values based on an expression
EQU	Test whether two values are equal
GEQ	Test whether one value is greater than or equal to another value
GRT	Test to see if one value is greater than a second value
LEQ	Test to see if one value is less than or equal to a second value
LES	Test whether one value is less than a second value
LIM	Test whether one value is between two other values
MEQ	Pass two values through a mask and test if they are equal
NEQ	Test whether one value is not equal to a second value

Figure 7-13 Relational instructions.

Equal To (EQU) Instruction

The EQU instruction is used to test if two values are equal. The values tested can be actual values or tags that contain values. An example is shown in Figure 7-14. Source A is compared to Source B to test to see if they are equal. In this example Source A is equal to Source B so the instruction is true and the output is on.

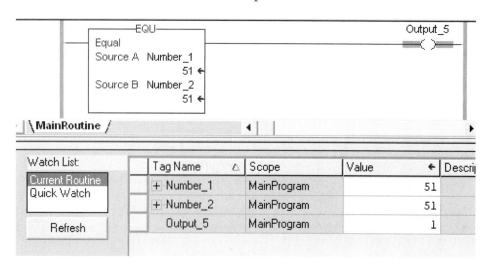

Figure 7-14 Use of an EQU instruction.

Greater Than or Equal To (GEQ) Instruction

The GEQ instruction is used to test two sources to determine whether Source A is greater than or equal to Source B. The use of a GEQ instruction is shown in Figure 7-15.

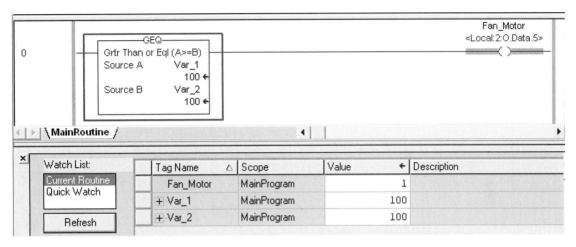

Figure 7-15 Use of a GEQ instruction.

Greater Than (GRT) Instruction

The GRT instruction is used to see if a value from one source is greater than the value from a second source. An example of the instruction is shown in Figure 7-16. If the value of Source A (Var_1) is greater than the value of Source B (Var_2), the output tag named Fan_Motor will be energized.

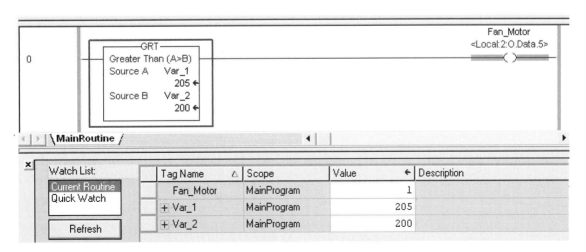

Figure 7-16 Use of a GRT instruction

Less Than (LES) Instruction

Figure 7-17 shows an example of a LES instruction. A LES instruction can be used to check if a value from one source is less than the value from a second source. If the value of Source A (Var_3) is less than the value of Source B (Var_2), output Fan_Motor will be energized.

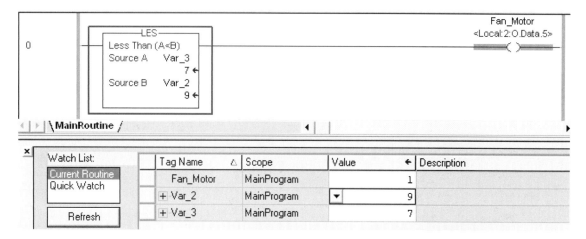

Figure 7-17 Use of a LES instruction.

Limit (LIM) Instruction

LIM instructions are one of the more useful comparison instructions. One LIM instruction can do the work of two comparison instructions. A LIM instruction is used to test a value to see if it falls between two other values. The instruction is true if the test value is between the low and high values.

The programmer must provide three pieces of data to the LIM instruction when programming. The first is a low limit, which can be a constant or an address that contains the desired value. The address will contain an integer or floating-point value. The second is a test value, which is a constant or the address of a value that is to be tested. If the test value is within the range specified, the rung will be true. The third value is the high limit, which can be a constant or the address of a value.

Figure 7-18 shows the use of a LIM instruction. If the value in Cycle_Timer.ACC is greater than or equal to the Low Limit value (0) and less than or equal to the High Limit value (10000), the rung will be true and the output Fan_Motor will be energized. Note that tags could have been used for the Low Limit and High Limit parameters.

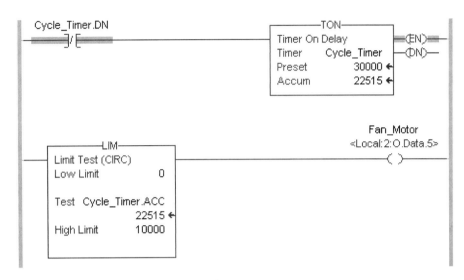

Figure 7-18 Use of a LIM instruction in RSLogix 5000.

Not Equal To (NEQ) Instruction

The NEQ instruction is used to test two values for inequality. The values tested can be constants or addresses that contain values. An example is shown in Figure 7-19. If Source A (Var_1) is not equal to Source B (5), the instruction is true and the output Fan_Motor (<Local:2:0.DATA.5>) is energized.

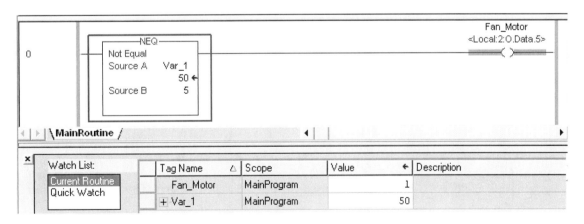

Figure 7-19 Use of a NEQ instruction.

Compare (CMP) Instruction

The CMP instruction performs a comparison on the arithmetic operations that the programmer specifies in the Expression. The execution of a CMP instruction is slightly slower than other comparison instructions. A CMP instruction also uses more memory than other comparison instructions. The advantage of the CMP instruction is that it allows the programmer to perform complex comparisons in one instruction. Figure 7-20 shows the use of a very simple comparison in a CMP instruction. Note that an EQU comparison instruction would have been quicker.

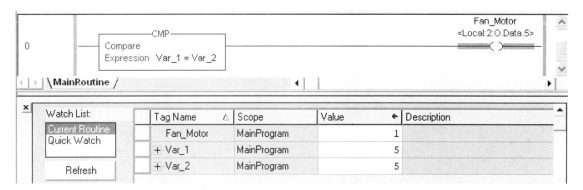

Figure 7-20 Use of a CMP instruction for a simple comparison.

Figure 7-21 shows the use of a CMP instruction to make a more complex comparison in a formula. If the comparison is true, the output coil will be energized.

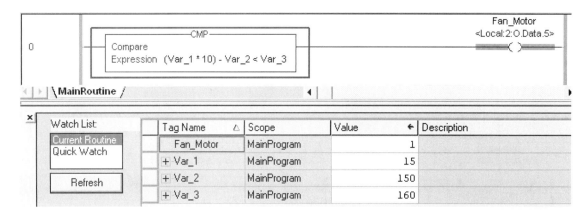

Figure 7-21 Use of a CMP instruction for a complex comparison in a formula.

LOGICAL INSTRUCTIONS

There are several logical instructions available. They can be very useful to the innovative programmer. They can be used, for example, to check the status of certain inputs while ignoring others.

AND Instruction

The AND instruction is used to perform an AND operation using the bits from two source addresses. The bits are ANDed and a result occurs and is stored in a third address. Figure 7-22 shows an example of an AND instruction. Source A (Var_1) and Source B (Var_2) are ANDed. The result is placed in the Dest tag Var_3. Examine the bits in the source addresses so that you can understand how the AND instruction produced the result in the Dest.

Source A	Source B	Result
0	0	0
1	0	0
0	1	0
1	1	1

Figure 7-22 Results of an AND instruction on the four possible bit combinations.

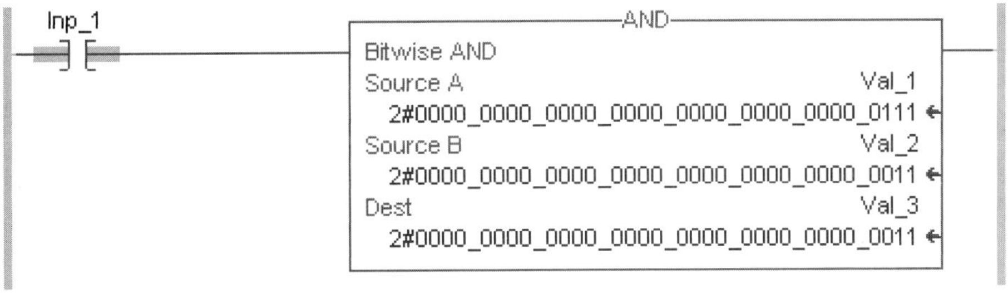

```
 Inp_1                                              -AND-
--] [--                  Bitwise AND
                         Source A                              Val_1
                            2#0000_0000_0000_0000_0000_0000_0000_0111
                         Source B                              Val_2
                            2#0000_0000_0000_0000_0000_0000_0000_0011
                         Dest                                  Val_3
                            2#0000_0000_0000_0000_0000_0000_0000_0011
```

Figure 7-23 Result of an AND instruction on two source addresses. The ANDed result is stored in Val_3.

NOT Instruction

NOT instructions are used to invert the status of bits. A 1 is made a 0, and a 0 is made a 1. Figure 7-24 shows the result of a NOT instruction on bits.

Source	Result
0	1
1	0

Figure 7-24 Result of a NOT instruction. Note that a 0 becomes a 1 and a 1 becomes a 0.

Figure 7-25 shows the use of a NOT instruction. If Sensor_1 is true, the NOT instruction executes. Every bit in the number in tag Var_1 is NOTed (1's complemented).

The result is stored in the Dest tag Var_2. Note that all 0s from the source were changed to 1s and all 1s were changed to 0s.

Figure 7-25 Use of a NOT instruction.

OR Instruction

Bitwise OR instructions are used to compare the bits of two numbers. Figure 7-26 shows how each Source A bit is ORed with each Source B bit.

Source A	Source B	Result
0	0	0
1	0	1
0	1	1
1	1	1

Figure 7-26 Result of an OR instruction on bit states.

See Figure 7-27 for an example of how an OR instruction functions. In this example Source A (Var_1) and Source B (Var_2) are ORed and the result is put in the Dest tag Var_3.

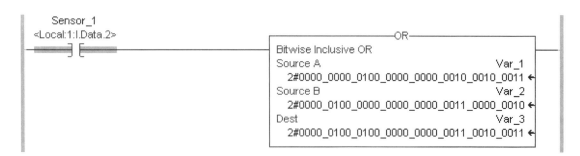

Figure 7-27 Use of an OR instruction.

TRIGONOMETRIC INSTRUCTIONS

There are a variety of trigonometric instructions available. Figure 7-28 shows some trigonometric instructions. The trigonometric instructions utilize radians in the calculations.

The radian is a unit of plane angle, equal to 180/π degrees, or about 57.2958 degrees. It is the standard unit of angular measurement in mathematics.

The radian is abbreviated as rad. For example, an angle of 1.5 radians would be written as 1.5 rad. Figure 7-28 shows how angles in degreees can be converted to rads and how rads can be converted to degrees.

> To convert from degrees to radians multiply by PI/180°
>
> For example $1° = 1 * \dfrac{PI}{180°} = 0.175$ rad
>
> To convert from radians to degrees multiply by 180/PI
>
> For example $1 \text{ radian} = 1 * \dfrac{180°}{PI} = 57.2958°$

Figure 7-28 Conversion of degrees to radians and radians to degrees. Note that there are CLX instructions to make these conversions.

The table in Figure 7-29 shows trigonometric instructions.

Instruction	Use
SIN	Takes the sine of the source value (in radians) and stores the result in the destination
COS	Takes the cosine of the source value (in radians) and stores the result in the destination
TAN	Takes the tangent of the source value (in radians) and stores the result in the destination
ASN	Takes the arc sine of the source value and stores the result in the destination (in radians)
ACS	Takes the arc cosine of the source value and stores the result in the destination (in radians)
ATN	Takes the arc tangent of the source value and stores the result in the destination (in radians)

Figure 7-29 Trigonometric instructions.

SIN Instruction

The SIN instruction takes the Sine of the Source value (in radians) and stores the result in the Dest. To find the Sine of 45 degrees, we would first need to convert 45 degrees into radians. From the formula in Figure 7-28, we could calculate that 45 degrees would be equal to 0.785375 radians. Figure 7-30 shows the use of a SIN instruction. Note that 0.785375 was entered for the Source value. This is 45 degrees in radians. The result of the SIN instruction is put into the Dest tag (Answer). The Sine of 45 degrees is 0.707.

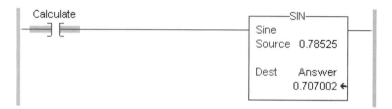

Figure 7-30 Use of a SIN instruction.

We could utilize math instructions to do the degrees-to-radians conversion. Figure 7-31 shows a CPT instruction that is used to convert degrees to radians. The result of the CPT instruction is then put into the Dest tag (Rads). The value of Rads was then used as the Source value in the SIN instruction. Note that there are CLX instructions available to do these conversions.

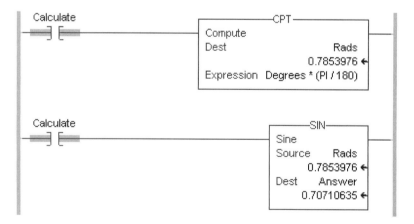

Figure 7-31 Use of a CPT and a SIN instruction. The CPT is being used to convert degrees to radians. The SIN instruction then finds the sine of the angle. The CPT instruction was used here to illustrate the use of a formula in logic.

MATH CONVERSION INSTRUCTIONS

There are several math conversion instructions available. The table in Figure 7-32 shows math conversion instructions.

Instruction	Use
DEG	Converts radians to degrees
RAD	Converts degrees to radians
TOD	Converts an integer to a BCD
FRD	Converts a BCD to an integer
TRN	Removes the fractional part of a value

Figure 7-32 Math conversion instructions.

There are many other math instructions available to the programmer. If you study the examples in the chapter, you should be able to use any of the other instructions.

QUESTIONS

1. Explain the terms *source* and *Dest*.
2. What is precedence in a math operation?
3. What has the highest precedence in a math operation?
4. Why might a programmer use an instruction that would change a number to a different number system?
5. Explain the following logic:

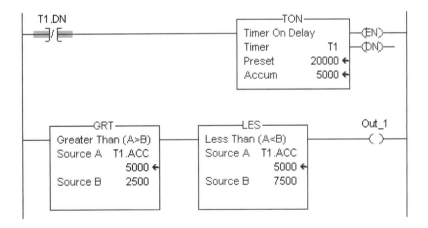

6. Explain the following logic:

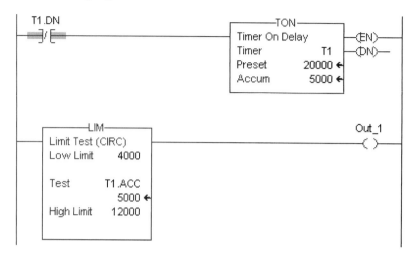

7. Write a rung of ladder logic that would compare two values to see if the first is less than the second. Turn an output on if the statement is true.
8. Write a rung of logic that checks to see if one value is equal to or greater than a second value. Turn on an output if true.
9. Write a rung of logic that checks to see if a value is less than 212 or greater than 200. Turn on an output if the statement is true.
10. Write a rung of logic to check if a value is less than 75 or greater than 100 or equal to 85. Turn on an output if the statement is true.
11. Utilize math instructions and any other instructions that may be helpful to program the following application:

    A machine makes coffee packs and puts 8 in a package. There is a sensor that senses each pack of coffee as it is produced. It would be desirable to show the number of packs that have been produced and the total number of packages of 8 that have been produced.

12. Write a ladder diagram program to accomplish the following:

    A tank level must be maintained between two levels. An ultrasonic sensor is used to measure the height of the fluid in the tank. The output from the ultrasonic sensor is 0 to 10 volts. This directly relates to a tank level of 0 to 5 feet. It is desired that the level be maintained between 4.0 and 4.2 feet. Output 1 is the inflow valve. The sensor output is an analog input to an analog input module. Utilize math comparison instructions to write the logic. Calculate the correct analog counts for the instruction. Use the tag names shown in the table.

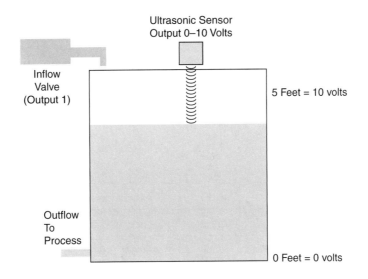

I/O	Tag Name	Description	Analog Counts
Ultrasonic Sensor	Level_Sensor	Analog Output (0–10 Volts)	0–32767
Inflow Valve (Output1)	Input_Valve	On or Off	
Start	Start	Momentary Normally Open Switch	
Stop	Stop	Momentary Normally Closed Switch	
Run	Run	BOOL	

13. Utilize math statements to convert Fahrenheit temperature to Celsius temperature.

    Hint: Tc = (5/9)*(Tf 32), where Tc = temperature in degrees Celsius and Tf = temperature in degrees Fahrenheit

    Utilize math instructions to convert radians to degrees.

14. Write a ladder logic routine for the following application. Utilize comparison instructions.

   This is a simple heat treat machine application. The operator places a part in a fixture, then pushes the start switch. An inductive heating coil heats the part rapidly to 1500 degrees Fahrenheit. When the temperature reaches 1500 degrees, the coil turns off and a valve that sprays water on the part is opened for 10 seconds to complete the heat treatment (quench). The operator then removes the part and the sequence can begin again. Note there must be a part present or the sequence should not start.

I/O	Type	Description
Part_Present_Sensor	Discrete	Sensor used to sense a part in the fixture
Temp_Sensor	Analog	Assume this sensor outputs 0–2000 degrees Fahrenheit
Start_Switch	Discrete	Momentary normally open switch
Heating_Coil	Discrete	Discrete output that turns the coil on
Quench_Valve	Discrete	Discrete output that opens the quench valve

CHAPTER 8

Special Instructions

OBJECTIVES

On completion of this chapter, the reader will be able to:

- Explain some of the special instructions that are available and their use.
- Utilize file instructions.
- Use proportional, integral, derivative (PID) instruction.
- Utilize communication instructions.
- Utilize sequencer instructions.
- Understand the use of special instructions so that new instructions can be quickly learned.

INTRODUCTION

ControlLogix has a wealth of instructions available to perform special functions. This chapter will sample a few of the many instructions that are available. Once you understand how to use a few of them, it will be easy to learn and utilize new ones.

FILE INSTRUCTIONS

Copy (COP) Instruction

COP instructions are used to copy values from one tag array to a different tag array.

One use of a COP command might be a process that can produce ten different kinds of cookies. The recipe for each cookie is different. Each recipe could be in located in a different array in memory. When the operator chooses the product, the correct recipe for that product could be copied into the operating parameters.

There are three entries for a COP instruction: Source, Dest, and Length. Study Figure 8-1. The Source is the tag name of an array that you want to copy. The Dest is the tag name of an array where you would like the data copied. Length is the number of values that you would like to copy. In Figure 8-2 (tag editor) the five (Length) values in Recipe_A[0] through Recipe_A[4] are copied to Production_Recipe[0] through Production_Recipe[4] if the rung becomes true.

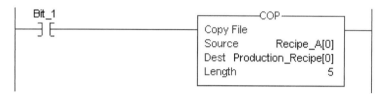

Figure 8-1 A COP instruction copies data from one place in memory to another.

Recipe_A	{...}
Recipe_A[0]	100
Recipe_A[1]	110
Recipe_A[2]	120
Recipe_A[3]	130
Recipe_A[4]	140
Production_Recipe	{...}
Production_Recipe[0]	0
Production_Recipe[1]	0
Production_Recipe[2]	0
Production_Recipe[3]	0
Production_Recipe[4]	0

Figure 8-2 Arrays in memory. A COP instruction could be used to copy the values from the array named Recipe_A into the array named Production_Recipe.

One use of this would be to load recipes for different products. By loading different parameters (numbers), a process can run differently. For example, the same machine might be able to produce different chemicals on the basis of different ingredients and process parameters. Different recipes (ingredients and process parameters) could be loaded by having multiple COP instructions in a program.

Move (MOV) Instruction

The MOV instruction can move a constant or the contents of one memory location to another location. This is a very useful instruction for many purposes. One example would be to change the preset value of a timer. There are two values that must be entered by the programmer: Source and Dest (see Figure 8-3). Source can be a constant or the address of the data to be moved. Dest is the tag name to move the data to. In the example in Figure 8-3 the number 212 was put in the tag named Cycle_Temp because Bit_1 is true.

Figure 8-3 A MOV instruction.

Note that a COP instruction copies data bit by bit, whereas the MOV moves a value. If a COP is used from a DINT to an INT array, the values may not be the same, but a MOV will always keep values the same even moving between dissimilar data types.

Masked Move (MVM) Instruction

An MVM instruction is similar to a MOV instruction except that the source data is moved through a mask before it is stored in the Dest. A mask is a very useful programming tool. It allows the programmer to selectively control which bits are used in instructions. Source, mask, and Dest must be entered for a MVM instruction. Study Figure 8-4. The values in the Source and Dest are shown in binary to make the example more understandable. The Mask value is shown in Hex (16#000000FF). The mask is an important concept in PLC programming. A mask value is used to control which bits are used in move, comparison, and other types of instructions. The value 000000FF in hex would be 0000 0000 0000 0000 0000 0000 1111 1111 in binary. For example, if we are moving a number from the tag named Number_1 and the first eight bits in the mask are 1s and the second half are 0s in the mask, only the first eight bits of the number in tag Number_1 will be moved to the Dest. The rest of the bits will not be changed in the Dest. They will remain in the same state they were before the move. The tag named Output_Number shows the result of this masked move. Note that only the first eight bit conditions were passed through the mask to the Dest result.

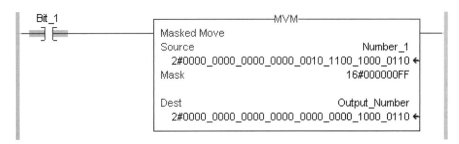

Figure 8-4 A MVM instruction.

File Fill (FLL) Instruction

The FLL instruction is used to fill a range of memory locations in an array with a constant or a value from a memory address. This could be used to put the same number in several or all of the memory locations in an array. For example, we could put 0s in all locations when we started a process running. For a FLL instruction the programmer must enter the Source, the Dest, and the Length (see Figure 8-5). The Source is a constant or a tag for a CLX controller. The Dest is the name of the tag array you would like to fill. The Length is the number of elements that you would like to fill. In this example the number 44 was the Source value, and it was put in all five of the array positions. Note that the array could be longer or shorter than five memory locations.

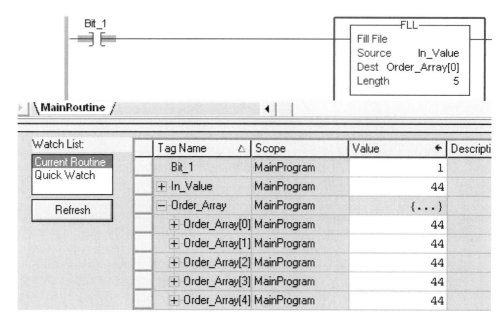

Figure 8-5 A FLL instruction. A FLL instruction copies data from one location in memory to multiple locations.

File Bit Comparison (FBC) and Diagnostic Detect (DDT) Instructions

These instructions can be used to compare large blocks of data. For example, they could be used for diagnostics to check current states against a table of what they should be. The FBC instruction compares values in a bit file (Source) with values in a Reference data file. The instruction then records the bit number of each mismatch in the Result array. The FBC instruction operates on contiguous memory.

The DDT instruction is similar to the FBC instruction except that when it finds a difference between the input file and the reference file it changes the reference file. The user could develop a list of conditions for the process under operation. These can then be compared to actual conditions for diagnostics and troubleshooting. The FBC instruction could be used to make sure the actual conditions are the same as the desired conditions.

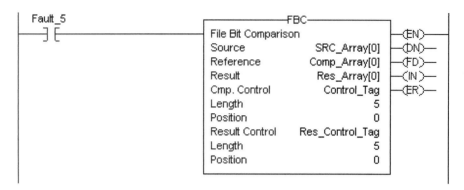

Figure 8-6 An FBC instruction.

The DDT instruction could be used to record the actual conditions in the reference file for diagnostics.

The difference between the DDT and FBC instructions is that each time the DDT instruction finds a mismatch, the instruction changes the Reference bit to match the Source bit. The FBC instruction does not change the Reference bit.

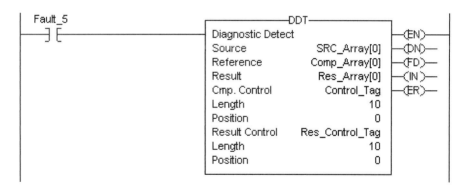

Figure 8-7 A DDT instruction.

MESSAGE (MSG) INSTRUCTIONS

The MSG instruction can be used to communicate between PLCs. A MSG instruction can be used to communicate with another CL controller, PLC5, PLC3, PLC2, as well as SLCs. Appendix D explains the use of an MSG instruction.

PROPORTIONAL, INTEGRAL, DERIVATIVE (PID) INSTRUCTION

The PID instruction is used for process control. It is used to control properties such as flow rate, level, pressure, temperature, density, and many other properties. The PID instruction takes an input from the process and controls the output. The input is normally from an analog input module. The output is usually an analog output. PID instructions are used to keep a process variable at a commanded setpoint.

A PID instruction typically receives the process variable (PV) from an analog input and controls a control variable (CV) output on an analog output to maintain the process variable at the desired setpoint.

Figure 8-8 shows a tank level example. In this example there is a level sensor that outputs an analog signal between 4 and 20 mA. It outputs 4 mA if the tank is 0 percent full and 20 mA if the tank is 100 percent full. The output from this sensor becomes an input to an analog input module in the PLC (see Figure 8-8). The analog output from the PLC is a 4–20-mA signal to a valve. The variable valve on the tank is used to control the inflow to the tank. The valve is 0 percent open if it receives a 4-mA signal and 100 percent open if it receives a 20-mA signal. The PLC takes the input from the level sensor and uses the PID equation to calculate the proper output to control the valve.

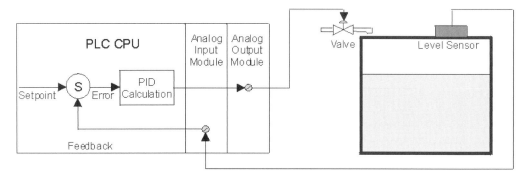

Figure 8-8 A level control system.

Figure 8-9 shows how the PID system functions. The setpoint is set by the operator and is an input to a summing junction. The output from the level sensor becomes the feedback to the summing junction. The summing junction sums the setpoint and the feedback and generates an error. The PLC then uses the error as an input to the PID equation. There are three gains in the PID equation. The P gain is the proportional function. It is the largest gain, and it generates an output that is proportional to the error signal. If there is a large error, the proportional gain generates a large output. If the

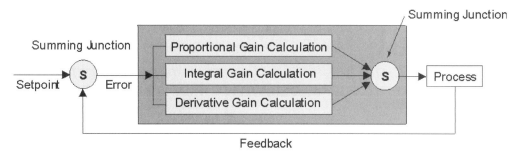

Figure 8-9 A block diagram of the use of a PID instruction.

error is small, the proportional output is small. The proportional gain is based on the magnitude of the error. The proportional gain cannot completely correct an error. There is always a small error if only proportional gain is used.

The I gain is the integral gain. The integral gain is used to correct for small errors that persist over time. The proportional gain cannot correct for very small errors. The integral gain is used to correct for these small errors over time.

The D gain is the derivative gain. It is used to help correct for rapidly changing errors. The derivative looks at the rate of change in the error. When an error occurs, the proportional gain attempts to correct for it. If the error is changing rapidly (for example, maybe someone opened a furnace door), the proportional gain is insufficient to correct the error and the error continues to increase. The derivative would see the increase in the rate of the error and add a gain factor. If the error is decreasing rapidly, the derivative gain will damp the output. The derivative's damping effect enables the proportional gain to be set higher for quicker response and correction.

As you can see from Figure 8-9 the output from each of the P, I, and D equations are summed and a control variable (output) is generated. The output is used to bring the process back to setpoint.

Refer back to Figure 8-8. As you can see in this figure the feedback from the process (tank level) is an input to an analog input module in the PLC. This input is used by the CPU in the PID equation, and an output is generated from the analog output module. This output is used to control a variable valve that controls the flow of liquid into the tank. Note that there are always disturbances that affect the tank level. The temperature of the liquid affects the inflow and outflow. The outflow varies also due to density, atmospheric pressure, and many other factors. The inflow varies because of pressure of the fluid at the valve, density of the fluid and many other factors. The PID instruction is able to account for disturbances and setpoint changes and control processes very accurately.

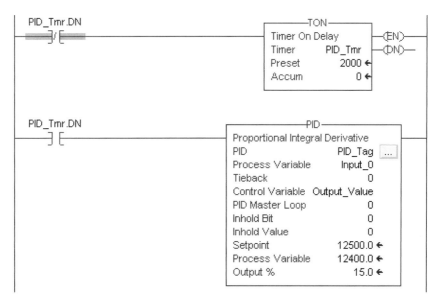

Figure 8-10 Use of a timer DN bit to perform a PID instruction every 2 seconds.

Figure 8-11 shows and describes the main PID parameters.

Operand	Type	Format	Description
Process variable	SINT, INT, DINT, REAL	Tag	Value that is being controlled. This is often the tag that specifies the actual input that is monitored for control.
Tieback	SINT, INT, DINT, REAL	Immediate or tag	Output of a hand/auto switch used to bypass the output of a controller. Enter 0 if you do not wish to use this parameter.
Control variable	SINT, INT, DINT, REAL	Tag	Value that is sent to the final control device. If you are using dead band, the control variable must be REAL or it will be forced to 0 when the error is in the dead band. This is normally the tag that specifies the actual output that is used to control the process.
PID master loop	PID	Structure	If the cascade control is being used with this PID is a slave loop, the name of the master PID loop is entered. Enter 0 if you do not want to use this parameter.
Inhold bit	BOOL	Tag	This is the current status of the inhold bit from a ControlLogix analog output channel to support bumpless restart. Enter a 0 if you do not want to use this parameter.
Inhold value	SINT, INT, DINT, REAL	Tag	This is the data read-back value from a ControlLogix analog output channel to support bumpless restart. Enter 0 if you do not want to use this parameter.
Setpoint			Displays the current value of the setpoint
Process variable			Displays the current value of the process variable.
Output %			Displays the current value of the output percentage.

Figure 8-11 Main PID parameters.

Figure 8-12 shows and describes parameters that can be configured by selecting the ellipses on the PID instruction.

Parameter	Enter
Setpoint	Value of the setpoint
Set output %	A percentage to output
Output bias	An output bias percentage
Proportional Gain (K_p)	The proportional gain
Integral gain (K_i)	The integral gain
Derivative time (K_d)	The derivative gain
Manual mode	Select either the manual mode or the software manual mode. Manual mode overrides software manual mode if both are selected.

Figure 8-12 Main PID parameters.

Figure 8-13 shows and describes additional PID configuration parameters.

Field	Enter
PID equation	Select independent gains when you would like the three gains (P, I, and D) to operate independently. Use dependent gain when you would like an overall controller gain to affect all three gains (P, I, and D)
Control action	Select either E=PV−SP (Error = Present Value − Setpoint) or E=SP−PV
Derivative of	Select PV or Error. PV is used to eliminate output spikes that result from setpoint changes. Error is used for fast response to setpoint changes when overshooting can be tolerated.
Loop update time	Update time for the instruction.
CV high limit	Enter a high limit for the control variable.
CV low limit	Enter a low limit for the control variable.
Dead-band value	Enter the desired dead-band value.
No derivative smoothing	Enable or disable.
No bias calculation	Enable or disable.
No zero crossing in deadband	Enable or disable.
PV tracking	Enable or disable.
Cascade loop	Enable or disable.
Cascade type	If cascade loop is enabled, select either slave or master.

Figure 8-13 Additional PID parameters.

Figure 8-14 shows and describes the alarms that can be set for a PID instruction.

Field	Specify
PV high	A value for a high alarm
PV low	A value for a low alarm
PV dead band	A value for an alarm dead band
Positive deviation	A value for a positive deviation
Negative deviation	A value for a negative deviation
Deviation dead band	A value for a deviation alarm dead-band level

Figure 8-14 Alarm parameters.

Figure 8-15 shows and describes parameters that can be used for scaling.

Field	Specify
PV unscaled maximum	The maximum PV value that is equal to the maximum unscaled value received from the analog input channel for the PV.
PV unscaled minimum	The minimum PV value that is equal to the minimum unscaled value received from the analog input channel for the PV.
PV engineering units maximum	The maximum engineering units.
PV engineering minimum	The minimum engineering units.
CV maximum	The maximum CV value corresponding to 100%.
CV minimum	The minimum CV value corresponding to 100%.
Tieback maximum	The maximum tieback value that equals the maximum unscaled value received from the analog input channel for the tieback value.
Tieback minimum	The minimum tieback value that equals the minimum unscaled value received from the analog input channel for the tieback value.
PID initialized	If scaling constants are changed during the run mode, turn this off to reinitialize the internal scaling values.

Figure 8-15 Scaling parameters for the PID instruction.

One way to execute a PID instruction is to place it in a task that is configured to be periodic. Set the loop update time (.UPD) equal to the periodic task rate. The PID instruction should be executed every scan of the periodic task.

If a periodic task is used, make sure that the analog input that provides the process variable is updated at a rate that is significantly faster than the rate of the task. If possible, the process variable should be sent to the processor at least five to ten times faster than the periodic task rate. This helps minimize the time difference between the actual samples of the process variable and execution of the PID loop. For example, if the PID loop is in a 500-ms periodic task, use a loop update time of 500 ms (.UPD = .5) and configure the analog input module to produce its data at least every 50 to 100 ms.

Another way to execute a PID instruction is to put the instruction in a continuous task and use a timer DN bit to trigger execution of the PID instruction. This method is slightly less accurate.

PID Timing

The PID instruction and the sampling of the process variable need to be updated at a periodic rate. The update time for a PID instruction is related to the process that is being controlled.

An update time of once per second or even longer is usually sufficient to obtain good control for slow processes such as temperature control loops. Faster loops, such as flow or pressure loops, may require an update time shorter than every 250 ms.

Some applications, such as tension control on an unwinder spool, may require a loop update time as fast as every 10 ms or less.

Derivative Smoothing

The calculation of the derivative term is enhanced by a derivative-smoothing filter. This digital smoothing filter helps to minimize large derivative term spikes that can be caused by noise in the PV.

The larger the value of the derivative is, the more aggressive the smoothing is. Derivative smoothing can be disabled if the process requires very large values of derivative gain (Kd > 10). Derivative smoothing can be disabled by selecting the No derivative smoothing option on the Configuration tab.

Dead-Band Parameter

Dead band enables you set an error range above and below the setpoint where the output does not change as long as the error remains within the specified range (dead band). The dead band is used to control how closely the process variable matches the setpoint without changing the output. The use of dead band also helps to minimize wear and tear on the final control device. The dead band extends above and below the setpoint by the value you specify. Enter zero to inhibit the dead band. The dead band has the same scaled units as the setpoint.

Zero Crossing

Zero crossing is a term that applies to dead-band control. When zero crossing is used, the instruction uses the error in its computation as the process variable enters the dead band until the process variable crosses the setpoint.

When the process variable crosses the setpoint, the error crosses zero and changes sign. While the process variable remains in the dead band, the output will not change. The dead band can be used without the zero-crossing feature by configuring it to the No zero crossing for dead band option.

The Control variable must be REAL if you are using the dead band, or it will be forced to 0 when the error is within the dead band.

Output Limiting

A percentage output limit can be set for the control output. If the instruction detects that the output limit has been reached, an alarm bit is set and the output is prevented from exceeding either the lower or the upper limit.

Feedforward or Output Biasing

Feedforward is an attempt to deal with known disturbances before they occur. In a feedforward application a disturbance is measured and fed forward to the control loop so that corrective action can be initiated before the disturbance can have an adverse effect on the system response. Feedforward control can respond quickly to known and measurable types of disturbances.

An example would be the cruise control on a car. Cruise control enables a car to maintain a steady speed. If the car begins to go up a steep hill, the car slows down below the set speed. The error in speed causes more gas to be sent to the engine, which returns the car to its original speed. A PID system would do a fine job of speed control in a cruise control.

A feedforward system would attempt to predict the decrease in speed. If the car could measure the slope of the road, it could open the throttle to send more gas as soon as an increase in slope was encountered. It could do this before any slowing in speed was sensed by the system. In this manner it would anticipate a speed decrease before it occurred and provide a feedforward corrrection.

A bias value is typically used when no integral gain is used. The bias value can then be adjusted to maintain the output required to keep the PV near the setpoint. You can provide a feedforward value by feeding the .BIAS value into the PID instruction's feedforward/bias value. The feedforward value represents a correction value fed into the PID instruction before the disturbance has had a chance to affect the process variable.

Cascading Loops

PID loops can be cascaded. Cascading is a control algorithm in which the output of one control loop provides the setpoint for another loop. Cascade is used to control difficult processes where minimal overshoot and quick stabilization are required.

Cascade control can be implemented by assigning the output in percent of a master PID loop to the setpoint of a slave PID loop. Study Figure 8-16. This is a level control system. The master PID loop is using level feedback and generating an error signal that is used as the setpoint to the slave PID loop. The slave loop uses flow feedback and the setpoint from the master to control the flow valve. The slave loop automatically converts the output of the master loop into the correct engineering units for the setpoint of the slave loop, on the basis of the slave loop's values for .MAXS and .MINS.

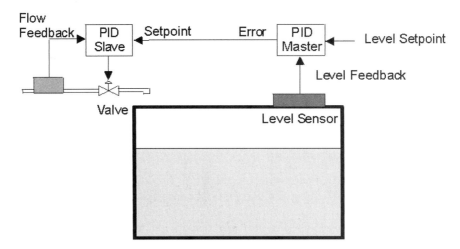

Figure 8-16 Cascade control.

SEQUENCERS

Many industrial processes are sequential. A sequential process is made up of a series of steps. Imagine an auto assembly line. This is a very sequential process. Many home appliances work sequentially. Washers and dryers are examples of sequential control.

Originally many industrial machines were controlled by a drum controller. A drum controller functions like a player piano, which is controlled by a paper roll with holes punched in it. The holes represent the program or notes to be played. The position across the paper roll indicates which note will be played. The position around the roll indicates when each note will be played.

A drum controller is the industrial equivalent. The drum controller is a cylinder with holes around the perimeter with pegs placed in the holes. There are switches that the pegs hit as the drum turns. The peg turns, closing the switch that it contacts and turning on the output to which it is connected. The speed of the drum is controlled by a motor. The motor speed can be controlled. Each step must, however, take the same amount of time. If an output must be on longer than one step, consecutive pegs must be installed.

Drum controllers have been around for centuries. Figure 8-17 shows an automated musical organ that uses a roller that has small pins around the circumference to play the notes of the song that is on the roller. This particular instrument has 20 possible notes. This one has simple valve and reed mechanisms that are each activated by the pins. The air passing through the reeds creates the notes. The rollers are called cobs. Each cob plays a different song. This one is from about 1889.

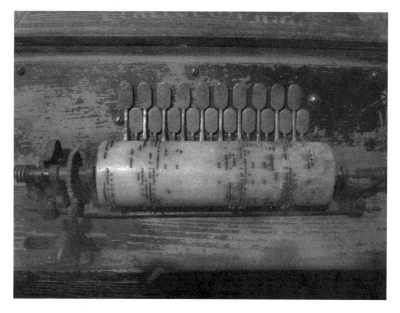

Figure 8-17 Music organ (drum controlled). Note the pegs that activate the valves as the drum is turned at slow speed.

The drum controller has some advantages. Drum controllers are easy to understand and program. They are easy to maintain. To program a drum controller, a simple chart is developed that shows which outputs are on in each step. Pegs are then installed to match the chart. The pegs control when each output is on.

Sequencer Instructions

Sequencer instructions can be used for processes that are cyclical in nature. Sequencer instructions can be used to monitor inputs to control the sequencing of the outputs. Sequencer instructions can make programming many applications a much easier task. The sequencer is very similar to a drum controller.

Processes which have defined steps can be easily programmed with sequencer instructions. There are three main instructions available for programming sequencers: sequencer output (SQO), sequencer input (SQI), and sequencer load (SQL).

Sequencer Output (SQO) Instruction

Imagine a sequential process that has five steps. The states of the outputs for every step are shown in Figure 8-18.

Look at step 0 in Figure 8-18. Step 0 in a sequencer is a special step. It is only true when the controller initially is put into run mode. The sequencer will be in step 0. When the sequencer is running and incrementing through the steps, it will not return to 0 after step 5. It will return to step 1. Notice that if the process has five steps, there are actually six steps in memory including step 0.

Outputs																Step
15	14	13	12	11	10	9	8	7	6	5	4	3	2	1	0	
																0
													On			1
													On	On		2
											On					3
								On	On			On				4
								On	On	On	On					5

Figure 8-18 Table showing which outputs are on in which steps.

This simple system would be a good application for a sequencer instruction. An SQO instruction would be used. Figure 8-19 shows a simple ladder diagram with a timer and an SQO instruction. In this example a timer is used to increment the SQO instruction through the steps.

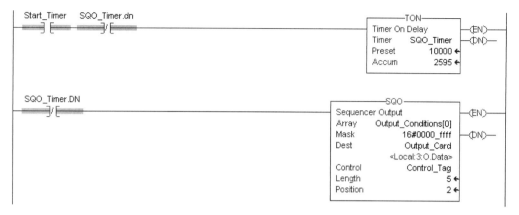

Figure 8-19 Simple SQO instruction ladder logic.

The first entry you need to make in the SQO instruction is the name of an Array that will hold the output states for each step. The tag name that is entered is the starting location in memory for the output conditions for each step. In this example the Array name is Output_Conditions[0].

The output conditions must be entered into the array in the tag editor. Study Figure 8-20. The output states for each step are entered at the starting location Output_Conditions[0]. In this application there are no outputs on in step 0. When the SQO instruction runs, it will only be in step 0 on initial startup. When the SQO instruction gets to step 5, it will go back to step 1 and continue from there.

The step number is shown by the Position value in the instruction. This process has five steps so the output states for the five steps of the process are located in elements Output_Conditions[1] through Output_Conditions[5].

Note that this is actually six elements in memory. The first element is for step (POS) 0. Step 0 is the start position for the application. Sequencers start at step zero the first time. When the SQO instruction reaches the last step, however, it will reset to step 1 (position 1).

− Output_Conditions	{...}
+ Output_Conditions[0]	2#0000_0000_0000_0000_0000_0000_0000_0010
+ Output_Conditions[1]	2#0000_0000_0000_0000_0000_0000_0000_0100
+ Output_Conditions[2]	2#0000_0000_0000_0000_0000_0000_0000_0110
+ Output_Conditions[3]	2#0000_0000_0000_0000_0000_0000_0000_1000
+ Output_Conditions[4]	2#0000_0000_0000_0000_0000_0000_0110_0100
+ Output_Conditions[5]	2#0000_0000_0000_0000_0000_0000_0111_1000

Figure 8-20 Output states in memory.

Next a mask value is entered. The mask can be a tag in memory or a constant value. If a tag name is entered for the mask value, the tag will hold the mask value.

If a constant is entered, the SQO instruction will use the constant as a mask for the steps. In our example a hex number was entered (see Figure 8-19). The value that was entered was 0000-FFFF. The eight hex digits represent 32 bits. Each bit corresponds to one output in the SQO instruction. If the bit is a 1 in the mask position, the output is enabled and will turn on if the step condition tells it to be on. If the bit in the step is a 1 and the corresponding mask bit is a 1, the output will be set. In this example the four hexadecimal Fs represent 16 binary 1s in memory. The four hexadecimal 0s represent 16 binary 0s in the mask value. The values in the output steps (Output_Conditions) will be passed through the mask value and then sent to the Dest. If the value in the mask is zero, the Dest bit is unchanged. A value of 1 in the mask means that the output state from the step will be sent to the Dest.

The Dest is usually the tag name of the outputs. In this example the outputs are in the tag named Output_Card, which is an actual output card that has 16 outputs.

Remember that the output states for each step are located in the array tag named Output_Conditions[1] through Output_Conditions[5].

When the SQO instruction is in Position 1 (step1), the third output will be on. In the next step the second and third outputs will be on.

A tag name must also be entered for Control. For this example the Control tag was named Control_Tag. An SQO instruction stores status information for the instruction in the control tag. An SQO, SQL, or SQI instruction uses multiple elements in memory. The control tag has several bits that can be used: enable (EN), done (DN), error (ER), and so on. The control element also contains the Length of the sequencer (number of steps) and the current Position (step in the sequence). The current Position in the sequence is in Control_Tag.POS in this example (see Figure 8-21).

Tag Name	Value
− Control_Tag	{...}
+ Control_Tag.LEN	5
+ Control_Tag.POS	2
Control_Tag.EN	0
Control_Tag.EU	0
Control_Tag.DN	0
Control_Tag.EM	0
Control_Tag.ER	0
Control_Tag.UL	0
Control_Tag.IN	0
Control_Tag.FD	0

Figure 8-21 Memory organization for a SQO instruction control memory.

Length is the number of steps in the process starting at position 1. Position 0 is the startup position. The first time the SQO instruction is enabled it moves from position 0 to position 1 when the instruction is toggled. The instruction resets to position 1 at the end of the last step.

Position is the location (or step) in the sequencer file from/to which the instruction moves data. In other words, Position shows the number of the step that is currently active. The value of Control_Tag.POS is 2 in Figure 8-21. This would mean that the SQO instruction would have moved the output data from Output_Data[2] to the output module when it incremented the Position from 1 to 2.

In Figure 8-22 a 10-second TON timer is used as the rung condition to the SQO instruction. The SQO instruction must see a transition from true to false to increment the Position. Note that the SQO_Timer.DN bit is used to energize the SQO instruction. Every time the timer ACC value reaches 10 seconds, the timer DN bit forces the SQO instruction to increment the .POS (Position) member of the Control tag and move to the next step. That step's outputs are turned on for 10 seconds until the timer DN bit enables the SQO instruction again. The SQO instruction then moves to the next step and sets any required outputs. Note that the timer DN bit also resets the accumulated time in the timer to 0 and the timer starts timing to 10 seconds again. When the SQO instruction is in the last step and receives the EN bit again, the SQO instruction will return to step (Position) 1 (not 0).

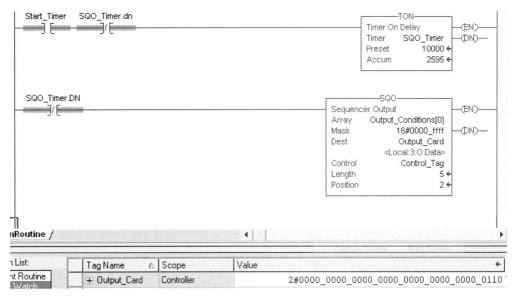

Figure 8-22 Ladder logic for an SQO instruction.

Sequencer Input (SQI)_Instruction

Every step was the same length of time in this example (10 seconds). In most industrial applications there are specific input conditions that need to be met before steps are incremented.

An SQI instruction can make it very easy to set up the input conditions to control moving from step to step. Figure 8-23 shows an SQI instruction.

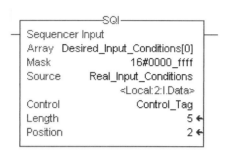

Figure 8-23 An SQI instruction.

Note that in the SQI instruction in Figure 8-23 we have to give the instruction the name of an Array that will hold the input conditions for each step. In this example Desired_Input_Conditions[0] was entered for the name of the Array. A Source also has to be entered. The Source could be the tag name of an input card. The states of the inputs on the card would then be compared to the current location (Position) in the Array of input conditions.

If they are equal, the instruction is true. The true condition can then be used as a rung condition to an SQO instruction. When the SQO instruction sees the transition, the Position would be incremented to the next Position (step). The SQO and SQI instructions share the same Control tag so the Position is incremented for both. Real_Input_Conditions entered as the Source tag.

The Length of this sequence is 5.

The desired input conditions to increment from step 0 to step 1 are entered into Desired_Input_Conditions[0] using the tag editor (see Figure 8-24). Remember that the sequencer will initially begin with step 0. In this example there are all 0s in Desired_Input_Conditions[0]. If all of the inputs are false, the instruction will increment to position 1.

The rest of the desired input conditions are entered as shown in Figure 8-24. Note that step 1 requires a 1 in the first position to increment to the next step, step 2 requires a 1 in the second position to increment, and so on.

Desired_Input_Conditions	{...}
+ Desired_Input_Conditions[0]	2#0000_0000_0000_0000_0000_0000_0000_0000
+ Desired_Input_Conditions[1]	2#0000_0000_0000_0000_0000_0000_0000_0001
+ Desired_Input_Conditions[2]	2#0000_0000_0000_0000_0000_0000_0000_0010
+ Desired_Input_Conditions[3]	2#0000_0000_0000_0000_0000_0000_0000_0011
+ Desired_Input_Conditions[4]	2#0000_0000_0000_0000_0000_0000_0000_0100
+ Desired_Input_Conditions[5]	2#0000_0000_0000_0000_0000_0000_0000_0101

Figure 8-24 Input condition table. Note that if we are presently in step 3, we would need real-world inputs 1 and 2 to be true in order to match our desired input conditions to increment to step 4.

Figure 8-25 shows the current states of the bits of the Source tag (Real_Input_Conditions). In this example the current state is that the second input is true. If the SQI instruction were in step 2, the real-world input states from Real_Input_Conditions would equal Desired_Input_Conditions[2] and the instruction would be true.

+ Real_Input_Conditions	2#0000_0000_0000_0000_0000_0000_0000_0010

Figure 8-25 Current input states in the tag named Real_Input_Conditions.

The mask value in an SQI instruction is just like the SQO instruction mask value. The mask can be a tag name in memory or one word (32 bits) in memory. In this example a hex constant was entered (0000-FFFF). The eight hex digits correspond to 32 binary bits. If a bit is a 1 in the mask, that input is enabled. The Source word (usually real-world input conditions) is passed through the mask value and compared to the current step (word) in the file. In this application we are only interested in the states of the first 16 inputs. The 0000-FFFF means the first 16 inputs are used and that the last 16 (0000 0000 0000 0000 1111 1111 1111 1111) are ignored.

Next the Source must be entered. The Source is often the tag name of the real-world inputs. In this example the Source tag name is Real_Input_Conditions. In this example it's an alias-type tag for an input module with 16 inputs. The states of these real-world inputs will be compared to the desired inputs in the present step. If they are the same, the SQI instruction is true; this causes the SQO instruction to increment to the next step by incrementing the Position value in the Control tag.

A tag name must also be entered for Control. Control is the location where the SQC instruction stores status information for the instruction. There are several bits that can be used: EN, DN, ER, found (FD), and so on. The Control tag also contains the Length of the sequencer (number of steps) and the current Position. Position is the location (or step) in the sequencer file from/to which the instruction moves data. In other words,

Position is the step that is currently active. In order to stay synchronized, the SQI and SQO instructions must use the same Control tag.

Figure 8-26 shows one method of programming a sequencer using an SQI and an SQO instruction. The SQI instruction is used to compare the real-world input conditions in the Source (Real_Input_Conditions) with the programmed step conditions in the Array tag named Desired_Input_Conditions. Note that the Position value in the instruction is currently 2 so the SQI instruction would be comparing the Source input conditions to Desired_Input_Conditions[2]. If they are equal, the SQI instruction is true, energizing the SQO instruction which then increments the Position value to 3 and outputs the states from Output_Conditions[3] to the Dest (Output_Card). Note that when the Position was incremented in the Control tag, it was incremented for the SQO and the SQI instructions because they share the same Control tag.

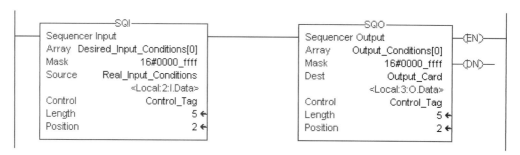

Figure 8-26 SQI and SQC instructions.

Resetting the Position Value of an SQO Instruction

Each time the controller goes from program to run mode, the SQO instruction clears (initializes) the .POS value. A RES instruction could also be used to clear the position value (POS = 0).

Sequencer Load (SQL) Instruction

The SQL instruction can be used to load reference conditions into a sequencer array.

SHIFT INSTRUCTIONS

A shift register is a storage location in memory. These storage locations can typically hold 32 bits of data, that is, 1s or 0s. Each 1 or 0 could be used to represent good or bad parts, the presence or absence of parts, or the status of outputs. Many manufacturing processes are very linear in nature. Imagine a bottling line. The bottles are cleaned, filled, capped, and so on. This is a very linear process. There are sensors along the way to sense for the presence of a bottle, there are sensors to check fill, and so on. All of these conditions could easily be represented by 1s and 0s.

Bit Shift Left (BSL) Instruction

Figure 8-27 shows an example of the use of a bit shift left (BSL) instruction. Array is the name of an Array tag name. Control is the name of the Control tag that the instruction will use to keep track of information. Source Bit is the tag name of the bit that will be shifted into the array. Length is the number of bits to be shifted. Figure 8-28 shows how a BSL operates.

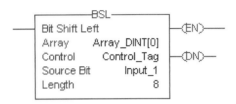

Figure 8-27 A BSL instruction.

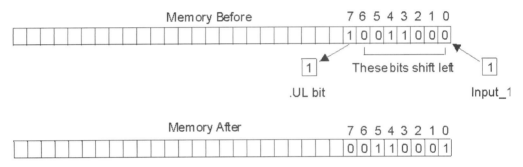

Figure 8-28 A BSL instruction on memory.

Shift registers essentially shift bits through a register to control I/O. Think of a bottling line. There are many processing stations; each station could be represented by a bit in the shift register. Each station should operate only if there is a part present that requires this station. As the bottles enter the line, a 1 is entered into the first bit. Processing takes place. The stations then release their product, and each product moves to the next station. The shift register also increments. Each bit is shifted one position. Processing takes place again. Each time a product enters the system, a 1 is placed in the first bit. The 1 follows the part all the way through production to make sure that each station processes it as it moves through the line. There is also a bit shift right (BSR) instruction. Note that there are many bit instructions available.

File Shifts

In addition to bit shift instructions, there are shift instructions that can shift whole elements in arrays. There are several shift instructions available. The table in Figure 8-29 shows some bit and file shift instructions.

Instruction	Operation
BSL	Bit shift left
BSR	Bit shift right
FFL	First-in, first-out (FIFO) load
FFU	First-in, first-out (FIFO) unload
LFL	Last-in, first-out (LIFO) load
LFU	Last-in, first-out (LIFO) unload

Figure 8-29 Bit and file instructions.

LOAD INSTRUCTIONS

FIFO Load (FFL) instruction

One use of load and unload instructions is order entry and order processing. Imagine a system where the operator enters orders into an array and the machine takes the orders one at a time and produces the products that were ordered. See Figure 8-30. When the operator enters an order it would be placed into the Source tag specified by the FFL instruction. When the FFL instruction is enabled, it loads the Source value into the position in the FIFO identified by the Position value. The instruction loads one value each time the instruction is enabled, until the FIFO is full. Typically, the Source and the FIFO are the same data type.

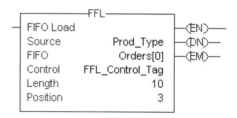

Figure 8-30 An FFL instruction.

FIFO Unload (FFU) Instruction

The FFU instruction is usually used in tandem with an FFL instruction. The FFL instruction is used to load the orders into the array, and the FFU instruction is used to take one order out of the array at a time in a first-in, first-out manner. The use of a FFU instruction is shown in Figure 8-31. Note that the same tag was used for the FIFO array and the Control tag in both instructions. Note also that the Length is the same in both. Because the same Control tag is used for both, they will work together. If the Position is incremented by the FFL instruction or decremented by the FFU instruction, it will change for both so they remain coordinated.

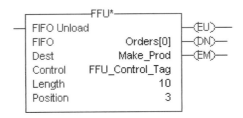

Figure 8-31 An FFU instruction.

JUMP INSTRUCTIONS

Jump-to-Subroutine (JSR) Instruction

Figure 8-33 shows an example of a JSR instruction. If Sensor_2 is true, the JSR instruction will execute the routine named Manual_Mode.

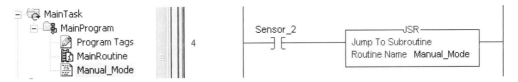

Figure 8-32 A Jump-to-Subroutine (JSR) instruction. Note in the left of the figure that Manual_Mode is a routine in the main program.

Jump (JMP) Instructions

JMP instructions can be used to jump to a different place in the logic. Figure 8-33 shows an example. Note the JMP instruction in the first rung. A label is used to specify the label to jump to. In this example the label name is Alt_1. If contact Sensor_1 is true in the first rung, the processor will jump to the rung that contains the label Alt_1. In this example it only jumps over one rung. A JMP can jump over multiple rungs.

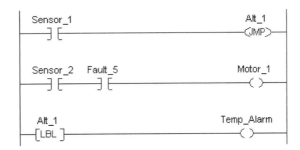

Figure 8-33 A JMP instruction and a label (LBL) instruction.

CONTROL INSTRUCTIONS

Master Control Reset (MCR) Instruction

MCR instructions are used in pairs. An MCR instruction can disable all rungs between the MCR instructions.

Figure 8-34 shows an example. When an MCR zone is enabled, the rungs in the MCR zone are scanned for their true or false conditions. When an MCR instruction is disabled, the controller still scans rungs within the MCR zone. Scan time is reduced because nonretentive outputs in the zone are disabled. The rung-condition-in is false for all the instructions inside of the disabled MCR zone.

Considerations for MCR Zones

MCR zones must be ended with an unconditional MCR instruction.

An MCR zone cannot be nested in another MCR zone.

Do not jump into an MCR zone. If the zone is false, jumping into the zone will activate the zone from the point to which you jumped into the zone to the end of the zone. An MCR instruction does not need to be used to end the zone if the MCR zone continues to the end of the routine.

The MCR instruction is not a substitute for a hardwired master control relay for emergency-stop capability. You should still utilize a hardwired master control relay.

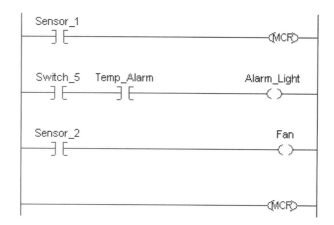

Figure 8-34 Use of MCR instructions.

Loop (FOR) Instruction

A FOR instruction can be used in any routine. A FOR instruction cannot be used to call the main routine.

When enabled, the FOR instruction repeatedly executes the routine until the Index value exceeds the Terminal Value. The FOR instruction cannot pass parameters to the routine. Each time the FOR instruction executes the routine, it will add the Step Size to the Index. An excessive number of repetitions can cause the controller's watchdog to time out; this causes a major fault.

Study Figure 8-35. When contact Sensor_1 is true, the FOR instruction repeatedly executes the routine named Loop_Routine and increments Tag_5 by 2 each time (Step Size). When Tag_5 is > 9 or a BRK instruction is enabled, the FOR instruction will quit executing the routine named Loop_Routine.

```
Sensor_1                    ┌───────FOR──────────────┐
──┤ ├──────────────────────│ For                     │
                            │ Routine Name Loop_Routine│
                            │ Index              Tag_5 │
                            │                       0 ←│
                            │ Initial Value          0 │
                            │ Terminal Value         9 │
                            │ Step Size              2 │
                            └─────────────────────────┘
```

Figure 8-35 A FOR (loop) instruction.

Return (RET) Instruction

The use of a RET instruction is shown in Figure 8-36. When an RET instruction is enabled, it returns to the FOR instruction. The FOR instruction increments the Index value by the Step Size and executes the subroutine again. If the Index value exceeds the Terminal Value, the FOR instruction completes and execution moves on to the instruction that follows the FOR instruction.

```
Sensor_2                         ┌──RET───┐
──┤ ├───────────────────────────│ Return  │──
                                 └────────┘
```

Figure 8-36 A RET instruction.

Break (BRK) Instruction

The BRK instruction can be used to interrupt the execution of a routine that was called by a FOR instruction. When enabled, the BRK instruction exits the routine and returns the CPU to the instruction that follows the FOR insruction. If there are nested FOR instructions, a BRK instruction will return control to the innermost FOR instruction. The use of a BRK instruction is shown in Figure 8-37.

Figure 8-37 A BRK instruction.

Get System Values (GSV) and Set System Values (SSV) Instructions

The GSV and SSV instructions get and set controller status data that is stored in objects. The controller stores status data in objects. There is no status file, as in the PLC-5 processor.

When enabled, the GSV instruction retrieves the specified information and places it in the Dest. When enabled, the SSV instruction sets the specified attribute with data from the Source.

Figure 8-38 shows an example of a GSV instruction. It is being used to acquire the proportional gain (PositionProportionalGain) for the X_axis. The instruction will place the value into the Dest tag (Present_PID_Gain_X).

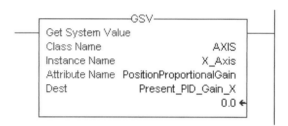

Figure 8-38 A GSV instruction.

The SSV instruction can be used to change controller data. Figure 8-39 shows the use of an SSV instruction to set the value of PositionProportionalGain for the X_Axis.

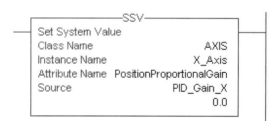

Figure 8-39 An SSV instruction.

When you enter a GSV/SSV instruction, the programming software displays the valid object classes, object names, and attribute names for each instruction. For the GSV instruction, you can get values for all the attributes. For the SSV instruction, the software displays only those attributes you can set.

Use the GSV and SSV instructions carefully. Making changes to objects can cause unexpected controller operation or injury to personnel.

EVENT Instruction

The EVENT instruction is used to execute an event-type task (see Figure 8-40). The name of the event-type task must be entered into the instruction. Each time the instruction executes, it triggers the event task that is specified by the instruction. When using an EVENT instruction, make sure the event task is given enough time to complete its execution before it is triggered again. If not, an overlap will occur. If an EVENT instruction is executed while the event task is already executing, the controller increments the overlap counter but it does not trigger the event task.

Figure 8-40 An EVENT instruction.

This chapter has covered a small sample of the many special instructions that are available. If you can understand and use the instructions in this chapter, you will be able to learn new instructions very easily. If you have a special need in an application, the odds are good that an instruction is available to meet that need. The instruction help file in RSLogix 5000 is a great source of available instructions and their use.

QUESTIONS

1. What instruction could be used to fill a range of memory with the same number?
2. What instruction could be used to move an integer in memory to an output module?
3. What would the Dest result be in the table below?

Source	0	0	1	1	0	1	0	1	0	1	0	1	0	1	1	0
Mask value	0	0	0	0	0	0	0	1	1	1	1	1	1	1	1	1
Dest																

4. What instruction could be used to move data if it desired to mask some of the bits?
5. What does PID stand for?
6. What does the proportional gain do?
7. What does the integral gain do?
8. What does the derivative gain do?
9. If an SQO's Control Tag's Length value is 5 and its Position value is 3, what values are being output to the output card, on the basis of the Array shown below?

Output_Conditions	{...}
+ Output_Conditions[0]	2#0000_0000_0000_0000_0000_0000_0000_0010
+ Output_Conditions[1]	2#0000_0000_0000_0000_0000_0000_0000_0100
+ Output_Conditions[2]	2#0000_0000_0000_0000_0000_0000_0000_0110
+ Output_Conditions[3]	2#0000_0000_0000_0000_0000_0000_0000_1000
+ Output_Conditions[4]	2#0000_0000_0000_0000_0000_0000_0110_0100
+ Output_Conditions[5]	2#0000_0000_0000_0000_0000_0000_0111_1000

10. Describe the operation of an SQO.
11. Describe the operation of an SQI.
12. What are at least two advantages of SQO and SQI instructions over a traditional mechanical drum controller?
13. What instruction might be used to skip several rungs of logic under certain conditions?
14. Which instruction should be used to execute logic five times?
15. Which instruction could be used to return from a routine to a FOR instruction?
16. Which instruction could be used to change a system value?
17. Which instruction can be used to execute a task?

CHAPTER 9

Structured Text Programming

OBJECTIVES

On completion of this chapter, the reader will be able to:

- Explain the basics of the structured text language.
- List at least three benefits of structured text.
- Understand structured text routines.
- Utilize structured text to develop routines.

INTRODUCTION

Ladder logic has been the overwhelming choice for PLC programming since PLCs were developed. Additional languages were specified by an international standard (IEC 61131-3). These languages are rapidly gaining in popularity and use. Structured text (ST) is one of the languages in IEC 61131-3. ST programming is more of a typical computer language than ladder logic. It is very similar to languages that are used in many industrial devices such as robots, vision systems, and so on. Structured text programming is also used within other PLC languages. The best way to learn the material in this chapter is to read the chapter and then work on the chapter questions before trying to write and test actual ST programs.

OVERVIEW OF STRUCTURED TEXT

Structured text language resembles C, Basic, Pascal, and even visual basic language. People are often most comfortable with the first programming language they learn or the one they have used the most. People who have used computer programming languages for other devices often believe that ST is the easiest language to use for programming PLC logic.

ST programs are written in short English-like sentences. This makes ST programs very readable and easy to follow and understand. ST programming is well suited for applications requiring complex mathematics or decision making. ST is also concise. An ST program for an application would be much shorter than a ladder logic program, and the ST logic would be easier to understand. Learning ST will help you program many industrial devices.

Benefits of ST

- People who have programmed a computer language can readily learn ST.
- Programs can be created in any text editor.
- Runs as fast as ladder logic.
- Is concise and easy to understand.
- Can be used for all or a portion of an application's logic.

FUNDAMENTALS OF ST PROGRAMMING

A short ST program is shown in Figure 9-1. This simple routine will be used to illustrate the sentence-like structure of ST programs. Don't worry about understanding all of the syntax. The rest of the chapter will cover ST programming in detail.

The logic in the Figure 9-1 program has two variables (tags), named Temp and Flow, and two outputs, named Pump and Green_Light. Temp, Flow, Pump, and Green_Light are all tags in a ControlLogix (CL) project. This routine controls the pump, flow, and a green_light.

The routine uses IF statements to make decisions. The first portion of logic compares the value of the Temp tag. If Temp is greater than or equal to 100 and less than 200, then Pump is turned on, the Flow variable (tag) is set to a value of 45, and Green_Light is turned on.

Else if (ELSIF) Temp is less than 100, Pump is turned off, Flow is set to 20, and Green_Light is turned off.

If neither of those conditions is true, then the Else is true and Alarm_Light is turned on.

```
IF Temp >= 100 & Temp < 200 THEN
   Pump := 1; Flow := 45; Green_Light := 1;
ELSIF Temp < 100 THEN
   Pump := 0; Flow := 20; Green_Light := 0;
ELSE
   Alarm_Light := 1;
END_IF;
```

Figure 9-1 ST logic.

ST logic is easier to understand than ladder logic. ST is not case sensitive. Tabs and carriage returns should be used to make programs clear and readable. They have no effect on the execution of the program, but they can really help make it more understandable. Indentation was used in the program in Figure 9-1 to make a program more understandable. It makes it much easier to follow the logic. Note that if this logic were created in a routine, it would have to be called from the main with a JSR instruction for the routine to actually run.

PROGRAMMING ST IN CONTROLLOGIX

A ST program is a routine within a ControlLogix project. If you right click the mouse button on the main program icon, it will give you the option of adding a new routine. Figure 9-2 shows the initial screen for creating a new routine. In this example the name Structured_Text_Stop_Light was entered. Next the language Type is chosen from the drop down menu in the Type box. In this example Structured Text was chosen.

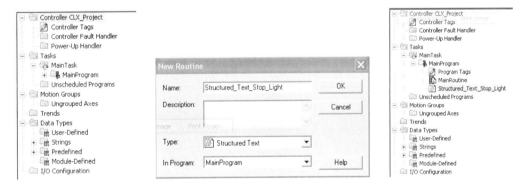

Figure 9-2 On the left is the Controller Organizer. If you right-click on MainProgram, you can create a new routine. In the center is the New Routine configuration screen. Note that Structured Text was chosen for the language Type. The figure on the right shows the Controller Organizer after the new routine was created. After the new routine is chosen, you can click on the name of the routine in the main program list and the screen shown in Figure 9-3 will appear.

The program code can be entered into the ST entry screen. A very simple one-statement routine was entered in Figure 9-3. Green_1 is a tag that was created in the tag editor. Green_1 is a discrete output. This simple logic would set the value for Green_1 to 1 (true) when this routine is run. The output assigned to Green_1 output would be turned on. The := is used to change the value of whatever tag is on the left to the value shown on the right. This is called an assignment statement. Almost all program lines must end with a semicolon (;). This will be explained later in the chapter.

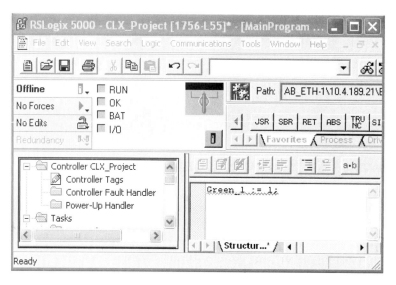

Figure 9-3 A simple program to turn output Green_1 on.

Note that to run, this routine would need to be called by the MainRoutine (see Figure 9-4). Note that a JSR instruction was used to run the structured text routine.

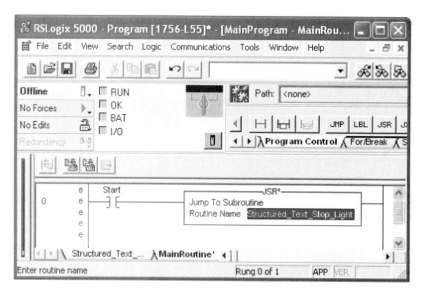

Figure 9-4 A JSR instruction being used to call (run) the ST routine. Note that in this example, the JSR will only be executed while the Start contact is true.

Note: A well-planned job is half done. It is wise to preplan the application and programming before you start writing the program. How many I/O will there be? What tags will be required, and what types will they be? What would appropriate names for all of the tags be? It will ease your task if you create the tags before you begin to write a program. You are not required to create tags first, but it will make programming and troubleshooting less frustrating.

Assignment Statements

Assignment statements are used to assign a value to a tag. The example below would assign a value of 115 to the tag named Temp.

Temp := 115;

The generic example of an assignment statement is shown below.

Tag := value or mathematical expression;

Tag on the left in the equation above represents a variable or tag that is being assigned a new value. A tag can be a Boolean (BOOL)-, single integer (SINT)-, an integer (INT)-, a double integer (DINT)-, or a real (REAL)-type tag. The DINT is the default integer type and is the one that should normally be used for integers. The := is the assignment operator. Note that you cannot just use the equal sign. You must use a colon followed by the equal sign. The expression on the right side of the equation is used to represent the value that will be assigned to the tag. An expression can be a constant value (a number), or it could be a tag of type BOOL, SINT, INT, DINT, or REAL. The last thing in the line is the semicolon. You must have a semicolon at the end of the line. In fact, almost all lines in a ST program must be terminated with a semicolon. The most common error in ST programming is forgetting the semicolon.

Study the examples below. The first example assigns a value of 212 to the tag named Temp. The second example uses a tag (variable) to assign a value to Temp. The third example shows that math statements can also be used in assignment statements. In this example Var_1 is being assigned the value of Var_2 multiplied by 2.

Temp := 212;

Temp := Var_1;

Var1 := Var_2 * 2;

Documenting Logic with Comments

Comments should be used in programs. They help make programs more understandable. They also help reduce the time and frustration of troubleshooting an application. They help a technician understand what the programmer intended. A comment can appear anywhere in a program line or on a line by itself. Figure 9-5 shows the format for various types of comments. Figure 9-6 shows examples of some of the ways comments can be used.

To Use a Comment	Format
On a line without instructions Or At the end of a line of ST	// Comment (* Comment *) /* Comment */
Within a line of ST	(* Comment *) /* Comment */
On more than one line	(* Start of comment ....end of comment *) /* Start of comment ....end of comment */

Figure 9-5 Table showing possible format for comments.

Example	
1	// This is an example of a comment at the beginning of a line.
2	IF Temp > 100 THEN // Comment at the end of a line
3	Temp := 212; /* Comment at the end of a line – different format */
4	Pump := 1; (* Comment at end of a line – different format*)
5	IF S1 (* Comment within a line) & S2 (* Comment within a line) THEN
6	IF Temp2 = 205 /* Comment within a line – different format */ THEN
7	(* Comment on more than one line. This is an example of a comment that takes up more than one line in a program *)
8	/* Comment on more than one line. This is an example of a comment that takes up more than one line in a program in a different format */

Figure 9-6 Examples of the use of comments.

ARITHMETIC OPERATORS

All of the standard arithmetic operators are available in ST programming. Study the examples shown below. In the first line the tag named Temp is assigned a value equal to Var1 plus 20. Line 2 assigns a value to the RPM tag equal to the Speed tag divided by 60. The third statement assigns the result of the tag named Cases multiplied by 12 to the Total_Cans tag.

Temp := Var1 + 20;

RPM := Speed/60;

Total_Cans := Cases * 12;

Figure 9-7 shows the arithmetic operators that can be used in ST programming. Math operators are most commonly used in assignment statements.

Instruction	Operator	Optimal Data Type
Add	+	DINT, REAL
Subtract	−	DINT, REAL
Multiply	*	DINT, REAL
Exponent (X to the power of Y)	**	DINT, REAL
Divide	/	DINT, REAL
Modulo	MOD	DINT, REAL

Figure 9-7 Arithmetic operators.

The programmer must be very careful to use correct number types when performing math operations. Integer math will not provide a decimal result. Study the example below. The tag Answer is a DINT type. Because it is an integer type, integer math will be done. You might expect the answer to be 2.5, but the answer would be 2. You would need to use a REAL type to get a decimal result. If the Answer tag is created as a REAL type, the answer would be 2.5.

Answer := 5/2;

Study the example below. You might expect the answer to be 2, but the answer would be 3. You must be careful when using integer math. If you want a decimal number as a result use REAL-type tags.

Answer := 5.1/2;

Modulo Instruction

The modulo instruction is a very interesting operator. It can be used to find the remainder of a division. The result of a modulo operation is the integer remainder of the division.

Answer := 5 MOD 2; //Answer = 1   (5 MOD 2 = 2 with a remainder of 1)
Answer := 7 MOD 3; //Answer = 1   (7 MOD 3 = 2 with a remainder of 1)
Answer := 17 MOD 3; //Answer = 2   (17 MOD 3 = 5 with a remainder of 2)
Answer := 13 MOD 5; //Answer = 3   (13 MOD 5 = 2 with a remainder of 3)

ARITHMETIC FUNCTIONS

There are also many arithmetic functions available to the programmer. Figure 9-8 shows which functions are available. Note that the chart in Figure 9-8 also shows the optimal data type to use for each function. The example below shows the use of the square root function. In this example, the function would calculate the square root of 515 and assign the result to the tag named Val. Note that a tag (variable) could have been used in place of the constant 515.

Val := SQRT(515);

For	Function	Optimal Data Type
Absolute value	ABS(numeric expression)	DINT, REAL
Arc cosine	ACOS(numeric expression)	REAL
Arc sine	ASIN(numeric expression)	REAL
Arc tangent	ATAN(numeric expression)	REAL
Cosine	COS(numeric expression)	REAL
Radians to degrees	DEG(numeric expression)	DINT, REAL
Natural log	LN(numeric expression)	REAL
Log base 10	LOG(numeric expression)	REAL
Degrees to radians	RAD(numeric expression)	DINT, REAL
Sine	SIN(numeric expression)	REAL
Square root	SQRT(numeric expression)	DINT, REAL
Tangent	TAN(numeric expression)	REAL
Truncate	TRUNC(numeric expression)	DINT, REAL

Figure 9-8 Arithmetic functions.

RELATIONAL OPERATORS

Relational operators are used to compare two values or strings and provide a true or false result. The result of a relational operation is a BOOL value. If the result of an operation is true, the result will be a 1. If the result of a relational operation is false, the result will be 0. These are used extensively for decision making. Figure 9-9 shows a table of relational operators.

Comparison Type	Operator	Optimal Data Type
Equal	=	DINT, REAL, string
Less than	<	DINT, REAL, string
Less than or equal	<=	DINT, REAL, string
Greater than	>	DINT, REAL, string
Greater than or equal	>=	DINT, REAL, string
Not equal	<>	DINT, REAL, string

Figure 9-9 Relational operators.

Relational operators can be used to make decisions. For example, an IF statement can make use of relational operators.

IF TEMP > 200 THEN

In this example the value of TEMP is evaluated to see if it is larger than 200; if it is, this evaluates to a value of 1 (true). If TEMP is less than 200, it would have a value of 0 and be false. The THEN would only be executed if the value of the operation is a 1 (true).

The value of a relational operation can also be assigned to a tag. In the statement below, if TEMP is greater than or equal to 200, the value of 1 will be assigned to the tag called STAT. If TEMP is less than 200, the value of 0 will be assigned to the tag named STAT.

STAT := (TEMP >= 200);

Relational operators can also be used to evaluate strings of characters or characters in strings. In the example below, the = relational operator is used to evaluate whether String_1 is equal to the second string (Password). If they are the same, the result will be a 1 (true). If the strings are not equal, the result will evaluate to a 0 (false).

IF String_1 = Password THEN

The example below would evaluate the first character in a string called String_1 to see if it is equal to 65. The letter A in ASCII is equal to 65. The String_1.DATA[] represents an array of characters. The 0 in square brackets represents the character we want to evaluate.

IF String_1.DATA[0] = 65 THEN

LOGICAL OPERATORS

Logical operators can also be used to check to see if multiple conditions are true. Figure 9-10 shows a table of logical operators.

Type	Operator	Type
Logical AND	&, AND	BOOL
Logical OR	OR	BOOL
Logical exclusive OR	XOR	BOOL
Logical complement	NOT	BOOL

Figure 9-10 Logical operators.

In the example below, Sensor_1 is evaluated. If the sensor is on, it would evaluate to a 1 (true) and the THEN would be executed.

IF Sensor_1 THEN

In the example below Sensor_1 is evaluated. In this case a NOT was used. So, if the Sensor_1 is a 0 (false), the THEN would be executed.

    IF NOT Sensor_1 THEN

In the next example, both must evaluate to true for the whole statement to be evaluated as true. Note that the & operator was used for the AND logical operator. AND or & can be used.

    IF Sensor_1 & (TEMP < 150) THEN

In the next example the OR logical operator was used. In this case if either Sensor_1 OR Sensor_2 are true, the statement will evaluate to a 1 and the THEN will be executed.

    IF Sensor_1 OR Sensor_2 THEN

An Exclusive-OR (XOR) is used in the next example. In this example only one of the two sensors can be true for the statement to be evaluated to true.

    IF Sensor_1 XOR Sensor_2 THEN

In the next example, the result (1 or 0) of the logical operation will be assigned to the tag called STATUS. If Sensor_1 and Sensor_2 are both true, the tag STATUS will be assigned a value of 1.

    STATUS := Sensor_1 & Sensor_2

PRECEDENCE

The table in Figure 9-11 shows the order in which math statements will be evaluated. This is very important. The wrong answer will be obtained if precedence is not carefully considered. Consider the example below. Normally math statements are evaluated from left to right if precedence is equal. In the example below precedence is not equal. Many people would say the answer is 21. They might add the 5 and the 2 (7) and then multiply by 3, getting a result of 21. The correct answer is 11. The multiplication has a higher precedence than the addition. You must multiply 2 * 3 (6) and then add the result to 5 (11).

    Answer := 5 + 2 * 3;

Another method to assure the proper order of calculation is to use parentheses.
    In the example below it was desired to do the addition first. It will not be done first because the multiplication operator has a higher precedence that the addition operator.

    Answer := Var1 + 17 * Temp;

It could be rewritten as shown below. In this example the parentheses assure that the addition will be done first. Parentheses have the highest precedence.

    Answer := (Var1 + 17) * Temp;

Order	Operation
1	( )
2	Function ( )
3	**
4	− (negate)
5	NOT
6	*, /, MOD
7	+, − (subtract)
8	<, <=, >, >=
9	=, <>
10	&, AND
11	XOR
12	OR

Figure 9-11 Table showing the order of precedence for arithmetic operators.

CONSTRUCTS

The definition of construct is to form by assembling or combining parts; to build. The table in Figure 9-12 shows the constructs that are available in structured text. A construct can also be thought of as a statement.

Construct	When to Use
IF THEN	If specific conditions are true
FOR DO	A specific number of times
WHILE DO	As long as a condition is true
REPEAT UNTIL	Until a condition is true
CASE OF	On the basis of a number

Figure 9-12 Constructs that can be used in ST.

An IF statement can be used to make a decision and then execute logic on the basis of the decision. The IF is followed by a test statement. The test statement is a Boolean expression that is evaluated to be true (1) or false (0). In the example below if the BOOL expression is true, all statements between the THEN and the END_IF will be executed. If the BOOL expression is false, processing would continue after the END_IF.

IF BOOL_expression THEN
   Motor_1 := 1;
   Temp := 150;
   Additional logic
END_IF

An ELSE may also be used. In the example shown below, if the tag Motor_On is true, the tag Red_Light will be set to 1. If Motor_On is false, the tag Green_Light will be set to 1. Note that semicolons are very important. There is no semicolon after the THEN or the ELSE, but every other line is terminated with a semicolon.

If Motor_On THEN
   Red_Light := 1;
ELSE
   Green_Light := 1;
END_IF;

ELSE IF (ELSIF) Statements

The example below shows the use of an ELSIF statement. Note the spelling of the ELSIF. In this example if the first IF is false, the ELSIF is evaluated. If Sensor_3 is true, Alarm will be assigned the value 1. Note also that you must have an END_IF statement.

IF Sensor_1 & Sensor_2 THEN
   Pump := 1;
   Heat_Coil := 0;
ELSIF Sensor_3 THEN
   Alarm := 1;
END_IF;

The example below adds a few new twists. An IF, an ELSIF, and an ELSE are used. In this example if the IF is false and the ELSIF is false, the ELSE will be executed and the Alarm_Light tag will be assigned the value 1. Note in this example more than one statement was put on a line. Each was separated by a semicolon. While this is permissible, it should not be used to excess. You should be careful that it does not make your program more difficult to understand. Note also that the statements were indented to make the program easier to understand. Finally note where semicolons are and are not used.

```
IF Temp > = 100 & Temp < 200 THEN
    Pump := 1; Flow := 45; Green_Light := 1; Alarm_Light := 0;
ELSIF Temp < 100 & Temp > 50 THEN
    Pump := 0; Flow := 20; Green_Light := 0; Alarm_Light := 0;
ELSE
    Alarm_Light := 1; Pump := 0; Flow := 0; Green_Light := 0;
END_IF;
```

FOR DO Statements

A FOR DO loop is used when we know how many times a loop should be executed. For example, if we needed to fill an array of ten integers with a number, we could use a FOR DO loop that would execute ten times. In the example below a tag named Temp was created and it was configured to be an array of ten tags named Temp (Temp[0] through Temp[9]). The short loop below will fill the array of ten tags (Temp[0] through Temp [9]) with the number 99.

```
FOR X := 0 to 9 by 1 DO
    Temp[X] := 99;
END_FOR;
```

In the next example the loop was incremented by 6 each time through. X in the loop is assigned a value of 0 the first time through the loop. The next time through the loop it is assigned a value of 6, then 12, then 18, then 24. The next time it would be 30, but 30 is larger than 24 so the loop would not execute again.

```
FOR X := 0 to 24 by 6 DO
    Additional ST Statements;
    Additional ST Statements;
END_FOR;
```

Note that when a loop is executed, the controller does not execute any other statement in the routine outside of the loop until the loop is completed. If the time it takes to complete a loop is greater than the watchdog timer for the task (default time is 500 ms), a major fault will occur. If this becomes a problem try a different type of construct such as an IF THEN.

WHILE DO Statement

A WHILE DO construct is used when you would like something to happen *while* a condition is true. In the following example we would like the loop to execute WHILE two things are true: the tag CNT is less than 50 AND the tag TEMP is less than 200. If CNT becomes larger than 49 or Temp becomes greater than 199, the loop will not run. Note that integer numbers are used.

```
WHILE ((CNT < 50) & (Temp < 200)) DO
    CNT := CNT + 3;
END_WHILE;
```

Remember: when a loop is executed, the controller does not execute any other statement in the routine (outside of the loop) until the loop is completed. If the time it takes to complete a loop is greater than the watchdog timer for the task, a major fault will occur. If this becomes a problem, try a different type of construct such as an IF THEN. Note that with the WHILE loop, the test is at the beginning of the loop. If the test is false, the loop will not even execute once.

REPEAT UNTIL Statement

The REPEAT UNTIL statement is used when we have something we want to do UNTIL certain conditions are true. In this example the REPEAT will execute UNTIL one of two conditions are true (note the use of the OR). In this example 2 is added to the value of the tag CNT every time the REPEAT is executed. The REPEAT will end when CNT is greater than 34 OR Temp is greater than 92. Note where semicolons are and are not used.

```
REPEAT
    CNT := CNT +2;
    UNTIL ((CNT > 34) OR (Temp > 92))
END_REPEAT;
```

Note that a REPEAT loop will execute at least once, because the test is at the end of the loop.

CASE OF Statement

The CASE OF construct is very useful. It can be used to perform different portions of the program depending on a value. For example, we might produce six different products on one machine. We could have the operator enter a number between 1 and 6 for the product that needs to be made. The CASE statement would then run the section of code to produce that product.

The value of the CASE variable (tag) determines which section of the program is executed. In the example below, the value of tag Var_1 is used to decide which part of the logic to execute. This example shows several ways in which more than one number can be used to select a section of code.

If Var_1 is equal to 1, Temp_1 will be assigned a value of 85 and Pump_1 will be turned on.

If the value of Var_1 is 2, Temp_1 will be assigned a value of 105 and Pump_1 will be turned on.

If Var_1 is equal to 3 or 4, Temp_1 will be assigned a value of 110 and Pump_1 will be turned on. Note the use of a comma. When a comma is used between values, each is valid. For example 1, 4, 7, or 9 could be used. In that case if Var_1 were equal to a 1 or 4 or 7 or 9, that portion of the program would be executed.

If the value of Var_1 is 5, 6, 7, or 8, Temp_1 will be assigned a value of 115 and Pump_1 will be turned on. Note the use of the periods between the 5 and the 8. That means that any numbers between the two specified numbers are also valid. In this case a 5, 6, 7, or 8 would cause that portion of the program to be executed.

In the next portion of the code, the comma and the period declarations have both been used. In this case, if Var_1 is a 9, 11, 12, 13, 14, or 15, this portion of code will be executed.

Note the use of the ELSE at the end. If Var_1 has a value that is different from any of the cases specified, the ELSE portion of code will be run.

```
CASE VAR_1 OF
   1:        Temp_1 := 85;
             Pump_1 := 1;
   2:        Temp_1 := 105;
             Pump_1 := 1;
   3,4:      Temp_1 := 110;
             Pump_1 := 1;
   5..8:     Temp_1 := 115;
             Pump_1 := 1;
   9,11..15: Temp_1 := 120;
             Pump_1 := 1;
ELSE
   Alarm := 1;
   Additional logic;
END_CASE;
```

TIMERS

Other traditional programming instructions can be used in ST. For example, some types of timers, like a retentive (TONR) time, can be used. To check which instructions are available and to see an example of their use, you can use the instruction help available in RSLogix 5000. Look up a particular instruction, and it will explain which languages it can be used in and will also give examples of the use of the instruction in the language. A TONR timer was created and given a tag name of Stop_light_Timer. Note that the tag type must be a *FBD Timer* type for ST. Figure 9-13 shows the tag and tag members for the timer.

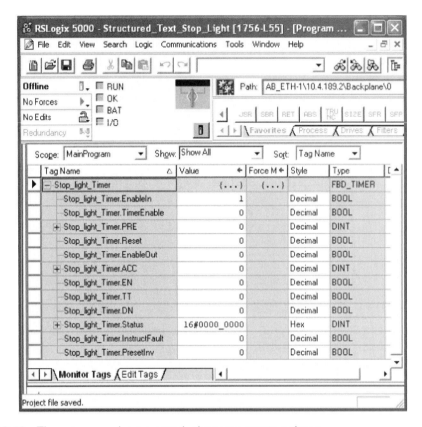

Figure 9-13 Timer tag members created when you create a timer.

The next example shows how a timer can be used in ST programming. Study the example. Note the use of the preset (.PRE), accumulated time (.ACC), and the timer enable (.TimerEnable). The first three lines are used to call the timer, set a PRE value, and enable the timer (start the timer timing). The next portion of code is IF statements that turn a green light on if the timer's ACC value is greater than 0 and less than 10000. The last IF statement resets the timer if the accumulated time has reached 30000 ms by making the timer enable 0 (false).

TONR(Stop_light_Timer);
Stop_light_Timer.PRE := 30000;
Stop_light_Timer.TimerEnable := 1;
IF (Stop_light_Timer.ACC > 0 & Stop_light_Timer.ACC < 10000) THEN
   Green_1 := 1;
ELSE
   Green_1 := 0;

END_IF;

IF Stop_light_Timer.ACC = 30000 THEN
Stop_light_Timer.TimerEnable := 0;
END_IF;

Figure 9-14 shows what this program would look like in RSLogix 5000. Note also the Watch list at the bottom of the screen. This enables the programmer to watch the values of tags during run mode.

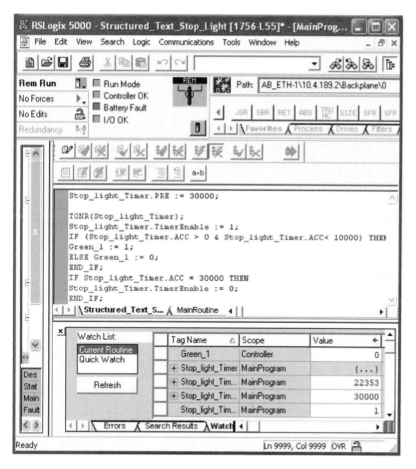

Figure 9-14 ST program example.

QUESTIONS

1. Write an assignment statement to:
   Assign the value of 5 to a tag named Count.
   Assign the value of the tag named T5 to a tag named Value.
   Multiply the value of a tag named count by a tag named Number and assign the result to a tag named Count.
2. Write one comment for each of the following. Make sure you use the correct format for a comment and make sure you explain the line. Use more than one format type.
   Temp := An_In_1/994.3;
   Var1 := 7;
   Total := ((5 + 8)/7) + 2;
3. Write ST for the following:
   Add 20 to a tag named TEMP and assign the result to Curr_Temp.
   Assign the value of a tag named Total to a tag named Amount.
   Multiply two tags and assign the result to a new tag.
   Motor_1 is a tag for digital output. Turn it on with an assignment statement.
4. What is the difference between DINT-type arithmetic and REAL-type arithmetic?
5. What is the result of the following statements?
   Answer := 4/2 /* Answer tag is a DINT */
   Answer := 5/2 /* Answer tag is a DINT */
   Answer := 5/2 /* Answer tag is a REAL */
   Answer := 5 MOD 2 /* Answer tag is a DINT */
6. Write a line of ST for each of the following:
   Find the square root of a number and assign the value to a tag.
   Find the tangent of a tag and assign it to another tag.
   Square a tag and assign the result to another tag.
7. Write an IF statement for each of the following:
   Temp is greater than 250.
   Var1 is greater than or equal to Var2.
   Var2 is less than Var1 multiplied by 6.3.
   If Sensor_1 is on, turn on Light_1.
   If Temp > 95, turn heater_1 off.
   If Temp is greater than 100 and Sensor_1 is true, turn Done_Light ON; otherwise turn Alarm ON.
8. When should a FOR DO loop be used?
9. Write a FOR DO loop to fill an array of temperatures with zeros.
10. Write a WHILE DO loop to turn Alarm_1 on until Temp is less than 150.
11. When should a REPEAT UNTIL loop be used?

12. Thoroughly explain the following ST:

    ```
    CASE Choice OF
    1:         Var1 := 65;
               Out_1 := 1;
    2,3:       Var1 := 85;
               Out_2 := 1;
    4:         Var1 := 95;
               Out_3 := 1;
    5,7...10:  Var1 := 115;
               Out_4 := 1;
    ELSE
       Statement;
       Statement;
    END_CASE;
    ```

13. Write ST to use a 60-second timer. An output should be on for the first half of the cycle and off for the second half of the cycle. It should continuously repeat.
14. Where can you find which CLX instructions are available for ST programming?
15. Write ST code for the following application:

    Develop a stoplight application. You must program both sets of lights. Make the overall cycle time 30 seconds.

Tagname	Description
Green_1	Green – East/West
Yellow 1	Yellow – East/West
Red_1	Red – East/West
Green_2	Green – North/South
Yellow_2	Yellow – North/South
Red_2	Red – North/South

CHAPTER 10

Sequential Function Chart (SFC) Programming

OBJECTIVES

On completion of this chapter, the reader will be able to:

- Explain what sequential function chart programming is.
- Explain types of applications that could benefit by the use of sequential function chart programming.
- Develop sequential function chart programs.

INTRODUCTION

Sequential function chart (SFC) programming is a very useful and friendly language. SFC is very useful for helping organize an application. This chapter will start with an overview of what a SFC program is and then explain each of the components of an SFC program in detail. It should be noted that the only way you will learn to program is to write programs. Utilize the questions at the end of the chapter to make sure you understand the concepts and then practice writing and testing your logic.

SFC PROGRAMMING

SFC programming is a graphically oriented programming language. SFC program looks like a flow diagram or a decision tree. If an application is sequential in nature, SFC programming is a natural choice. In SFC programming an application is broken into logical steps. For example, when making a pot of coffee, there are some definite steps that are followed.

1. Measure and add coffee.
2. Measure and add water.
3. Turn the pot on.
4. Remove completed coffee and turn the pot off.
5. Remove the filter and grounds and clean the pot.

SFC programming could be used to program this system. The coffee-making process could be broken into five steps. Figure 10-1 shows an example of what it might look like. Step 1 has an *action* that the operator must perform. The action is to measure and add coffee. There is a decision at the end of the step. If coffee has been added, the process continues at step 2; if not, processing remains in step 1. Step 2 has one action associated with it (measure and add water). At the end of step 2 is another decision. Step 3 also has an action and a decision. Steps 4 and 5 have two actions associated with each. Note that every step has a decision point after it that determines when processing for the step is done and the next step should be started.

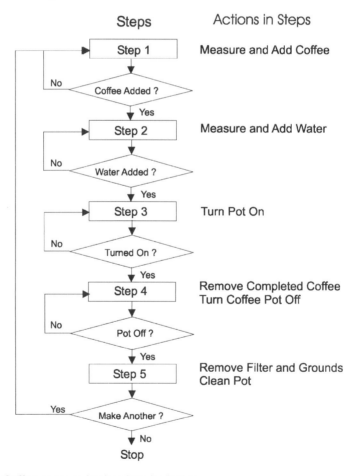

Figure 10-1 Coffee process broken into logical steps.

An SFC program consists of three main components: steps, actions, and transitions. Carefully study the typical SFC program in Figure 10-2. In this example, the first two step names were left with their default name: Step_000 and Step_001. The remaining three steps were given names that reflect the purpose of the steps: Normalize, Assemble, and Paint. There are transitions between steps (Tran_000, Tran_001, and Tran_002).

Study the transition between Step_000 and Step_001. A transition is a condition to determine when processing moves from one step to the next by evaluating to true or false. If the transition is true, processing will move to the next step. The first two steps are linear. In other words the first one is processed, then the second, and then the next. The next three steps are concurrent steps. All three are being processed at once. Note the actions that are attached to steps (Action_000, Action_001, and Action_002). Actions can use ST to make decisions and control I/O. Note that actions can be hidden to make a program less cluttered. Note also that each action has a qualifier that controls when an action starts and stops. These will be covered later in this chapter.

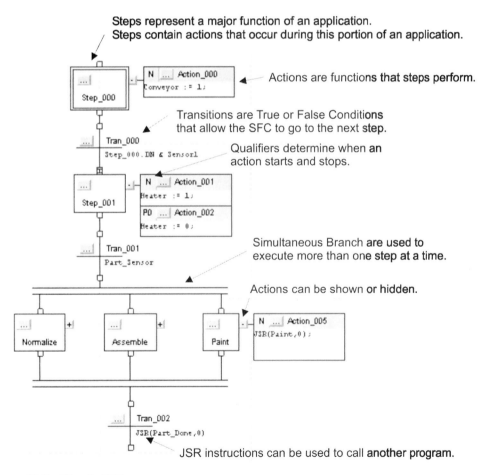

Figure 10-2 Simple SFC program.

SAMPLE APPLICATION

For this application imagine a system that is used to heat parts to a specified temperature.

- The operator puts a part in the fixture and pushes a start switch.
- The heating coil is turned on for 50 seconds.
- When 50 seconds is done, the operator removes the completed part, places another part in the fixture, and starts a new cycle.

Figure 10-3 shows two examples of simple SFC programs that could control the heating application. Step 1 is named Wait_For_Start_Cycle. The processor will stay in this step until the transition after the step is true. In this example the transition condition is the state of the start switch (Start_Switch). When the operator pushes the start switch, the transition is true and the processor will execute the next step (Heat_Cycle).

Heat_Cycle is a timed step. Parameters were set in the step to make it 50-second step. An action was added to the step. The action turns on the heater output (Heat_Output). The heater output was programmed to be nonretentive so it will shut off when the processor leaves this step. When the step has reached 50 seconds, the Heat_Cycle.DN bit will be true. Note that this bit was used for the next transition. In the example on the left the transition was then wired to the first step, so the process can be executed again when the start switch is pushed by the operator. The example on the right is almost the same except it will only run once. A stop was programmed at the end of the sequence.

Steps are the logical groupings that we break our application into; for example, the first step in our coffee-making process is measuring and adding coffee. The step is the organizational unit. Steps can perform timing and counting functions and can have actions associated with them.

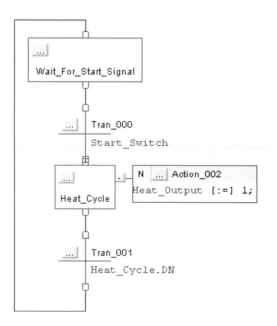

CHAPTER 10—SEQUENTIAL FUNCTION CHART (SFC) PROGRAMMING

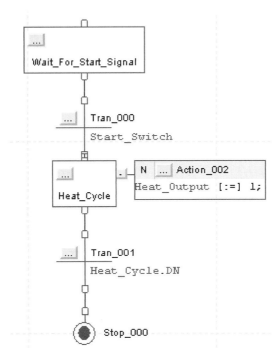

Figure 10-3 Simple heating cycle SFC programs. The one on the top will repeat. The one on the bottom will run once and stop.

Actions can be thought of as the inputs and outputs we use to accomplish the tasks in a step. Actions in a step are repeated until the transition to the next step becomes true. Transitions are BOOL statements that must be true to move to the next step.

ORGANIZING THE EXECUTION OF THE STEPS

Linear Sequence

A linear sequence is used to execute one or more steps in a linear fashion. Figure 10-4 shows a bottling operation that is linear.

Figure 10-4 A linear step bottling process.

Figure 10-5 shows a linear SFC. One step is executed continuously until the transition becomes true, then the next step in the sequence is executed continuously until the next transition is true, and so on.

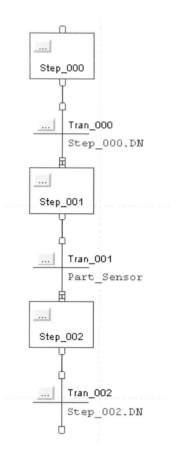

Figure 10-5 Linear steps in an SFC.

Wiring (Connecting) Steps

Wires are used to connect steps. Wiring is done by dropping elements on the attachment tabs or by clicking on the tab of one element and dragging it to the tab of the element you want to connect to. Tabs turn green when you can connect. You can connect a step to a previous point in an SFC (see Figure 10-6). This enables you to loop back to repeat steps or return to the beginning of an SFC and start over.

CHAPTER 10—SEQUENTIAL FUNCTION CHART (SFC) PROGRAMMING

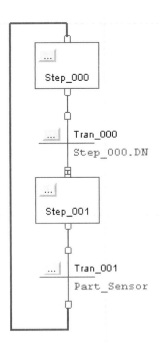

Figure 10-6 Looping back in an SFC.

Concurrent (Simultaneous) Processing

If there were more than one person making coffee, they could do some steps at the same time. One person could be putting a filter and coffee into the coffee machine while another could be measuring water and pouring it into the pot. Figure 10-7 shows an example of concurrent coffee processing.

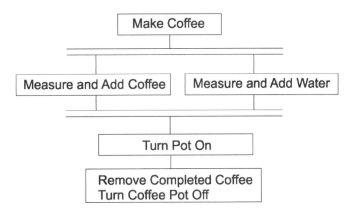

Figure 10-7 Concurrent processing.

A simultaneous branch is used to execute two or more steps or groups of steps at the same time. Figure 10-8 shows an example of a simultaneous branch. All paths must finish before continuing on in the SFC program. A single transition is used to enter a simultaneous branch and a single transition is used to end a branch. The SFC program will check the end transition, after the last step in each path has executed at least once. If the transition is false, the last step is repeated.

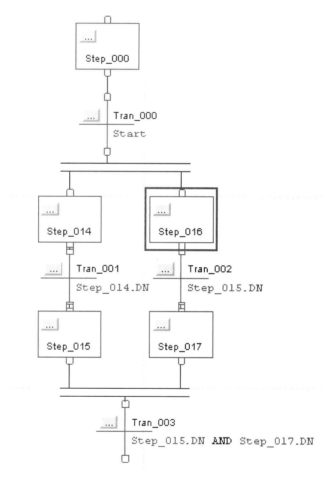

Figure 10-8 A simultaneous branch. Note the parallel horizontal lines at the top and bottom of the branch.

Selection Branching

A selection branch is used to choose between paths of steps depending on logic conditions. Figure 10-9 shows a selection branch SFC. Note that each path begins with a transition. The SFC program will check the transitions that start each path from left to right. It will take the first path that is true. If no transitions are true, the previous step will be repeated. The software will let you change the order in which transitions are checked. It is acceptable for a path to have no steps and only a transition. This could be used to skip an entire selection branch under certain conditions. In Figure 10-9 if the first three transitions are false and Tran_015 is true, processing would go through the selection branch to the Pack step after the branch.

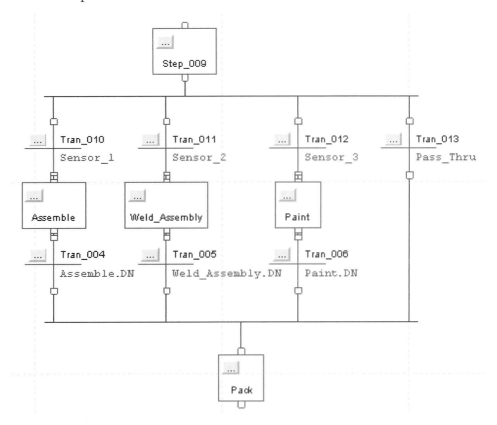

Figure 10-9 A selection branch SFC. Note the *single* horizontal lines at the beginning and end of the selection branch.

At this point you should understand the basics of what an SFC is. You should understand what steps, actions, and transitions are used for. Next, steps, actions, and transitions will be examined in more detail.

STEPS

Steps have a lot of functionality built into them. To identify steps in a process, it may be helpful to look for physical changes in the process. A physical change might be something such as a pressure that is reached, a temperature that is reached, a new part that is now in position, or a choice of which product or recipe to use, and so on. This physical change can represent the end of a step. The step will consist of the actions that occur before that change. Be careful not to have too many steps or too few steps. Too many or too few may make the program confusing. Make steps meaningful.

When a step is created, several tag members are automatically created that are associated with the step's tag name. Figure 10-10 shows an example of members that were automatically created for a step tag named Step_000.

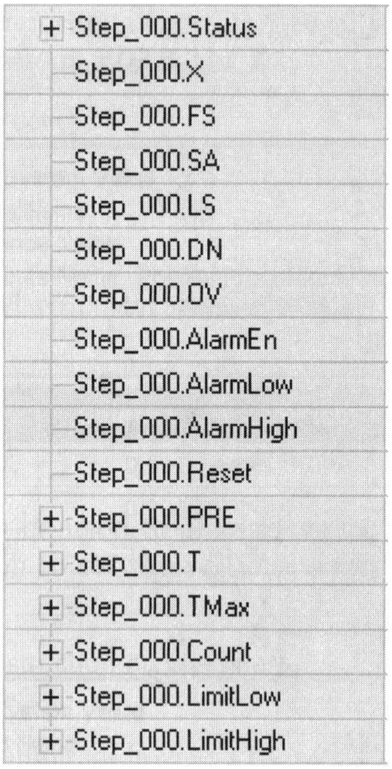

+ Step_000.Status
Step_000.X
Step_000.FS
Step_000.SA
Step_000.LS
Step_000.DN
Step_000.OV
Step_000.AlarmEn
Step_000.AlarmLow
Step_000.AlarmHigh
Step_000.Reset
+ Step_000.PRE
+ Step_000.T
+ Step_000.TMax
+ Step_000.Count
+ Step_000.LimitLow
+ Step_000.LimitHigh

Figure 10-10 Step_000 tag members.

Let's consider a few of the available tag members.

Step_000.DN would be the done bit. If we set a PRE value in milliseconds for the step, the DN bit for the step will be true when the step's time gets to the PRE value which is found in Step_000.T.

Step_000.AlarmHigh is a bit that would be set to true when the step's accumulated time (Step_000.T) gets to Step_000.LimitHigh value. Any of the tag members can be used in logic.

You can change the name of the step by right clicking on the step and choosing the Rename option. When you rename a step, all of the step tag members are renamed to the new name. One of the tags is a DN bit.

Figure 10-11 is a table that shows all of a step's members and their use. These step members (tags) can all be used in logic. To use tag members, the name of the step is followed by a period (.) and the member name.

To	Tag Member	Data Type	Function
Length of time a step has been active	T	DINT	When the step becomes active, the T value is reset and then begins to increment in milliseconds.
Length of the timer preset	PRE	DINT	Enter the length of time you want for the timer PRE value.
Timer DN Bit	DN	BOOL	The DN bit is set when the timer reaches the PRE value. The DN bit will stay on until the step becomes active again.
If a step did not execute long enough	LimitLow	DINT	Enter a value in milliseconds. If the step becomes inactive before the timer (T) reaches the LimitLow value, the AlarmLow bit turns on. The AlarmLow will stay on until you reset it. To use the alarm function, make sure the alarm enable (AlarmEn) is checked.
	AlarmEn	BOOL	To use the alarm bits, you must turn (check) on the alarm enable (AlarmEn) bit in the step.
	AlarmLow	BOOL	If the step becomes inactive before the timer (T) has reached the LimitLow value, the AlarmLow bit will be turned on. The bit will stay on until you reset it. To use this alarm function, you must turn on (check) the alarm enable (AlarmEn) bit in the step.
If a step is executing too long	LimitHigh	DINT	Enter a value in milliseconds. If the step becomes inactive before the timer (T) reaches the LimitHigh value, the AlarmHigh bit turns on. The AlarmHigh will stay on until you reset it. To use the alarm function, make sure the alarm enable (AlarmEn) is checked.
	AlarmEn	BOOL	To use the alarm bits, you must turn (check) on the alarm enable (AlarmEn) bit in the step.
	AlarmHigh	BOOL	If the step becomes inactive before the timer (T) has reached the LimitHigh value, the AlarmHigh bit will be turned on. The bit will stay on until you reset it. To use this alarm function, you must turn on (check) the alarm enable (AlarmEn) bit in the step.
Do something while the step is active (including the first and last scan)	X		The X bit is set the entire time a step is executing. It is recommended to use an action with an N (nonstored) qualifier to do this.

Continued

Do something once when the scan becomes active	FS	BOOL	The FS bit is on during the first scan of a step. It is recommended to use an action with a P1 Pulse (rising edge) to do this.
Do something while the step is active	SA	BOOL	The SA bit is on while a step is executing except for the first and last scan.
Do something one time on the last scan of the step	LS	BOOL	The last scan (LS) bit is on during the last scan of a step. This bit should only be used if you do the following: In the controller's properties box, SFC execution tab, set the Last Scan of Active Step to Don't Scan or Programmatic Reset.
Determine the target of an SFC Reset (SFR) instruction	Reset	BOOL	An SFC Reset (SFR) instruction resets the SFC to a step or stop that the instruction specifies. The Reset bit indicates to which step or stop the SFC will go to begin executing again. Once the SFC executes, the Reset bit is cleared.
The maximum time that a step was active during any execution	TMax	DINT	This is normally used for diagnostics. The controller will clear this value only when you select the Restart Position of Restart at initial step and the controller changes modes or experiences a power cycle.
Determine if the timer (T) value rolls over to a negative value	OV	BOOL	This is used for diagnostics.
Find out how many times a step has been active	Count	DINT	This is the number of times a step has been active, not the number of scans. The count is incremented each time a step becomes active. It will only be incremented after a step goes inactive and then active again. The count only resets if you configure the SFC to restart at the initial step. Under this configuration it will be reset when the controller is changed from program to run mode.

Figure 10-11 Table showing the tag members for a step.

Using the Preset Time of a Step

The PRE value of a step can be used to control how long a step executes. Figure 10-12 shows the properties screen for a timed step. In this example the heating step (Step_000) needs to be run for 40 seconds (40000 ms). Note that the PRE value for a step is entered into the Preset member. Figure 10-13 shows the step and transition. The DN bit of the step is used as the transition to quit executing Step_000 and move to the next step. Note that the Step_000 DN bit was used (Step_000.DN) in the transition. When the step's accumulated time equals 40000, the Step_000.DN bit will be true, making the transition true, and the program will move to the next step.

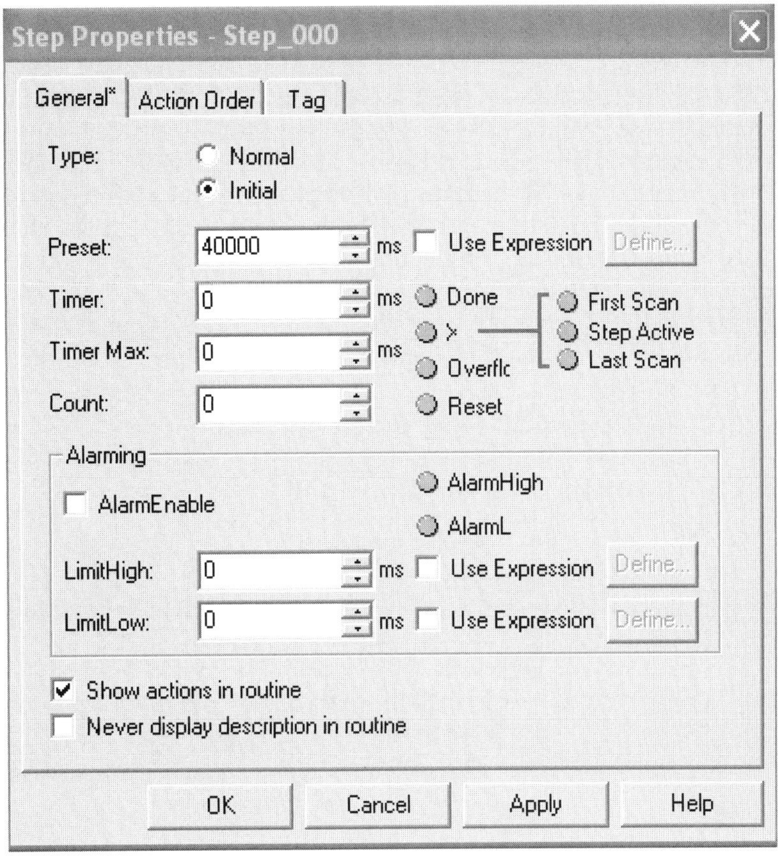

CHAPTER 10—SEQUENTIAL FUNCTION CHART (SFC) PROGRAMMING

Figure 10-12 Step properties.

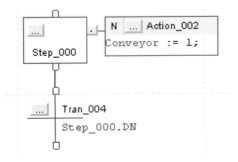

Figure 10-13 A timed step.

Use of a Step's Alarm Members

The next example makes use of the Step's AlarmHigh member. The AlarmHigh member is one bit. Study Figure 10-14. To use the alarm bits in a step you must turn on the Alarm Enable (AlarmEn) parameter bit in the step (refer back to Figure 10-12). This is done in the check box labeled AlarmEnable. You must also enter a value for LimitHigh. If the step's timer (T) has reached the LimitHigh value, the AlarmHigh bit will be turned on. The bit will stay on until you reset it. In this example the step was supposed to be finished successfully in less than 10 seconds (10000 ms). If it is finished, the transition on the left of the selection branch will be true and the step named Process will be executed. If the tasks in the step are not completed in less than 10 seconds, the step's AlarmHigh bit (INIT.AlarmHigh) will be set, the transition on the right will be true, and the step named Shutdown will be executed.

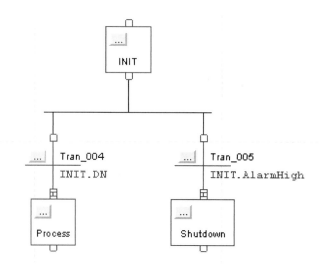

Figure 10-14   Use of a Step's AlarmHigh member (bit).

Turning Devices Off at the End of a Step

Devices can be turned off at the end of a step through program logic or automatically. There are three choices for how steps are scanned. The choices are found in Controller Properties under the SFC execution tab (see Figure 10-15). The three choices that control how the last scan of an active step is handled are Automatic reset, Programmatic reset, and Don't scan.

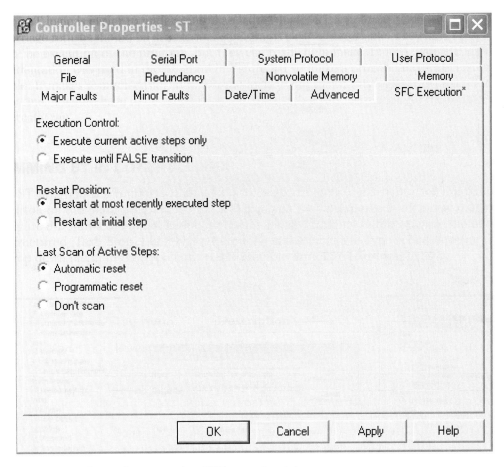

Figure 10-15 Controller properties, SFC execution tab.

Automatic Reset Option
Automatic reset may be the most straightforward. If you check the Automatic reset option choice (see Figure 10-15), square brackets in assignment statements as shown in Figure 10-16, outputs will be turned off when leaving the step. If the square brackets were not used, the output would remain on.

Output_5 [:=] 1;

The square brackets would make this action nonretentive, and Output_5 would be turned off when the step ends. Remember that you must choose Automatic reset in the Controller Properties for this to work.

Programmatic Reset Option

If the Programmatic reset option is chosen, you can use the last scan of a step to change the state of devices. Figure 10-16 shows an example of Programmatic reset. Note that the step's LS bit is used in the IF statement. During all scans except the last scan, the output is on. On the last scan of the step, the output would be turned off.

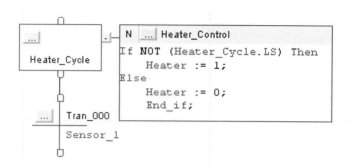

Figure 10-16 Use of the LS bit to control a device.

Don't Scan Option

The Don't scan is the default option for scanning. If this option is chosen, all data keep their current value when they leave a step. The programmer must use assignment statements or other instructions to change any data that need to be changed at the end of a step. A falling edge pulse (P0) action can be used. The P0 action should be the last action in a step. Only P and P0 actions are executed in the last scan if the Don't scan option is chosen. Figure 10-17 shows an example of the use of a P0 in Action_002 to turn a device off at the end of a step.

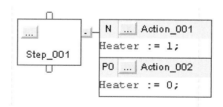

Figure 10-17 Use of a P0 action to turn an output off at the end of a step.

These have been only a few examples of what can be done with the step tag members. Study the table of step tags and their uses in Figure 10-11 for additional possibilities.

ACTIONS

Actions are used to perform functions such as turning outputs on or off in a step. Actions are added to steps by right clicking on a step and then choosing Add Action. Two actions have been added to a step in Figure 10-18.

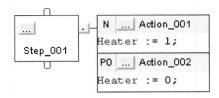

Figure 10-18 Actions in a step.

Action Tag Structure

Action tag members are created when you create an action. Figure 10-19 is a table that shows the action tag members and their uses.

To	Check or set this member of the structure	Type	Details	
Determine when the action is active.	Q	BOOL	The status of the Q bit depends on whether the action is a Boolean or a non-Boolean action.	
			Type of Action	State of Q Bit
			Non-Boolean	On while the action is active, but off at the last scan of the action
			Boolean	On the entire time the action is active including the last scan of an action
			Use the Q bit to determine when an action is active.	
	A	BOOL	The A bit is true the entire time an action is active.	
Determine how long an action has been active in milliseconds.	T	DINT	When an action becomes active, the timer (T) value resets and then starts to count up in milliseconds. The timer will count up until the action goes inactive, regardless of the PRE value.	
Use a time-based qualifier such as L, SL, D, DS, or SD.	PRE	DINT	Enter a time limit, or delay, in the preset (PRE) member. The action starts or stops when the timer (T) reaches the PRE value.	
Determine how many times an action has become active.	Count	DINT	Count is not a count of the scans of the action. The count is incremented each time the action becomes active. Count will increment only when the action goes inactive and then active again. The count will only reset if the SFC program is configured to restart at the initial step. With that configuration, it resets when the controller changes from program to run mode.	

Figure 10-19 Action tag member table.

There are two types of actions, non-Boolean and Boolean.

Non-Boolean Actions

Non-Boolean actions contain the logic for an action. They use ST to execute instructions or call a subroutine. ST can be used in actions for assignment statements, logic, or instructions. Figure 10-20 shows an example of an assignment statement in an action.

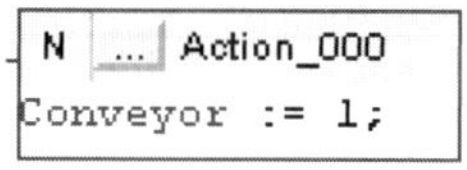

Figure 10-20 A retentive action.

Non-Boolean actions can also be used to call other subroutines. Figure 10-21 shows an example of a transition being used to execute a JSR. This subroutine could be another language or another SFC. JSRs are also commonly used in actions to call other subroutines.

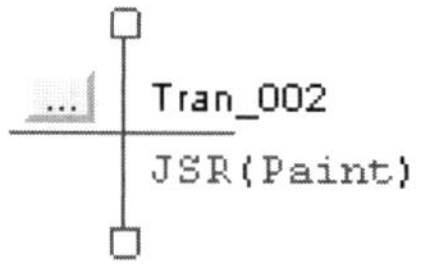

Figure 10-21 ST to call another routine.

Boolean Actions

Boolean actions can also be used. Figure 10-22 shows how a Boolean action is configured. Note the checkmark in the Action Properties screen. The Q bit for this action is used in logic in Figure 10-23. The Q bit for this action will be true when the action is active.

The use of the Q bit is explained in the table in Figure 10-19. Study the Q bit. If the action is set up as Boolean, the Q bit will be true the entire time the action is active, including the last scan. If the action is non-Boolean, the Q bit will be true while the scan is active until the last scan. The Q bit will be set to false in the last scan. This can be very helpful.

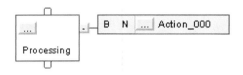

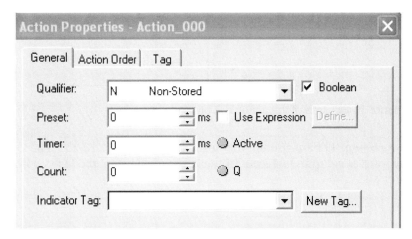

Figure 10-22 A Boolean action. Note that Boolean was checked in the Action Properties screen.

Figure 10-23 A Boolean action being used to call a subroutine named Processing_Cycle.

A Boolean action contains no logic for the action. It is used to set a bit in an action's *member*. To use the actual action, other logic must monitor the bit in the action tag and execute when the bit is set. If you use Boolean actions, you will normally have to manually reset the assignments and instructions that are associated with the action. There is no link between the Boolean action and the logic to perform the action so the Automatic reset option does not affect Boolean actions.

Figure 10-24 shows an example of a Boolean action. When Step_002 is active, the Boolean Action_007 executes. When the action is active, the Q bit is true for the action. The Q bit is true while the step is active until the last scan when it turns false. In this example, when the step becomes active, the Q bit will be true and Heater_Output will be set to 1. During the last scan of the actions, the Q bit becomes false and Heater_Output is set to 0.

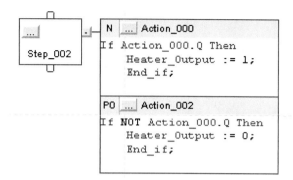

Figure 10-24 Use of the action's Q bit.

A step's Q bit is used in the next example (Figure 10-25). The Q bit is always true if the step is not in its last scan. In the last scan of a step, the Q bit is false (0).

Figure 10-25 Use of an action's Q bit in a ladder diagram.

The Order of Execution for Actions

Actions are executed for a step from top to bottom. The order of the actions for a step can be changed in the step's properties. Figure 10-26 shows how the order of actions can be changed. The programmer simply selects an action and moves it up or down into the desired execution order. Figure 10-27 shows a step and its actions before and after the order of execution was changed.

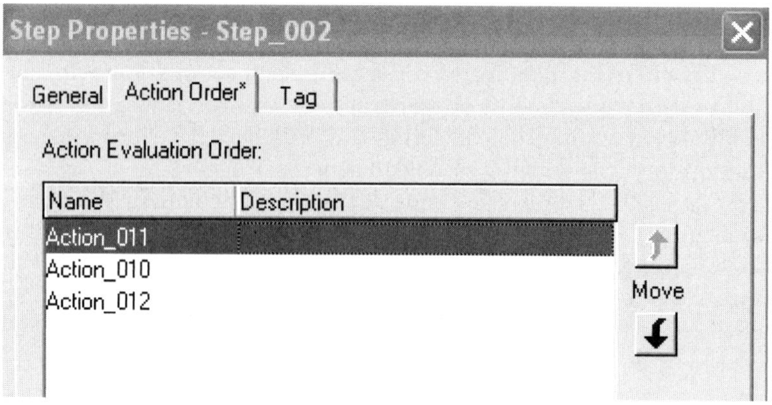

Figure 10-26 The Step Properties setup screen with the Action Evaluation Order chosen.

CHAPTER 10—SEQUENTIAL FUNCTION CHART (SFC) PROGRAMMING

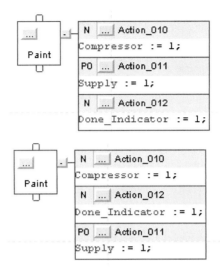

Figure 10-27 Order of execution for actions before and after the order was changed.

Using an Action to Call a Subroutine

Figure 10-28 shows the use of an action to call a subroutine named temp. Note the tag names between the parenthesis. The name of the subroutine is temp. We will be sending one parameter to the subroutine (Setpoint), and one will be returned from the subroutine (Current_Temp). Figure 10-29 shows the setup screen for the JSR call. The programmer entered Setpoint for the input parameter and Current_Temp for the return parameter. Note that you do not have to send or return values. In this example, the JSR instruction (action) sends the value of the tag Setpoint to the subroutine and the subroutine returns the value of the tag Current_Temp.

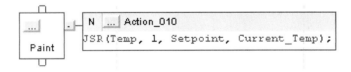

Figure 10-28 An action used to call a subroutine.

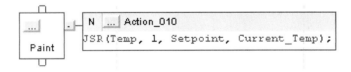

Figure 10-29 Configuration screen for the JSR action in Figure 10-25.

Qualifiers for Actions

Study the table shown in Figure 10-30. There are several action qualifiers that can be used to control how an action executes.

- N is a nonstored action. The N action will stop when the step is deactivated.
- The P1 qualifier executes once when the step is activated.
- S is the stored qualifier. An S action will remain active until a reset action turns off this action.
- A D qualifier causes the action to activate a specific time after a step has been activated and deactivate when a step is deactivated.
- A P qualifier will execute once when the step is deactivated.
- An R qualifier is used to reset (deactivate) a stored step.

If the Action is to	And	Use	Type
Start when a step is activated.	Stop when the step is deactivated.	N	Nonstored
	Execute only once.	P1	Pulse (Rising edge)
	Stop before the step is deactivated or when the step is deactivated.	L	Time limited
	Stay active until a reset action turns off this action.	S	Stored
	Stay active until a reset action turns off this action or a specific time expires, even if the step is deactivated.	SL	Stored and time limited
Start a specific time after the step is activated and the step is still active	Stop when the step is deactivated.	D	Time delayed
	Stay active until a Reset action turns off this action.	DS	Delayed and stored
Start a specified time after the step is activated, even if the step is deactivated before this time.	Stay active until a reset action turns off this action	SD	Stored and time delayed
Execute once when the step is activated.	Execute once when the step is deactivated.	P	Pulse
Start when the step is deactivated.	Execute only once.	PO	Pulse (Falling edge)
Turn off (reset) a stored action. S Stored SL Stored and time limited DS Delayed and stored SD Stored and time delayed		R	Reset

Figure 10-30 Table showing qualifiers for actions.

TRANSITIONS

Transitions are physical conditions that must happen or change before going to the next step (see Figure 10-31). A transition uses the state of Boolean logic (true or false) to determine whether processing should move to the next step. In Figure 10-32 a tag named Start was used for the Boolean state of the transition.

Transition State	Value	Result
True	1	Go to the next step.
False	0	Continue to execute the current step.

Figure 10-31 A transition table.

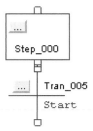

Figure 10-32 A transition. Note that the step will continue to execute until the transition is true.

A transition can be a Boolean or can be a JSR instruction to call another routine. Figure 10-33 shows some examples of Boolean transitions. The first example is just a Boolean tag. It could be a sensor's state, for example. If it is true (1), the transition will be true. The second example uses an AND to see if both Boolean tags are true. Both must be true for the transition to be true. The third example uses a Boolean operator (>) and a Boolean tag in an expression. If the expression is evaluated as true, the transition will be true.

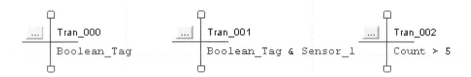

Figure 10-33 Use of Boolean expressions as transitions.

Figure 10-34 shows the use of a JSR instruction in an action to call a subroutine. In this example the action in Figure 10-34 calls subroutine Heater_On. The Heater_On subroutine is ladder logic that simply turns on an output named Heater_Output (see Figure 10-35).

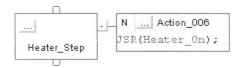

Figure 10-34 Step whose action calls a subroutine named Heater_On.

Figure 10-35 Heater_On subroutine. When this subroutine is called, it turns on Heater_Output.

Figures 10-36 and 10-37 show examples of the use of a JSR instruction in a transition. The JSR instruction in Figure 10-37 is used to call the routine named ST_Decision. The subroutine logic in the ST_Decision routine is used to control the transition state.

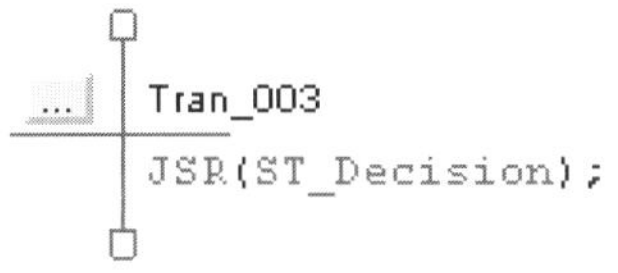

```
    Tran_003
    JSR(ST_Decision);
```

Figure 10-36 This transition calls a ST routine named ST_Decision.

The ST_Decision routine uses ST to decide the state of the transition. Note the use of the End of Transition (EOT) in the last line of the routine to return the state of BOOL-type tag to the transition. An EOT must be used at the end of a subroutine to return a 1 (true) or 0 (false) for the transition.

```
If Sensor_1 & Sensor_2 Then
    Trans_Tag := 1;
Else
    Trans_Tag := 0;
End_if;
EOT(Trans_Tag);
```

Figure 10-37 The ST_Decision routine. An EOT is used to return a value of 0 or 1 to the transition shown in Figure 10-36.

In the example shown in Figure 10-38, A JSR instruction is used to call a subroutine named Part_Done. Part_Done is a ladder logic subroutine. The Part_Done program is shown in Figure 10-39. Sensor_1, State_2, and State_3 must be true to make

Boolean_Tag true. If Boolean_Tag is true, the transition will be true. Note the use of the EOT to return the state of Boolean_Tag to the transition.

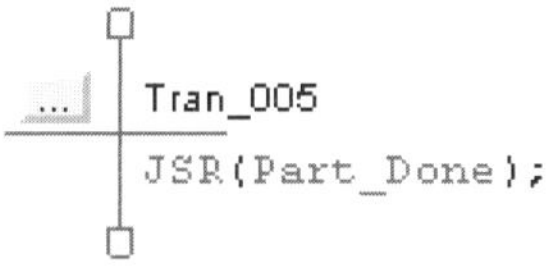

Figure 10-38 A transition that calls subroutine Part_Done.

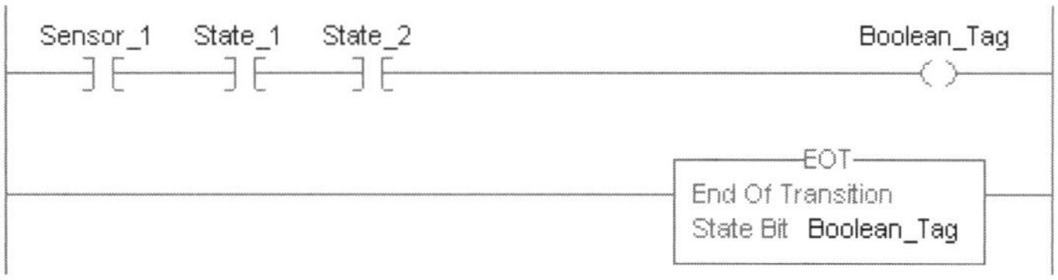

Figure 10-39 Subroutine Part_Done. If Sensor_1 AND State_1 AND State_2 are true, the EOT will return a true to the transition and the next step will be executed.

Keeping Outputs on During Multiple Steps

Figure 10-40 shows one method to turn a device on and keep it on for multiple steps. A regular assignment statement could be used to turn a device on as shown in Step_1's action. It will remain on until an assignment statement in an action in a different step turns it off. In this example it will be turned off by the action in Step_2.

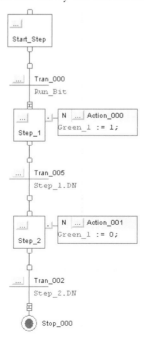

Figure 10-40 Keeping an output on for more than one step.

The example in Figure 10-41 shows the use of a simultaneous branch to keep an output on during multiple steps. This method may make the logic easier to understand.

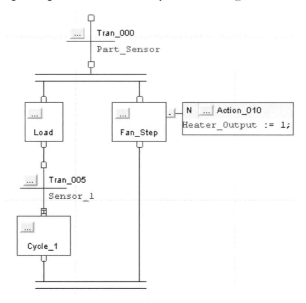

Figure 10-41 A simultaneous branch. Note the branch on the right keeps the Heater_Output on for the whole length of time that the simultaneous branch is active; in fact it will not turn off at the end of the branch either.

Another method of keeping actions active for multiple steps is by using Stored(S)-type actions. Figure 10-42 shows an example of an S-type action. An S-type action is used to keep the action active. Note that a reset action only turns off the desired action; it does not automatically turn off the devices in that action. You must use another action after the reset action to turn off the device.

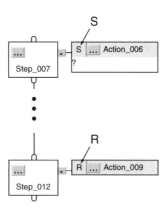

Figure 10-42 Set and reset actions.

Ending an SFC Program

Once an SFC program ends the last step, it does not automatically restart at the first step. If you would like to automatically go back to an earlier step in an SFC program, you would wire the last transition to the top of the step you want to execute next. If you would like to stop after the last scan and wait for a command to restart the SFC program, you would use a stop element. Figure 10-43 shows a SFC program with a Stop element at the end. When the SFC program reaches a stop element, the X bit of the stop element will be set to 1. Stored actions remain active. Execution of the SFC program stops. If a stop element is used in one path of a simultaneous branch, only that path's SFC program will stop scanning; the rest of the SFC program will execute.

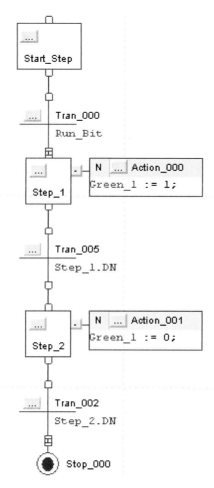

Figure 10-43 Use of a stop element.

Restarting a SFC Progam after a Stop

If another SFC program calls the subroutine, it will be reset to the initial step and execute if Automatic reset was chosen for the scan option. Figure 10-44 shows an example of calling another SFC program as a subroutine. If the Programmatic reset option or the Don't scan option was chosen, to restart an SFC program after a stop, you must use a SFC reset or logic to clear the X bit of the stop element.

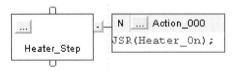

Figure 10-44 Use of a JSR instruction to call and execute another SFC program.

If no other SFC program calls the routine as a subroutine, use a SFC reset (SFR) instruction to restart the SFC program at the required step or use logic to clear the X bit of the stop element.

PROGRAMMING A SIMPLE SFC

To begin programming, right-click on MainProgram and add a new routine (see Figure 10-45). In this example the routine was given the name SFC_Routine.

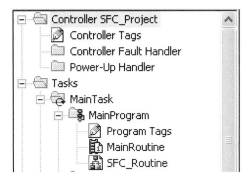

Figure 10-45 SFC Routine added.

Study Figure 10-46. This program has three steps. The program will stay in the Start_step until transition Tran_000 is true. The transition in this case is a Boolean tag named Run_bit. When you try your program, you can force this bit true to move to the second step.

Step_1 is used to turn the green light on for 30 seconds. The Preset in the step properties must be set to 30000 milliseconds (30 seconds). When the step time has reached

30000, the step's DN bit (Step_1.DN) will be set to 1. This will make the next transition true, and the next step will be executed. Step_2 turns the green light off. Then the program ends at the stop.

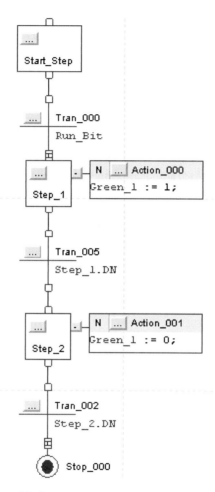

Figure 10-46 SFC program with three steps.

Adding Program Elements

Programming an SFC essentially consists of dragging and dropping program elements (steps and transitions) and then configuring the program elements and adding actions.

To begin the program, select the step and transition icon from the toolbar (see Figure 10-47). Drag it to the programming screen. Next select another step and transition icon from the toolbar and drag to the programming just under the first transition until you see a green dot. When you see the green dot, you can release the mouse button and the second step will be connected to the first transition. Add the third step and transition. Then select the stop icon from the toolbar and drag and drop it below the last transition.

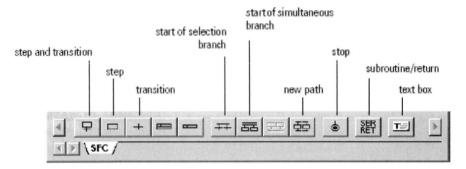

Figure 10-47 The SFC toolbar. (Courtesy of Rockwell Automation, Inc.)

Right-click on the first step and select properties or click on the ellipses button on the step to get the properties screen. This is the first step (initial step) so you must choose the Initial checkbox (see Figure 10-48).

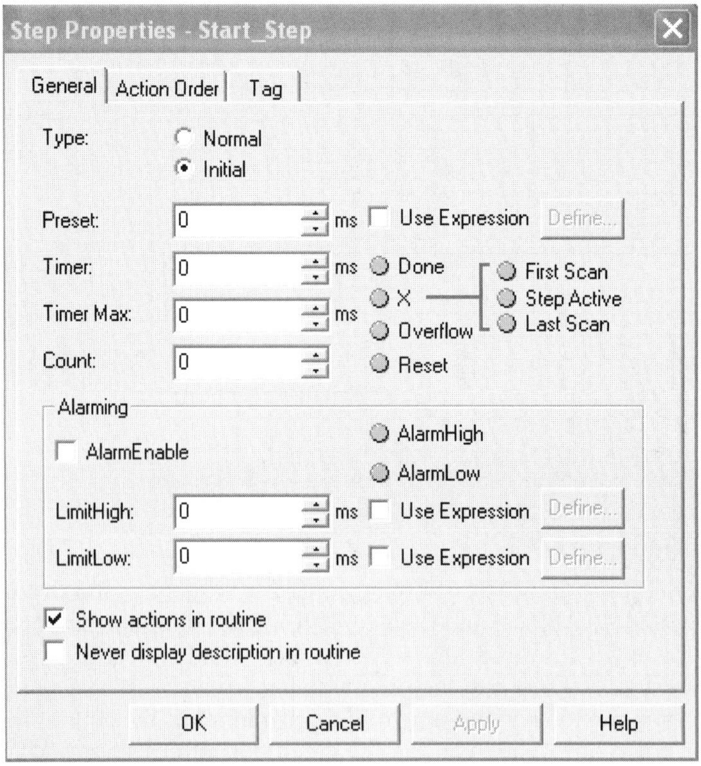

Figure 10-48 Step properties for the initial step.

As shown in Figure 10-49, right-click on Step_1, select properties, and enter a Preset of 30000 milliseconds (30 seconds). Next you can right-click on Step_001 and choose add action. Double-click in the bottom of the action, and you can enter your ST for the action as shown in Figure 10-50. Step_2 has one action to turn Green_1 off. This step will not need a preset time. You can add an action for Step_2 and add the action.

Figure 10-49 Step properties. Note that a Preset of 30000 milliseconds was used to create a 30-second preset.

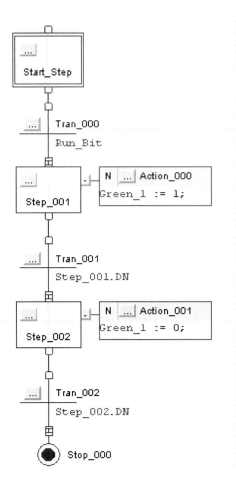

Figure 10-50 ST for the action.

At this point the program would still have errors because the Green_1 tag has not been created. This program has three step tags, two action tags, one tag in the actions and three tags in transitions that were created automatically when the elements were created. Right-click on the step tag name for the first step. In this example the tag is named Start_step. Right click on the tag and choose Edit tag to rename the tag. This tag type should be SFC step. Do the same for the other step tags. Next rename the tags for the transitions. These should be BOOL type. Lastly create the Green_1 tag. This could be an alias for a real-world output or a bit.

Programming a Simultaneous Branch

Simultaneous branches are programmed as shown in Figure 10-51. Click on the start of simultaneous branch button on the toolbar (see Figure 10-47), then drag it to where you want it. Next add paths to the branches. Click the first step of a path that is to the left of where you want to add the path, then click on the horizontal line of the simultaneous branch. After you have added the steps, the simultaneous branch can be wired to the preceding transition by clicking the bottom pin of the transition and then the horizontal line of the branch. A green dot will show the valid connection point.

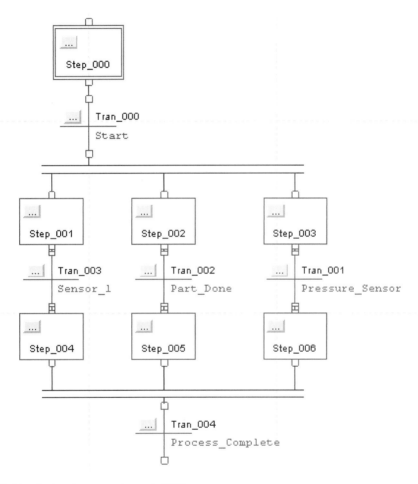

Figure 10-51 A simultaneous branch SFC.

Ending a Simultaneous Branch

To end a simultaneous branch, the last step of each path in the simultaneous branch is selected. Each branch must end with a step, not a transition (you will be connecting the last *step* of each branch to the simultaneous branch end, not transitions). This can be done by clicking and dragging the pointer around all of the desired steps. Or you may click on the first step and then press and hold the shift key down while clicking on the rest of the desired steps. Then you will click on the simultaneous branch end button in the SFC toolbar. It is located just to the right of the simultaneous branch icon in Figure 10-47. A transition can be added to the branch end.

There is another way to program the end to simultaneous branches. Wire from the connection point on the end of each step to a connection point on the end of the next step, and they will be joined with a simultaneous branch.

Programming a Selection Branch

Figure 10-52 shows a selection branch program. To add a branch, click on the start of selection branch button on the SFC toobar (see Figure 10-47) and drag the branch to the desired location. To add a path, select (click) the first transition of the path that is to the left of where you want to add the new path. Then click the start of selection branch button. Add the rest of the paths. To wire the selection branch to the preceding step, click the bottom of the step and then click on the horizontal line of the branch. A green dot shows where the connection point is.

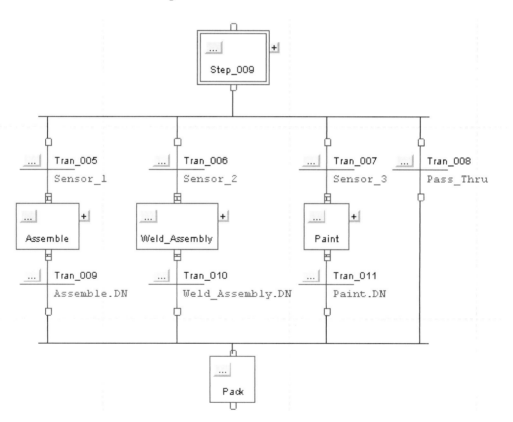

Figure 10-52 Programming a selection branch.

Ending a Selection Branch
Figure 10-51 also shows how to end a selection branch. First you must select all of the last transitions for each path (select the first transition and then hold the shift key down as you select the rest of the transitions). Then click on the end selection branch button. The other way to end a selection branch is to wire the connectors of each of the transitions together, and the end selection branch will be created automatically.

Note that a step follows a selection branch end.

Setting the Priorities for a Selection Branch
A selection branch evaluates the transitions in a selection branch from left to right. The first branch transition that is true will be executed. You may also change the execution priorities for each branch. To change priorities, right-click on the horizontal line that starts the selection branch and then choose Set Sequence Priorities. Figure 10-53 shows the priorities screen. Uncheck the Use default priorities check box. Select one sequence at a time and you can move them into the order you would like. Click OK when complete.

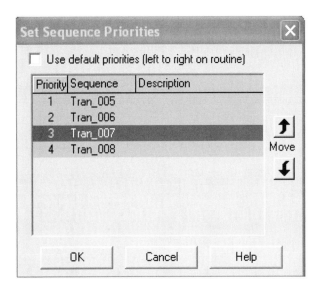

Figure 10-53 Sequence priority screen for a selection branch.

DOCUMENTATION OF SFC PROGRAMS

There are several methods to document an SFC program. Descriptions can be added for tags just like in ladder logic or any of the languages. Comments can be added in ST. Figure 10-54 shows an example of a comment in an action using ST.

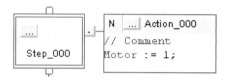

Figure 10-54 Comment added to an action in ST.

Text boxes can be added to program elements to document the program. In Figure 10-55 a text box was added to a step to help explain the purpose of the step. To add a text box, select the text box tool from the toolbar. A text box will appear that you can move around. Select the pushpin on the text box and you can attach it to the desired element by dragging the wire to the element.

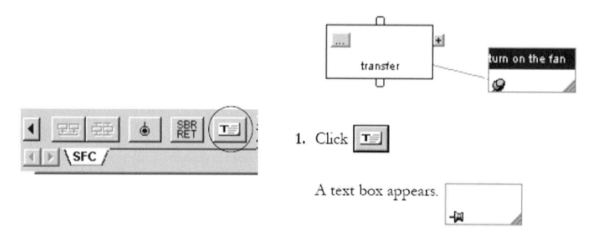

Figure 10-55 Adding text boxes. (Courtesy of Rockwell Automation, Inc.)

QUESTIONS

1. What do the letters SFC stand for?
2. What types of applications is SFC programming best suited to?
3. What is a step?
4. What is an action?
5. How can the order of action execution be changed?
6. What is a transition?
7. On paper, write a start step and a transition. Use a bit named Start for the transition.
8. Write a three-step program on paper. The first step is a start step that will wait until the run bit is true to move to the second step. Step 2 should turn on an output that turns a a motor on. The step should execute for 30 seconds. The last step should turn off the motor output. Explain how the step will be set up to run the motor for 30 seconds and what will be used for the transition.
9. Write a SFC routine in RSLogix 5000 from question 7 and make it continuously repeat after the run bit is turned on. The output should be on for 30 seconds and off for 5 seconds.
10. Explain how the order of execution is determined for selection branches.
11. Explain the following logic:

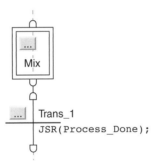

12. Explain the following logic:

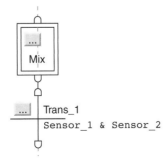

13. Explain the following logic:

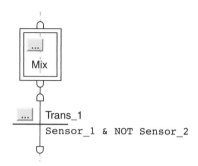

14. Explain the following logic:

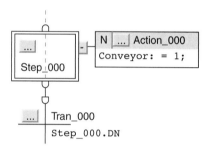

15. Explain the following logic:

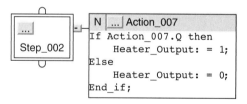

16. Explain the following logic:

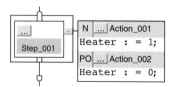

17. Explain the following logic:

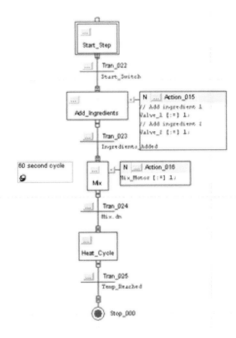

18. Explain the following logic:

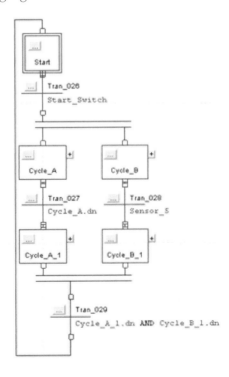

19. How can comments and text boxes be added to an SFC routine?
20. Write logic that uses ladder logic to determine the state of a transition.
21. Write logic in an action to call a subroutine.
22. Develop an SFC routine on paper or in a CLX to accomplish the following:

Step 1	Wait for Start_Switch.
Step 2	Turn a discrete valve named Product_A_Valve on. Turn on the output named HEAT. This step should run until a tag named Level_Fill_Sensor becomes true. When the step is done, the valve should be turned off but the heater output should remain on.
Step 3	Turn on Valve_2, and turn on the output named Mix_Motor. This step should continue until Level_2_Sensor becomes true. When the step ends, you must turn Valve_2 off.
Step 4	This step should continue until the Temp_tag reaches 150 degrees.
Step 5	Then turn Mix_motor and the heater off. Turn on the drain valve until Tank_Empty_Sensor is true. Run a subroutine named Bottle_Routine.
	Return to Step 1.

23. Write an SFC routine for the following application:

This is a simple heat treat machine application. The operator places a part in a fixture then pushes the start switch. An inductive heating coil heats the part rapidly to 1500 degrees Fahrenheit. When the temperature reaches 1500, the heating coil turns off and a valve is opened for 10 seconds to spray water on the part to complete the heat treatment (quench). The operator then removes the part and the sequence can begin again. Note there must be a part present or the sequence should not start. For simplicity assume the analog temperature sensor outputs a value that is equal to the actual temperature.

I/O	Type	Description
Part_Present_Sensor	Discrete	Sensor used to sense a part in the fixture
Temp_Sensor	Analog	Assume this sensor outputs a value that exactly corresponds to 0–2000 degrees Fahrenheit
Start_Switch	Discrete	Momentary normally open switch
Heating_Coil	Discrete	Discrete output that turns coil on
Quench_Valve	Discrete	Discrete output that turns quench valve on

CHAPTER 11

Function Block Diagram Programming

OBJECTIVES

On completion of this chapter the reader will be able to:

- Explain what function block programming is.
- Explain the types of applications that are appropriate for function block programming.
- Develop function block routines.

INTRODUCTION

Function block diagram (FBD) programming is one of the IEC 61131-3 languages. FBD is a powerful and friendly language once you have learned the basics. A function block can take one or more inputs make decisions or calculations, and then generate one or more outputs. A function block can output information to other function blocks. There are many types of function blocks available to perform various tasks. Function block programming can simplify programming and make a program more understandable. In ControlLogix the user develops a function block routine and uses a JSR instruction to run the routine from the main routine or another routine.

FBD programming is very useful for applications where there is extensive information/data flow. Process control typically involves more data flow and calculations than discrete manufacturing applications.

ADD Function Block

When you use a function block instruction in a routine, a tag is created for the function block that has several tag members. Figure 11-1 shows an ADD function block. The name of the function block (ADD_01) is the default name that is automatically created when you put the ADD function block into the routine. You can use the default tag name or you can change the tag name. The function block tag members are used to store configuration and status information about the instruction. Each function block tag has several tag members that can be used in logic. Figure 11-2 shows the tag and tag members that are created when you use a function block. In this example, five tag members were created that all use the name ADD_01 followed by a period (.) and a member name. These tag members can be used in logic.

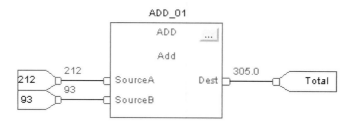

Figure 11-1 An ADD function block.

The EnableIn tag member (ADD_01.EnableIn) could be used in logic to enable this function block. The ADD_01.SourceA member would be the first of two values to be added. The tag member ADD_01.SourceB would be the second value to be added. Tag member ADD_01.EnableOut can be used to determine whether the result of the instruction is actually output. ADD_01.Dest is the output of the instruction. It is the result of the addition and would be put into the tag member Total. You do not need to use all of these tags. You could simply input a value to the SourceA input and a value to the SourceB input, and the instruction would output the sum of the two values to the Dest output line of the instruction.

Tag Name	Value
─ ADD_01	{...}
ADD_01.EnableIn	1
ADD_01.SourceA	0.0
ADD_01.SourceB	0.0
ADD_01.EnableOut	0
ADD_01.Dest	0.0

Scope: MainProgram   Show: Show All   Sort:

Figure 11-2 The tag members for an ADD function.

CHAPTER 11—FUNCTION BLOCK DIAGRAM PROGRAMMING   261

By right clicking on the function block, its parameters can be set (see Figure 11-3). Note that in this example, SourceA, SourceB, and the Dest were enabled. Function block instructions allow the programmer to determine which inputs and outputs will be used. Note that in this example the EnableIn was not checked to be used, but its value is 1 so the instruction is enabled.

Vis	Name	Value	Type	Description
1 ☐	EnableIn	1	BOOL	Enable Input. If False, the...
1 ☑	SourceA	0.0	REAL	Source A value
1 ☑	SourceB	0.0	REAL	Source B value
0 ☐	EnableOut	0	BOOL	Enable Output.
0 ☑	Dest	0.0	REAL	Dest value

Figure 11-3 Properties of an ADD function block.

Function Block Elements

Function blocks are used to take inputs, do some processing, and then provide one or more outputs. There are several methods to get information into and out of a function block. If you want to use tag-type information for an input to a function block, you would use an input reference (IREF). If you want to put the output from a function block into a tag, you would use an output reference (OREF) (see Figure 11-4).

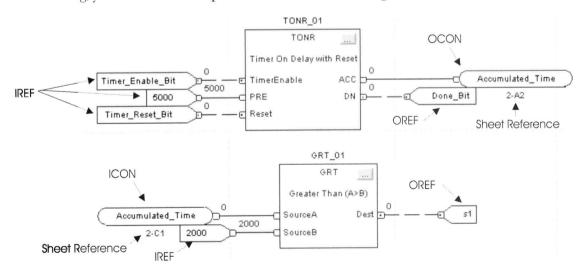

Figure 11-4 IREFs, OREFs, an OCON, and an ICON.

Function blocks can be connected to other function blocks by wiring their outputs to the input of another function block. If there are many function blocks on a sheet, the wiring can make the logic confusing. The other way to connect the output from a function block to the input of a function block is to use output connectors (OCONs) and input connectors (ICONs) (see Figure 11-4). ICONs and OCONs are used as connectors between function blocks. An OCON is an output connector and an ICON is an input connector. You cannot have an OCON without using an ICON of the same name. The table in Figure 11-5 shows what OCONs, ICONs, IREFs, and OREFs are used for.

Need	Element to Use
To send a value to an output device or a tag.	Output reference (OREF)
To receive a value from an input device or a tag.	Input reference (IREF)
To transfer data between function blocks. Note they can be on the same sheets or on different sheets.	Output wire connector (OCON) and input wire connector (ICON)
To send data to several places in a routine.	Single output connector (OCON) and multiple input wire connectors (ICONs)

Figure 11-5 Purpose of references and connectors.

When you use an IREF or an OREF you must create a tag or assign an existing tag to the element. You may use any of the tag data types for an IREF or an OREF.

Order of Execution

The order of execution is controlled by the way elements are wired together and by indicating feedback wires if they are required. The location of a block does not affect the order in which blocks are executed. Figure 11-6 shows an example of a simple FBD and the symbols. Note that the wire type indicates which type of data is being shared. A BOOL value would be a dashed line, and a solid line would indicate a SINT, INT, DNT, or REAL value. If function blocks are not wired together, it does not matter which block executes first as there is no data flow between the blocks. If blocks are wired sequentially, the execution order moves from input to output. The data must be available before a controller can execute a block. In Figure 11-6, the second function block (GRT_02) must execute before the third function block (BAND_02) because the output of the second function block is an input to the third function block.

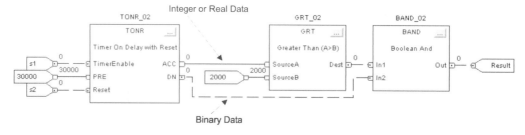

Figure 11-6 Simple FBD.

In Figure 11-7 there are two groups of blocks on one sheet. Execution order is only important for the blocks that are wired together.

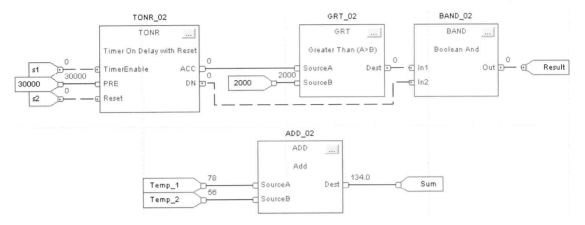

Figure 11-7 Execution of blocks that are not connected.

Feedback

Feedback to a block is done by wiring an output pin from a block to an input pin on the same function block. The input pin would receive the value of the output that was produced on the last scan of the function block. Study Figure 11-8. In Figure 11-8 the DN bit (TONR_02.DN) is used to reset the timer.

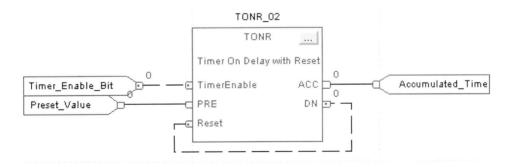

Figure 11-8 Feedback loop.

A controller cannot determine which function block to execute first for function blocks that are in a loop. The programmer must identify which block should be executed first by marking an input wire with the Assume Data Available Marker (see Figure 11-9). The arrow indicates that this data serves as the input to the first function block in the loop. Only one input to a function block in a loop should be marked. To add the Assume Data Available Marker, select the wire, right-click the mouse, and select the Assume Data Available choice.

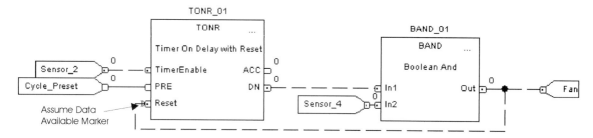

Figure 11-9 Assume Data Available Marker.

Figure 11-10 shows that if there is more than one connection between function blocks, they must all either be marked with the Assume Data Available Marker or none must be marked. The top example in Figure 11-10 is incorrect. The bottom example is correct.

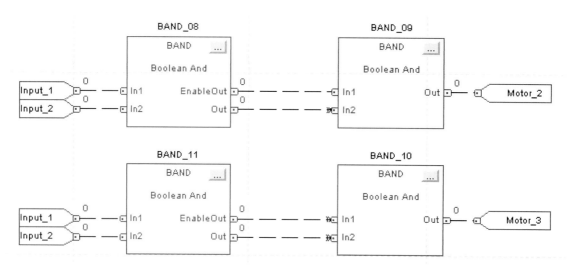

Figure 11-10 Use of the Assume Data Available Marker.

The Assume Data Available Marker can be used to create a one-scan delay between blocks (see Figure 11-11). In this figure, the first block is executed and then the second block uses the data that was generated in the previous scan of the function block routine.

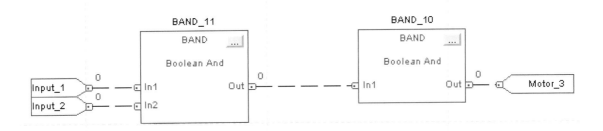

Figure 11-11 Use of Assume Data Available Marker to create a one-scan delay.

IREFs are used to provide input data to a function block instruction. The data in an IREF is latched (won't change) during the function block scan. IREF data is updated at the beginning of a function block scan.

Summary of the Execution of a Function Block Scan

1. The processor latches the data in all IREFs.
2. The processor executes the function blocks in the order determined by their wiring.
3. The processor writes outputs to the OREFs.

Connectors

ICONs and OCONs are used to transfer data between output and input pins (see Figure 11-12). They can be used to pass information between function blocks instead of wires when the elements you want to connect are on different sheets, when a wire might be hard to route on a sheet, when you want to provide the data to several points in a routine, or when you wish to pass data to another sheet of FBD. Note that in Figure 11-12 the output Accumulated_Time from the TONR_01 function block is used as the input Accumulated_Time to the GRT_01 function block. Note also that if SourceA (Accumulated_Time) were greater than SourceB (2000), the output (Dest) would be true. Accumulated_Time is not greater than 200 in this example so the output is false. Note that the input lines to the GRT function block are solid and the output line is dashed. The dashed line means that the output is a discrete value (1 or 0).

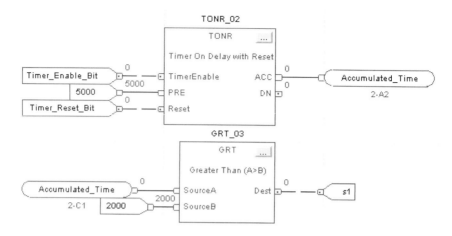

Figure 11-12 Use of connectors instead of wires.

Using Connectors
Each OCON must have a unique name. Connector names follow tag name rules, although they are not tags. Each OCON must also have at least one corresponding ICON. In other words, there must be at least one ICON with the same name as the OCON. Multiple ICONS can be used for the same OCON. This enables you to use an output value (OCON) in multiple places in your routine as an ICON.

MATHEMATICAL FUNCTION BLOCKS

There are many types of function blocks available. Mathematical function blocks are one type. Let's examine a few function blocks to see how a typical function block works.

ADD Function Block

An ADD function block can be used to add two numbers or the values of tags or numbers. Figure 11-13 shows an example of the use of an ADD function block. In this example two constants (numbers) were used as the inputs. Here 212 was added to 93 and the result was put into the output reference tag named Total. You can see that the result (305.0) is also shown above the output pin.

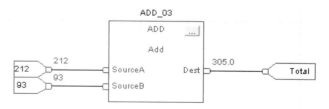

Figure 11-13 An ADD function block.

SUB Function Block

A SUB function block can be used to subtract two numbers or the values of tags or numbers. Figure 11-14 shows an example of the use of a SUB function block. In this example two constants (numbers) were used as the inputs. Here 93 was subtracted from 212 and the result was put into the output reference tag named Total. You can see that the result (119.0) is also shown above the output pin.

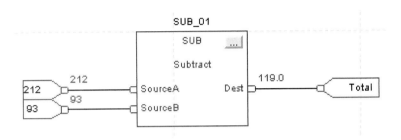

Figure 11-14 A SUB function block.

MUL Function Block

A MUL function block is shown in Figure 11-15. There are two inputs and one output. In this example, 212 was multiplied by 93 and the answer (19716.0) was output to the output reference tag named Total.

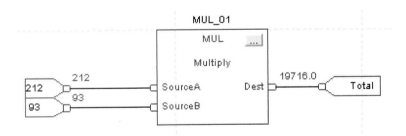

Figure 11-15 A MUL function block.

DIV Function Block

A DIV function block is shown in Figure 11-16. There are two inputs and one output. In this example, 212 was divided by 93 and the answer (2.2795699) was output to the output reference tag named Total.

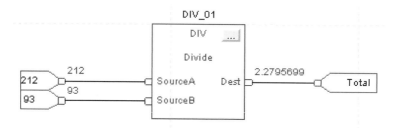

Figure 11-16 A DIV function block.

Boolean AND (BAND) Function Block

A BAND function block can be used to compare two or more discrete inputs (see Figure 11-17). If all are true, the discrete output of the function block will be true. If any or all are false, the output will be false.

This is a very useful instruction. There are many times in an application when we need to do something if exact input conditions are met. For example, if Sensor_1 is true AND Sensor_2 is true AND Sensor_3 is true AND Sensor_4 is true and we want to output a true from the instruction, BAND is the perfect instruction.

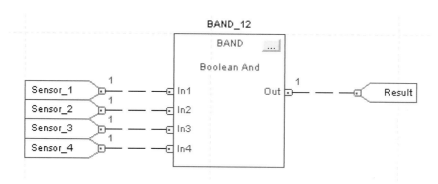

Figure 11-17 A BAND function block.

Boolean OR (BOR) Function Block

A BOR function block can be used to compare two or more discrete inputs. If any are true, the discrete output of the function block will be true. Figure 11-18 shows an example of a BOR function block. Note that there are four discrete inputs in this example. If one, some, or all of the four inputs are true, the output will be set to true. In this example only input 3 is true and the output is set to true. Note that if you right-click on a BOR function block, you can reduce or increase the number of inputs.

CHAPTER 11—FUNCTION BLOCK DIAGRAM PROGRAMMING 269

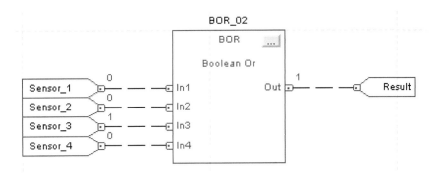

Figure 11-18 A BOR function block.

Figure 11-19 shows additional Compute/Math function blocks that are available. These function blocks include add, subtract, multiple, divide, modulo, square root, negate, and absolute. The MOD instruction is used to find the remainder of a division. The NEG instruction is can be used to change the sign of the Source and places the result in the Dest. The absolute instruction (ABS) takes the absolute (positive) value of the Source and places the result in the Dest.

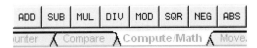

Figure 11-19 Compute/Math function block instructions.

TRIGONOMETRIC FUNCTION BLOCKS

Figure 11-20 shows trigonometric function blocks. Trigonometric function blocks include sine, cosine, tangent, arc sine, arc cosine, and arc tangent.

Figure 11-20 Trigonometric function block instructions.

STATISTICAL FUNCTON BLOCKS

Figure 11-21 shows statistical function blocks that are available. A MAVE instruction is a moving average instruction. The MSTD instruction can be used to calculate the moving standard deviation for a process. The MIN instruction is actually a MINC instruction. It means minimum capture. The MINC instruction finds the minimum of an input signal to the instruction over time. The MAX instruction is actually the MAXC instruction. The MAXC instruction finds the maximum of an input signal over time.

Figure 11-21 Statistical function block instructions.

Moving Average (MAVE) Instruction

The MAVE instruction calculates a time, average value for the In signal. This instruction optionally supports user-specified weights. It is available in function block and ST programming. An example of the use of this instruction is to monitor the size of product that is being made. This instruction could look at a moving average so that the correct adjustment could be made on the basis of the average size of a number of products that are made rather than just the last one made. This can make adjustments more accurate.

Initializing the Averaging Algorithm

Certain conditions, such as instruction first scan and instruction first run, require the instruction to initialize the moving average algorithm. Figure 11-22 shows an example of a MAVE instruction.

Each scan, the instruction places the input value from the In_Value tag in the StorageArray named Values. The most current input is put in the first element (Values[0] in this example) of the array named Values. The instruction calculates the average of the values in the StorageArray, optionally using the weight values in array weight, and places the result in Out. Note that a new value is input every scan that this instruction is true.

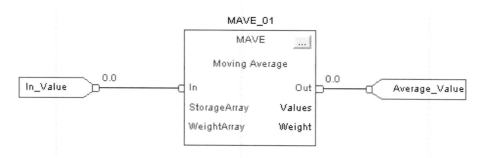

Figure 11-22 Use of a MAVE instruction.

Minimum Capture (MINC) Instruction

The MINC instruction finds the minimum of the input signal over time (see Figure 11-23). A good example of this might be to record the lowest temperature in a process during one day of operation. This instruction is available in function block and ST programming. The parameters for a MINC instruction are shown in Figure 11-24.

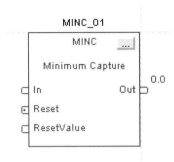

Figure 11-23 A MINC function block.

Inputs/Outputs	Data Type	Description
EnableIn	BOOL	If the Enable Input (EnableIn) is cleared, the instruction does not execute and outputs are not updated.
In	REAL	This is the analog signal input to the instruction. Any float is valid.
Reset	BOOL	This is a request to reset the control algorithm. This instruction sets Out = ResetValue as long as Reset is set. Any float is valid.
ResetValue	REAL	This is the reset value for the instruction. This instruction sets Out = ResetValue as long as Reset is set. Any float is valid.
EnableOut	BOOL	Enable output.
Out	REAL	This is the calculated output of the algorithm.

Figure 11-24 Parameters for a MINC instruction.

There is also a maximum capture (MAXC) instruction available to capture a maximum value from an input. An example of its use is to record the highest temperature during a day of production.

MATHEMATICAL CONVERSION FUNCTION BLOCK INSTRUCTIONS

Figure 11-25 shows some mathematical conversion function blocks. A DEG instruction can be used to convert radians to degrees. A RAD instruction is used to convert degrees to radians. A TOD instruction can be used to convert an integer to a BCD value. A FRD (convert to integer) instruction can convert a BCD value to an integer. A truncate (TRN) instruction is used to truncate an integer or a real value.

Figure 11-25 Mathematical conversion function block instructions.

Scale (SCL) Instruction

The SCL instruction converts an unscaled input value to a floating-point value in engineering units. These are very useful for converting the counts from an analog value to a number that makes more sense to an operator, for example, scaling the counts from an encoder on a motor to an actual speed in RPMs. Figure 11-26 shows the use of a SCL instruction. In this example the input raw values will be between 0 and 600. The SCL instruction will scale the input to a value between 0 and 60.

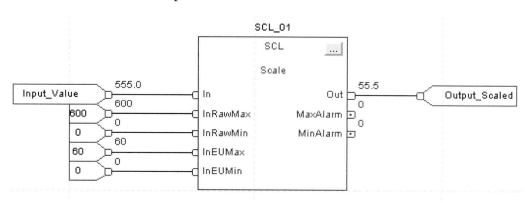

Figure 11-26 A SCL function block instruction.

FUNCTION BLOCK TIMERS

Figure 11-27 shows a timer on delay with reset (TONR) function block. It should look fairly familiar. It has the same basic inputs, parameters, and outputs as a TONR ladder logic timer. This timer was named TONR_01. TONR_01 is the default name and it can be changed. The TimerEnable input is used to enable the timer. In this example an input reference uses a tag named Start to enable the timer. A constant value (30000) was used in an input reference for the PRE value. The time base for CLX timers is milliseconds so this timer would be a 30-second timer. The accumulated time of the timer is being output to an output reference tag named ACC. The timer's DN bit (TONR_01.DN) is being used to reset the timer.

CHAPTER 11—FUNCTION BLOCK DIAGRAM PROGRAMMING   273

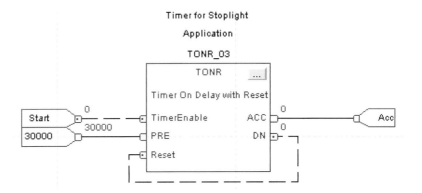

Figure 11-27 A TONR function block. Note that the DN bit was wired to the Reset input of the timer to automatically reset the timer when the DN bit becomes true. Note also that the connection is a dashed line. This means the connection passes a discrete value.

FUNCTION BLOCK COUNTERS

Figure 11-28 shows a function block counter. This counter can be used to count up and to count down.

Note that there is a count-up enable input and a count-down enable input. There is also a preset input (PRE) and a Reset input. The counter outputs include the present count (ACC) and a DN bit. The DN bit is set if the ACC is equal to the PRE.

Imagine a manufacturing cell where a component enters the cell to be worked on. As it enters the cell we might want to add it to the count of parts ready to be worked on. We could use a sensor (Part_In_Sensor in this example) to sense the part coming in as an input to the count-up input of the counter. As a part is finished and leaves the cell, we could have a sensor (Part_Out_Sensor) sense it leaving and use it as an input to the count-down input of the counter. The counter's accumulated value would always contain the number of parts actually in the cell. If we set the PRE value to 2, we could use the DN bit to warn the operator when there are only two parts left in the cell.

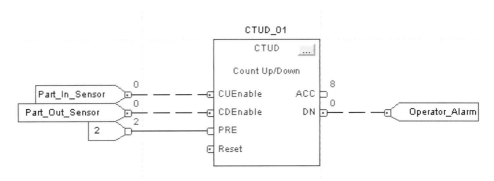

Figure 11-28 A count-up and count-down counter.

Figure 11-29 shows the timer and counter instructions that are available in function block programming. The TONR instruction is a nonretentive timer that accumulates time when TimerEnable is set. The TOFR instruction is a nonretentive timer that accumulates time when TimerEnable is cleared. The RTOR instruction is a retentive timer that accumulates time when TimerEnable is set. The CTUD instruction counts up by 1 when CUEnable transitions from clear to set. The instruction counts down by 1 when CDEnable transitions from clear to set.

Figure 11-29 Function block timers and counters.

PROGRAMMING FUNCTION BLOCK ROUTINES

To start a function block program, right-click on the main program (see Figure 11-30). Then you will add a new routine. Choose Function Block for the type of routine and give it a name. In this example the routine was given the name Stop_light_FB. It is shown below the MainProgram in the Controller Organizer in Figure 11-30.

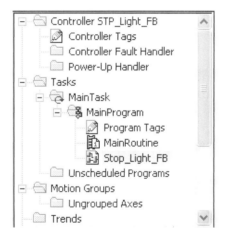

Figure 11-30 Program list in RSLogix 5000. Note the routine named Stop_light_FB. The ICON shows that it is a function block routine.

The next example will use timer (TONR) and limit (LIM) function block instructions. Figure 11-31 shows the use of a TONR timer function block. The TONR has a preset of 30000 ms (30 seconds). It has accumulated a count of 28385 at this point in time (about 28 seconds).

The LIM function block will be used to check to see if a value is between a low and a high limit. If the input to the LIM function block Test input is between the LowLimit and HighLimit, the output from the LIM function block will be true. In this example the output will only be true if the input is between 0 and 10000. In this example the output from the LIM function block is a real-world output.

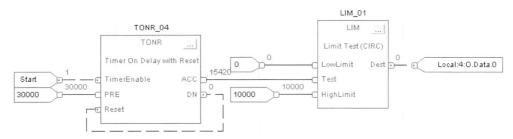

Figure 11-31 Two function blocks wired together.

Study Figure 11-32. This is the same application as shown in Figure 11-31, but programmed slightly differently. The two function blocks in this example are not wired together. An OCON and an ICON are used to pass information between the two function blocks. Note that the input to the LIM instruction is an ICON named ACC in this example. ICON ACC gets its data from the OCON output from the TONR function block. Note that these two function blocks are not connected but work together. The TONR output OCON (ACC) provides input to the TEST input of the LIM through the ICON (ACC). If ACC is between 0 and 10000, the LIM function block's output will be set to true. The numbers and letters under the ACC output show the page and are of the page where this output is used. Note that the output is an OCON named ACC. This application would perform exactly as the one shown in Figure 11-31. Note also that the use of an OCON and an ICON would enable these two function blocks to be on different sheets of the function block routine.

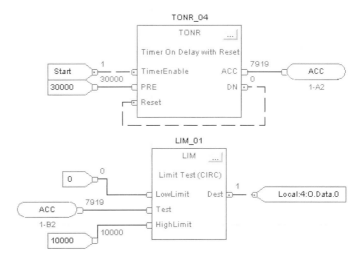

Figure 11-32 TONR and LIM function blocks.

In RSLogix 5000 a function block routine can be broken into multiple sheets. Sheets are like separate pages of a program (see Figure 11-33). This helps you organize your program and make it easier to understand. Sheets do not affect the order in which the function blocks execute. When a function block routine executes, all sheets execute. It is a good idea to use one sheet for each device that is to be programmed. In Figure 11-33 there are four sheets (pages) of an FBD program. In this example each sheet controls one device. Note that this is about as simple as it gets. Normally there might be multiple function blocks on a page to perform different but related tasks.

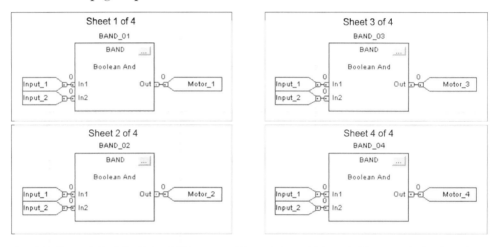

Figure 11-33 Multiple sheets in a function block routine.

ADDITIONAL FUNCTION BLOCKS

Select (SEL) Function Block

The SEL function block uses a digital input to select one of two inputs. This instruction is only available in function block programming. An example is shown in Figure 11-34.

The SEL function block selects In1 or In2 on the basis of SelectorIn. If SelectorIn is set, the instruction sets Out = In2 (see Figure 11-34). If SelectorIn is cleared, the instruction sets Out = In1. In this example 0 was input to SelectorIn so the value at In1 (150) is output to the OREF named Result.

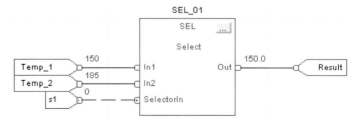

Figure 11-34 A SEL function block.

PROGRAM/OPERATOR CONTROL OF FUNCTION BLOCKS

Several function blocks support program/operator control. Program/operator control enables the programmer to control these instructions alternatively from a program or from an operator interface device. When an instruction is in program control, the instruction is controlled by the program inputs to the instruction. When an instruction is under operator control, the instruction is controlled by the operator inputs to the instruction. Program or operator control is determined by the inputs shown in the table in Figure 11-35.

Input	Description
.ProgProgReq	A program request to go to program control
.ProgOperReq	A program request to go to operator control
.OperProgReq	An operator request to go to program control
.OperOperReq	An operator request to go to operator control

Figure 11-35 Table showing program/operator control inputs and options.

If both ProgProgReq and ProgOperReq are set, the instruction will be under operator control.

You can determine whether the instruction is in program or operator control by looking at the ProgOper output. If ProgOper is set (1), the instruction is in program control. If ProgOper is set (0), the instruction is in operator control.

Program request inputs take precedence over operator request inputs. This enables the user to use the ProgProgReq and ProgOperReq inputs to lock an instruction in the desired mode. For example, assume you always want an instruction to operate in operator mode. You do not want the program to control the running or stopping of the instruction. To do this, you would input 1 into the ProgOperReq. This would prevent the operator from putting the instruction into program control by setting the OperProgReq from an operator input device. Let's examine one instruction that utilizes program/operator control.

Enhanced Select (ESEL) Function Block

The ESEL instruction (see Figure 11-36) lets you select one of as many as six inputs or the highest, lowest, median, or average of the inputs and send the result to the output. The SelectorMode input (see Figure 11-37) value determines whether the instruction will select the highest, lowest, median, or average of the inputs. In the example shown in Figure 11-36, 4 is the input to the SelectorMode input so the average of the input values will be output by the instruction. There are also inputs to determine if the instruction is under program or operator control. This instruction is available in function block and in ST.

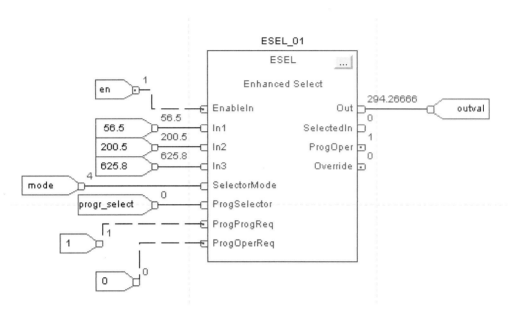

Figure 11-36 An ESEL instruction.

Value	Description
0	Manual select
1	High select
2	Low select
3	Median select
4	Average select

Figure 11-37 Selector modes.

Switching between Program Control and Operator Control

The following list states how the ESEL instruction changes between program control and operator control.

1. You can lock the instruction in operator control mode by leaving ProgOperReq set.
2. You can lock the instruction in program control mode by leaving ProgProgReq set while ProgOperReq is cleared.

Multiplex (MUX) Instruction

The MUX instruction can be used to select one of eight inputs to send to the output on the basis of the selector input. On the basis of the selector value, the MUX instruction sets Out equal to one of the eight inputs. The number of inputs can be reduced. An example of the use of this might be a process where we have different potential temperatures, depending on the product that needs to be produced.

Figure 11-38 shows the use of a MUX instruction to choose which one of eight input values should be sent to the output. This MUX instruction selects between In1, In2, In3, In4, In5, In6, In7, and In8, on the basis of the selector. The instruction sets Out = In, which becomes an input parameter for MUX_01. For example, if Select_Value = 2 (value into the selector input), the instruction sets Out = Analog_Input2, 6.7 in this example.

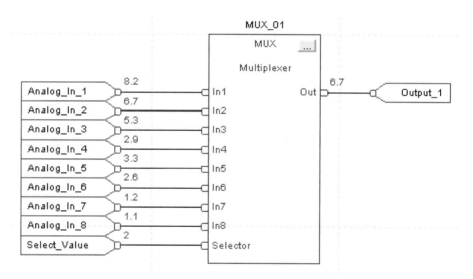

Figure 11-38 A MUX instruction.

ADD-ON INSTRUCTIONS

Add-On instructions are custom instructions you can create yourself. They are like personalized instructions. Add-On instructions can be used to create new instructions for sets of commonly used logic. Add-On instructions are not necessarily function block instructions. You develop the logic in your choice of logic. You can also provide documentation for the instruction so that it appears to be a standard instruction.

These are some benefits of Add-On instructions.
- If there are algorithms that are used multiple times in the same project or in different projects, it may make sense to use the code inside an Add-On instruction to make it modular and easier to reuse. It may also make the logic easier to understand, by hiding some of the complexity behind the code.
- Add-On instructions allow a programmer to reuse the work invested to develop algorithms.
- Add-On instructions can provide consistency between projects by reusing commonly used control algorithms.
- Add-on instructions allow the programmer to put complicated algorithms inside of an Add-On instruction, and then provide an easier-to-understand interface by making only essential parameters visible.
- The use of Add-On instructions reduces documentation development time by automatically generating instruction help.
- The proprietary code the programmer develops can be put inside the Add-On instruction and Source Protection can be used to prevent others from viewing or changing the code. This can help keep the code proprietary. This is very important for companies that manufacture and sell automation equipment.
- An Add-On instruction can be used across multiple projects.
- Once an Add-On instruction is defined in a project, it behaves like the standard instructions already available in the RSLogix 5000 software. Add-On instructions appear on the instruction toolbar and in the instruction browser.

Developing an Add-On Instruction

Note you must have at least RSLogix version 16 to develop Add-On instructions.

The General tab contains information from when an Add-On instruction is first created. The General information tab can be used to update the information. Also note that the description, revision, revision note, and vendor information is copied into the custom help for the instruction. The programmer is responsible for defining how the revision level is used and when it is updated. Revision levels are not automatically managed by the software.

Parameters

The Parameters define the instruction interface and also how the instruction appears when used in logic. The parameter order that you develop defines the order that the parameters appear on the instruction. Figure 11-39 shows the Parameters input screen.

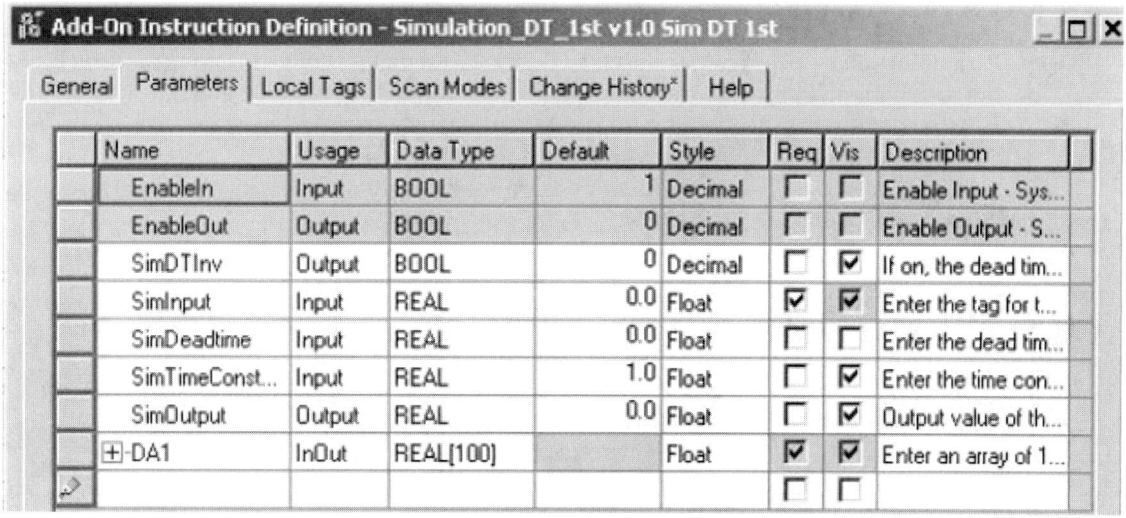

CHAPTER 11—FUNCTION BLOCK DIAGRAM PROGRAMMING   281

Figure 11-39 Parameters screen. (Courtesy of Rockwell Automation, Inc.)

Local Tags

Local Tags are hidden members and are not visible outside the instruction. They cannot be referenced by other programs or routines. They are private to the instruction. This can be very important. It can make the instruction appear to be simple and easy to understand. It also may be of benefit to hide proprietary logic. Figure 11-40 shows the Local Tags entry screen.

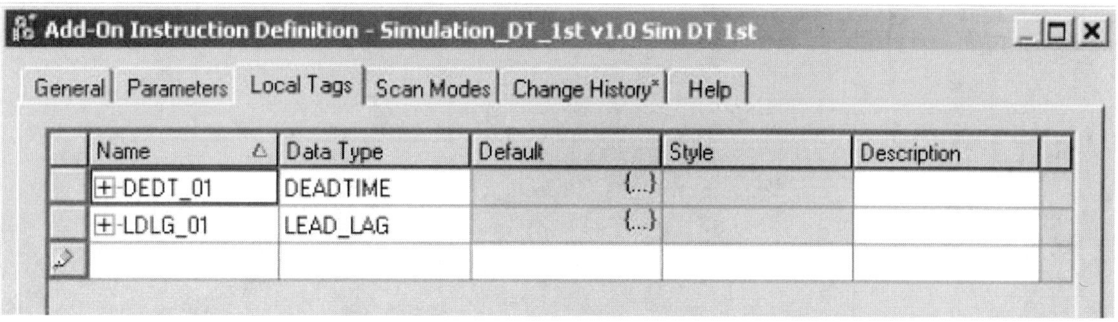

Figure 11-40 Local Tags screen. (Courtesy of Rockwell Automation, Inc.)

Data Type

Parameters and Local Tags are used to define the data type that is used when executing the instruction. The software builds the associated data type. The software orders the members of the data type that correspond to the parameters in the order that the parameters are defined. Local Tags are added as hidden members.

Logic Routine

The Logic routine of the Add-On instruction defines the primary functionality of the instruction. It is the code that executes whenever the instruction is called. Figure 11-41 shows the interface of an Add-On instruction and its primary Logic routine which defines what the instruction does.

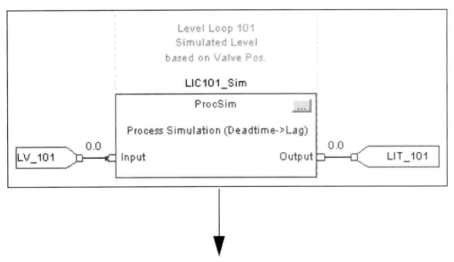

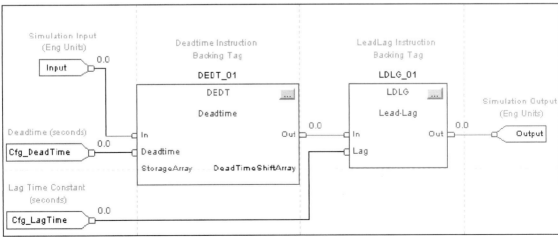

Figure 11-41 A Logic routine for an Add-On instruction. (Courtesy of Rockwell Automation, Inc.)

Optional Scan Mode Routines

Scan mode routines can also be defined for Add-On instructions. Figure 11-42 shows the Scan Modes configuration screen.

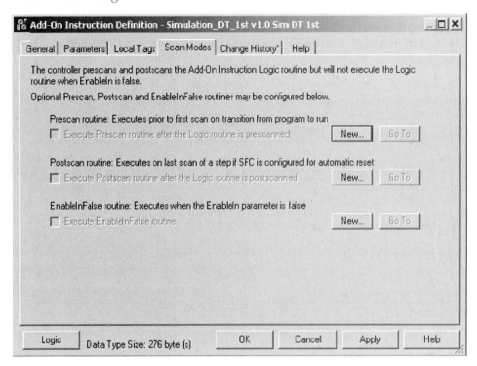

Figure 11-42 Scan Modes screen. (Courtesy of Rockwell Automation, Inc.)

Figure 11-43 shows the Change History entry screen. The Change History tab displays the creation and latest edit information that is tracked by the software. The By fields are used to show who made the change on the basis of the Windows user name at the time of the change.

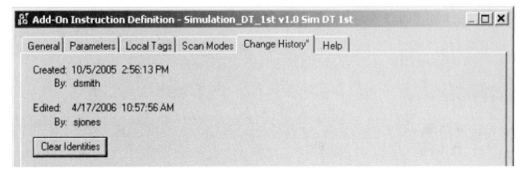

Figure 11-43 Change History screen. (Courtesy of Rockwell Automation, Inc.)

Help

Figure 11-44 shows the Help tab screen. The name, revision, description, and parameter definitions are used to build the Help instruction. This is done automatically. The Extended Description Text is used to provide additional Help documentation for the instruction. The Instruction Help Preview shows how the instruction will appear in the various languages, on the basis of parameters defined as required or visible.

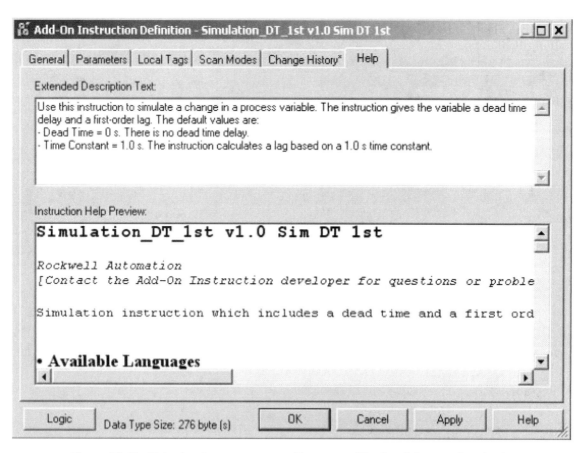

Figure 11-44 Help development screen. (Courtesy of Rockwell Automation, Inc.)

Available Languages

Add-On instruction routines have a choice of ladder diagram, FBD, or ST. Once the Add-On instruction has been created, it can be called from any of the RSLogix 5000 languages. An Add-On instruction written in one language can be used as an instruction through a call in another language.

Creating an Add-On Instruction

Figure 11-45 shows part of the Project Explorer window. To create an Add-On instruction, select Add-On Instructions and then New Add-On Instruction.

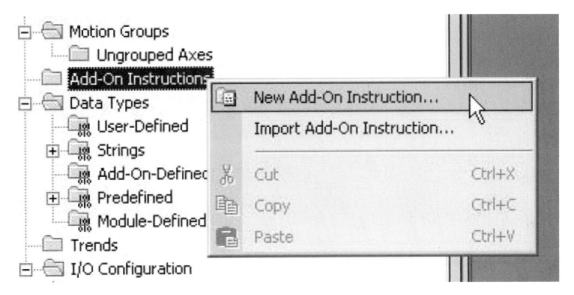

Figure 11-45 Screen to create a New Add-On Instruction. (Courtesy of Rockwell Automation, Inc.)

After you choose New Add-On Instruction, the window shown in Figure 11-46 appears. You must enter a name for the new instruction. You may enter a description. Note that some of the information on this screen will be used to automatically document this instruction as well as develop the Help file instructions for this instruction. You must choose the language that will be used. You may enter a Revision as well as text and a note. You may also enter a name or information in the Vendor area. When you have finished, select OK.

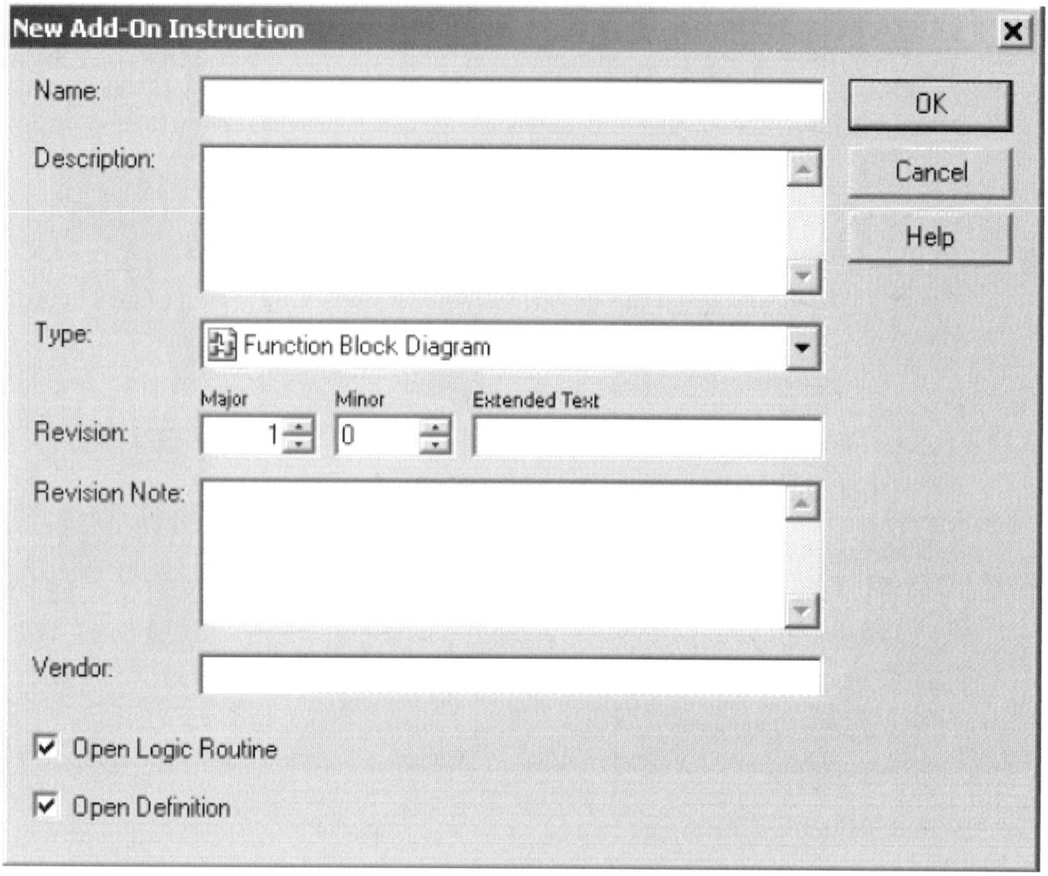

Figure 11-46 Information screen for a New Add-On Instruction. (Courtesy of Rockwell Automation, Inc.)

Next the logic can be developed. Choose the name of the instruction from the Add-On Instructions list. In the example in Figure 11-47 there are four Add-On Instructions shown.

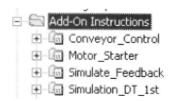

Figure 11-47 Add-On Instructions portion of the Controller Organizer screen. (Courtesy of Rockwell Automation, Inc.)

In Figure 11-48 an Add-On instruction named Motor_Starter was selected. To develop the logic routine for the instruction, you would choose the Logic choice and then develop the logic.

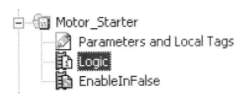

Figure 11-48 Routines and tags under the Motor_Starter Add-On instruction. (Courtesy of Rockwell Automation, Inc.)

Instructions can be stored so that they are easy to access when developing new applications. Source protection can also be applied to hide or protect proprietary code.

This chapter has only covered a small number of the available function block instructions. If you become comfortable with these, you will readily learn others that you need to use. You can bet that there is an instruction or a combination of instructions that will solve any application need you have. If it is a recurring need, you can even create your own Add-On instruction for future use.

QUESTIONS

1. *True or False:* Function block programming is one of the languages that IEC 61131-3 specifies.
2. What does the acronym IREF stand for?
3. What does the acronym OREF stand for?
4. Can an IREF be a tag? A number?
5. Can an OREF be a tag?
6. What does the acronym ICON stand for?
7. What does the acronym OCON stand for?
8. What is the difference between an OREF and an OCON?
9. What are ICONs and OCONS used for?
10. What is the Assume Data Available indicator used for?
11. What is a sheet in a function block routine?
12. What are two ways you could get information from an instruction on one sheet to an instruction on another sheet?
13. What is an Add-On instruction?

14. List three advantages of Add-On instructions.
15. Thoroughly explain the logic below. Make sure you explain the types of inputs and outputs to each instruction as well as what the logic does.

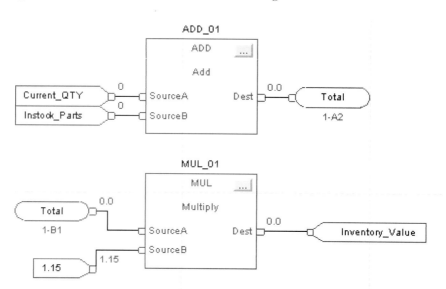

16. Thoroughly explain each instruction and the logic below. Make sure you explain the types of inputs and outputs to each instruction as well as what the logic does.

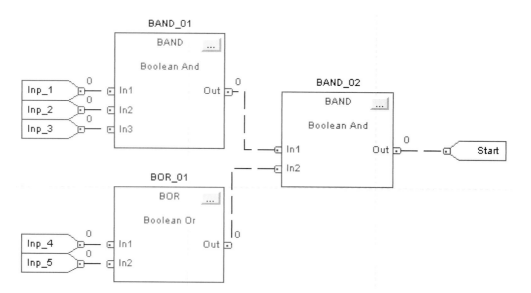

17. Thoroughly explain each instruction and the logic below. Make sure you explain the types of inputs and outputs to each instruction as well as what the logic does.

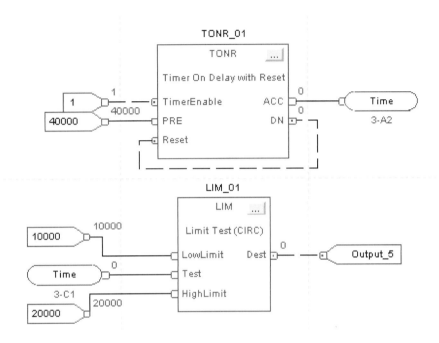

18. Thoroughly explain each instruction and the logic below. Make sure you explain the types of inputs and outputs to each instruction as well as what the logic does.

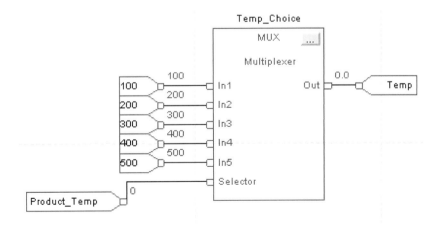

19. Thoroughly explain each instruction and the logic below. Make sure you explain the types of inputs and outputs to each instruction as well as what the logic does.

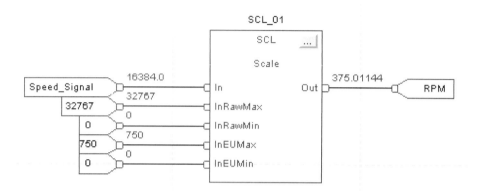

20. Write a function block routine for the following application. You may do it on paper or on a controller.

    This is a simple heat treat machine application. The operator places a part in a fixture, then pushes the start switch. An inductive heating coil heats the part rapidly to 1500 degrees Fahrenheit. When the temperature reaches 1500, turn the coil off and open the quench valve which will spray water on the part for 10 seconds to complete the heat treatment (quench). The operator then removes the part and the sequence can begin again. Note there must be a part present or the sequence should not start. Note: you may want to use a small ladder diagram program for the start/stop logic. It will simplify the task.

I/O	Type	Description
Part_Present_Sensor	Discrete	Sensor used to sense a part in the fixture
Temp_Sensor	Analog	Assume this sensor outputs 0–2000 degrees Fahrenheit. To keep it easy assume that the sensor is analog and will output the number 1500 when the temperature reaches 1500
Start_Switch	Discrete	Momentary normally open switch
Heating_Coil	Discrete	Discrete output that turns coil on
Quench_Valve	Discrete	Discrete output that turns quench valve on

CHAPTER 12

Industrial Communications

OBJECTIVES

On completion of this chapter the reader will be able to:

- Describe the typical levels of an industrial network.
- Explain terms such as serial, synchronous, asynchronous, multi-drop, full duplex, half duplex, deterministic, and so on.
- Describe the typical use of DeviceNet, ControlNet, and SERCOS.
- Describe token passing and CSMA/CD.

INTRODUCTION

There are three general categories of industrial networks: device networks, control networks, and information networks. Figure 12-1 shows an illustration of the three levels.

The device level is the lowest level. The device-level network is used with industrial devices such as sensors, switches, safety devices, drives, motors, valves, and so on. There are many industrial device-level networks available. DeviceNet is one of the more common device-level networks.

The control level would be the networks that industrial controllers are on. This level would include the PLCs, operator I/O devices, drives, robot controllers, vision systems, and so on. Communication on the control level includes sharing I/O and program data between controllers. In ControlLogix systems this would include sharing producer/consumer data between controllers as well as special purpose cards such as motion controllers communicating over specialized networks like SERCOS to drives. Communication at the control level can often affect the safety of the system and personnel. ControlNet is one of the more popular protocols at the control level.

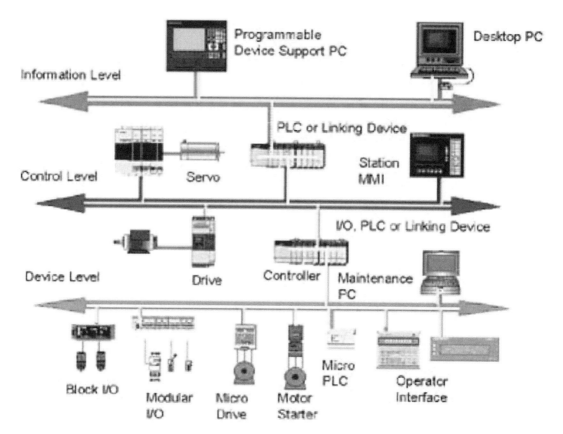

Figure 12-1 Levels of industrial networks. (Courtesy of Rockwell Automation, Inc.)

The information level typically is composed of the company's business networks and computers. These would include financial, sales, engineering, management information systems, Internet, intranet, email, scheduling, and so on. These systems and computers typically utilize an Ethernet network. The rest of the chapter will examine the types of communication networks from the device level up.

SERIAL COMMUNICATIONS

Serial communications is a term that means that communication takes place in a linear fashion. In serial communications a message is broken into individual characters and each character sent one bit at a time. A coding system named ASCII is typically used. ASCII uses a unique binary number to represent every letter, number, and special character. There is a 7-bit ASCII and an 8-bit extended ASCII coding system. A 7-bit ASCII has 128 possible different letters, numbers, and special characters. An 8-bit ASCII has

256 available. In serial communications each character is sent as its ASCII equivalent. In 7-bit ASCII the letter A is coded to be 1000001 (see Figure 12-2).

Serial communications can be synchronous or asynchronous. In the asynchronous communications mode individual characters are sent one at a time. Each character that is sent is delineated by a start bit and a stop bit. Asynchronous communications are typically used for low-speed simple data transmission.

When devices communicate, the sender and receiver must have a way to extract individual characters or blocks (frames) from the whole message. Imagine the message as being a very long list of 1s and 0s (bits). The receiving device has to be able to break the bits into logical groupings to make the message understandable.

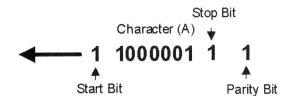

Figure 12-2 Asynchronous mode transmitting the letter A.

When a character is transmitted in asynchronous communications more than 7 bits are used. Extra bits are added before each character transmission so that the receiving device is warned that a message is coming and was not corrupted and after so that the receiver knows that the character has been sent.

The first bit that is sent is called a start bit (see Figure 12-3). This bit alerts the receiver that a message is about to be sent. In effect it tells the receiver to pay attention. The next 7 bits (8 if 8-bit ASCII is used) are the ASCII equivalent of the character that is being sent. The next bit is used for parity. Parity is used to check the received message for errors. There are several choices for the way that parity is used for checking. The parity choices include odd or even, mark or space, or none.

Start Bit	Data Bits	Parity Bit	Stop Bit
1 bit	7 or 8 bits	1 bit, can be odd, even, mark, space, or none	1, 1.5, or 2 bits

Figure 12-3 How a typical ASCII character is transmitted.

Study the example shown in Figure 12-3. This example uses odd parity. There is an even number of 1s in the character A (2), so the parity bit is set to 1 to make the total number of 1s odd. If the character had an odd number of 1s, the parity bit would be 0. The receiving device uses the parity bit as a check to see if the character may have been corrupted during transmission. The receiving device counts the number of 1s in the character and checks the parity bit. If the total number of 1s plus the parity bit is odd, the receiver assumes that the message was received accurately. Although this method is somewhat crude, it is quite effective. Note that it is not perfect; two or more bits could change state and the parity bit could still be correct but the message would be wrong.

Synchronous Communications

Synchronous communications are much faster than asynchronous communications. They synchronize the sending device and receiving device with a signal or a clock that is encoded into the data stream. The sending device and the receiving device synchronize with each other before any data is sent. Synchronous communications utilize a special bit-transition pattern in the signal that maintains the timing between sender and receiver. Synchronous communications are used in the more complex communications protocols.

RS-232 Communications

RS-232 is the most common asynchronous serial communications mode. The RS in the standard's name means *recommended standard*. The RS-232 was designed to use a 25-pin plug. It specifies a function for each of the 25 pins. The standard did not require that any of the pins must be used. Some devices use only three pins (see Figure 12-4). Some devices utilize more than three pins to do electric handshaking. Handshaking ensures the devices coordinate their communications.

Handshaking is cooperative. The first device alerts the other device that it has a message it would like to send. It accomplishes this by setting pin 4 high. Pin 4 is the request to send (RTS) pin. The receiving device sees the RTS pin high, and if it is ready to receive a message, it sets the clear to send (CTS) pin high. The first device then knows that the other device is ready to receive a message. Some devices have the ability to electrically handshake; some do not.

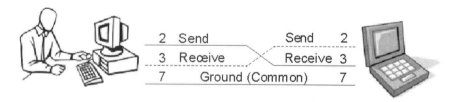

Figure 12-4 Simple RS-232 wiring scheme showing the simplest of RS-232 connections.

RS-422 and RS-423

RS-422 and RS-423 were designed to improve on the weaknesses of RS-232. RS-422 and RS-423 permit longer-distance and higher-communication speeds. RS-422 uses a balanced serial method of communications. RS-232 has only one common wire, so the transmit and receive lines must share the same wire. Sharing a common wire can accentuate noise problems. RS-422 has separate common wires for the transmit and receive lines. Having a separate wire for transmit and receive makes the method more noise immune. The balanced mode is less susceptible to external interference and exhibits lower cross talk between the transmitted and received signals. Cross talk is defined as being the bleeding of one signal onto another.

Noise problems reduce the potential speed and distance of communications. Communications are always a trade-off between speed and distance because of noise. RS-422 has much higher speed and longer distance than RS-232 because the balanced mode is used. RS-422 features speeds of 10 Mb for distances of over 4000 feet. RS-232 is only 9600 baud and 50 feet.

Termination

As speed and length increase, the reflection of the signal can become a factor. Resistors can be added to each end of the cable to terminate the line. The termination resistors are used to match the impedance of the device to the impedance of the cable. When they are matched, the signal won't reflect back into the line. For RS-422, a 100-ohm resistor is used on both ends of the cable.

RS-485

RS-485 is a derivation of the RS-422 standard. The standard is officially now known as EIA/TIA-485. It is commonly still referred to as RS-485 however.

RS-485 is a multidrop protocol. *Multidrop* means that multiple devices can be connected on the same network. The standard limits the number of stations to 32. This allows for up to 32 stations with transmission and reception capability, or 1 transmitter and up to 31 receiving stations. The maximum distance for RS-485 is 1200 meters. The total number of devices and maximum distance can be extended if repeaters are used.

RS-485 uses twisted pair communications wiring. Twisted pair wiring is a pair of wires that are twisted around each other. Having the two wires twisted around each other reduces the possibility of noise interference changing the message. An RS-485 network can use two sets of twisted wire or one set. RS-485 can be half duplex or full duplex. If two sets of twisted wire are used, the communications can be full duplex. Full duplex means that simultaneous, two-way conversation can be done.

Full-duplex mode allows communication simultaneously in both directions. Full duplex is like a two-lane highway with one lane for each direction. Cars can travel on each road in different directions at the same time.

A half-duplex system allows communication in both directions, but in only one direction at a time. A walkie-talkie is a good example of a half-duplex system. If a person wants to talk, he or she hit the call button and then talk. When done, he or she says the keyword *Over* and waits for the receiver to respond. Only one person can talk over the two walkie-talkies at a time.

Another example of half duplex is road construction. When there is road construction, there are often workers with flags at each end of the construction. They only allow traffic flow in one direction at a time.

A ground wire is also used in RS-485 so RS-485 wiring is typically involves three or five wires. The ends of the RS-485 communications lines are terminated with resistors.

DF1 PROTOCOL

DF1 can be used to talk with Rockwell Automation PLCs from computers or other devices with a serial communications port.

DF1 can be used in a peer-to-peer mode or in a main-sub mode. Peer-to-peer communication is a mode in which any device may talk to any other device. They are all peers. Peer-to-peer mode uses full-duplex protocol. The main-sub mode is a method of communication where the main device controls the communication and the other devices (subs) simply react to commands that they receive from the main device. The main device can communicate with several subdevices. Up to 255 devices can communicate in a DF1 main-sub mode. But only one can be the main. The subs do not speak unless the main requests information.

The main device is used continuously to poll the sub devices. Essentially they are polled to see if they have any data to transmit to the main. If the sub device has a message to transmit, it sends it. The main device receives the message and then polls the next subdevice, and so on. The main device maintains a list of active sub devices. If a sub device does not respond to a poll command from the main, the main removes it from the list of active sub devices. The main will repoll the sub device at a later time to see if it is active again. The main-sub mode of communication uses the half-duplex protocol.

MODBUS

Modbus is one of the oldest field buses. Modicon developed the standard in 1978. It was developed to exchange data between PLCs and other devices. Modbus is based on a serial master-slave system (RS-232/485). Modbus can utilize simple or peer-to-peer communications. Modbus protocols include ASCII/RTU, Modbus Plus, and Modbus/TCP. Modbus/TCP is used to communicate on an Ethernet. The ASCII mode and Remote Terminal Unit (RTU) mode are similar. RTU has a few differences in the format of the frame and it can transmit more data in the same amount of time than ASCII. The maximum transmission distance for ASCII/RTU modes is 350 meters. Five wires are used.

ASCII and RTU modes use a master-slave mode. Only the master device can initiate communication. The master can communicate with individual slave units or with all slave units simultaneously. The latter can be called a broadcast message. Slave devices can only respond to the master device's requests. Slaves respond by transmitting the requested data or by performing the requested operation.

In Modbus ASCII/RTU mode, the frame of the message is

> The device address field
> The function code field
> The data bytes field
> The error-check field

This frame format is used by both the master and slaves.

Address Field

There are two characters in the address field. Slave device addresses can be between 0 and 247. If the master sends a message, it puts the slave device's address in this field. When a slave responds, it puts its own address in this field so that the master knows which slave replied. If the master uses 0 as the address, all slaves react to the message. That is, 0 is used as a broadcast address for the master to talk to all slave devices.

Function Code Field

The function code has two characters that tell the slave devices what function to perform, such as turning outputs on, reading the status of inputs, reading the values of memory addresses, writing values to memory, and so on.

The valid numbers for function codes are 1 to 255. Some of the functions are universal for all equipment. Some of the function codes are specific to Modicon. Some of the function codes are reserved for future use.

Data Field

The data field has information that the slave devices need to process the command from the master. For example, if a read function were sent, the data field would tell the slave which input statuses should be read. When the slave sends its response, it puts the data that was requested in the data field. If there was an error, the slave sends an error code in the data field.

Error Check Field

The error check field in ASCII mode contains the result of the longitudinal redundancy check (LRC) calculation. This calculation is an arithmetic algorithm that is performed on the message to be sure the message that was received is exactly the same as the message that was sent. RTU mode uses a cyclical redundancy check (CRC). A CRC is an arithmetic algorithm that is performed on the message to be sure the message that was received is exactly the same as the message that was sent.

The last portion of a Modbus ASCII message is the carriage return and line feed characters. When the receiving device receives these two characters, it is the end of the message.

Message Formats for Modbus ASCII and RTU

Figures 12-5 and 12-6 show the communication formats for ASCII and RTU communication.

Modbus ASCII Message Format

Start Field	Address Field	Function Field	Data Field	Error Check Field	End
1 character, a colon (:) character	2 characters	2 characters	X characters (dependant on message length)	2 characters	2 characters, CR and LF

Figure 12-5 ASCII message format.

Modbus RTU Message Format

Start Field	Address Field	Function Field	Data Field	Error Check Field	End
4-character delay time	8 bits	8 bits	N * 8 bits (dependant on message length N)	16 bits	4-character delay time

Figure 12-6 RTU message format.

Modbus Plus

Modbus Plus makes peer-to-peer communication possible. With Modbus Plus any device can initiate communication with any other device. Although any device can initiate communication, on a message level it is still master-slave. The device that initiates communication is the master and the device it wants to communicate with responds like a slave. Modbus Plus can interface with up to 32 devices up to 1500 meters. A total of three repeaters can be used to extend the distance to 6000 meters and the number of devices to 64.

DATA HIGHWAY PLUS

Rockwell Automation's Data Highway Plus (DH+) is a proprietary protocol. A DH+ network can connect up to 64 devices. DH+ is a token-passing bus network. A bus can be thought of as a network with a long backbone cable. The backbone cable is often called the trunk line. Devices are simply connected to the trunk cable.

Token Passing

Devices must gain access to the network to be able to communicate. Token passing is one of the access methods. In the token-passing access method, only the device that has control of the token can talk. The token is passed from device to device until one of them wants to talk. The device then takes control of the token and is free to talk. The token is simply a bit pattern in the message.

A device that would like to gain access waits for a free token. When a free token arrives, the device that would like to talk sets the token busy bit, adds the information field, adds the actual message to be sent, and adds a trailer packet. The header packet has the address of the device that the message is being sent to.

As the message moves through the network, every device checks the address in the header to see if it is being talked to. If not, it ignores the message and sends it on.

The message arrives at the addressed device, and the device copies the message. The receiving device sets bits in the trailer field to show that the message was received. The receiving device regenerates the message and sends it back out on the network. The device that originally sent the message receives the message back and notes that the message was received. The device then frees up the token bit and sends it out for other devices to use.

With token passing, access times for a device are predictable. This can be very important in a manufacturing environment. The access times in a token-passing access method are called deterministic because actual access times can be calculated on the basis of the actual bus and nodes.

In general in a DH+ network, a single device cannot keep the token more than 38 ms. A device can keep the token for up to 100 ms in special circumstances.

WIRELESS COMMUNICATION

The Wireless Ethernet Compatibility Alliance (WECA) was formed in 1999 by Aironet (now Cisco), 3Com, Lucent Technologies (now Agere), Harris Semiconductor (now Intersil), Nokia, and Symbol Technologies. It was formed to certify interoperability of wireless local area network products based on the IEEE 802.11 specifications. It was renamed the Wi-Fi Alliance in October 2002. Wi-Fi product certification began in March 2000. Equipment passing these interoperability tests can use the Wi-Fi logo.

IEEE 802.11 is a standard for wireless networking; 802.11b is the most common. The 802.11b can operate at speeds of up to 11 Mb/second. The 802.11g can operate at speeds up to 54 Mb/second. It is backward compatible with the original 802.11b standard.

Wireless communications are attractive because wiring does not have to be run between devices. Running wire is labor intensive and expensive. Wired devices are difficult and expensive to move once they have been installed. Wire is also susceptible to electrical noise. Wireless local area networks (WLANs) are finding a home everywhere from the factory floor to the office.

There are four main types of network topologies available.

Peer to Peer

Each device communicates directly with another device. This is typically used for a small and simple network.

Multipoint to Point

Remote devices communicate their data to a central location. This is often used for data collection applications.

Point to Multipoint

One device talks to many devices simultaneously. The master device (point) broadcasts a message and all devices receive it.

Mesh

In a mesh network, each device passes the message to its neighboring device until it reaches the destination device. If a neighboring device is damaged, another neighbor is used.

Spread Spectrum Technology

Spread Spectrum Technology (SST) is used for wireless communications. SST was developed by the U.S. military during World War II to prevent jamming of radio signals. It also helped make them harder to intercept.

In SST the bandwidth of the transmitted signal is much wider than the bandwidth of the information. In contrast, a radio station uses a narrow bandwidth for transmission. The transmitted signals utilize virtually all of the bandwidth.

In spread spectrum communications the message is modulated across a wide bandwidth. SST spreads data transmission over many different frequencies. This ensures that interference on a single frequency cannot prevent the data from reaching its destination. A special code determines the transmitted bandwidth. Authorized receivers use the code to extract the message from the signal. The transmission looks like noise to unauthorized receivers. This makes SST very noise immune.

The Federal Communications Commission (FCC) dedicated three frequency bands for commercial use: 900 MHz, 2.4 GHz, and 5.7 GHz. There is very little industrial electrical noise in these frequencies.

DEVICE-LEVEL NETWORKS

Device-level networks are sometimes called field buses or industrial buses. Industrial networks create standards to allow different-brand field devices to communicate and be used interchangeably no matter who manufactured them.

Networks minimize the amount of wiring that needs to be done. Imagine a large system that has hundreds of I/Os and no device network. Now imagine the time and expense of wiring each and every I/O device back to the controller. This would be hundreds of wires and might involve long runs of wire. Conduit has to be fabricated and mounted for the wires that need to be run. If an industrial device bus had been used, only a single communications cable would have to be run. The cable would be the trunk line (or bus), and all devices could then connect directly to the bus. In many cases they would just be screwed onto connectors on the trunk line. The cabling system often eliminates the need for conduit or cable tray also. Multiple devices can even share one connection with the bus. I/O blocks are available that allow multiple I/O points to share one connection point to the bus.

Field bus devices gain a cost advantage when one considers the labor cost of installation, maintenance, and troubleshooting. The cost benefit received from using distributed I/O is really in the labor saved during installation and startup. There is also a material savings when one considers all the wire that does not have to be run. There is a tremendous savings in labor because only a fraction of the number of connections needs to be made and only a fraction of the wire needs to be run. A field bus system is also easier to troubleshoot. If a problem occurs, only one twisted pair cable needs to be checked. A conventional system might require the technician to sort out hundreds of wires.

Field Devices

Drives, sensors, valves, actuators, and starters are examples of I/Os that are called field devices. Field devices can be digital or analog.

DEVICENET

DeviceNet is intended to be a low-cost method to connect devices such as sensors, switches, valves, bar-code readers, drives, operator display panels, and so on, to a simple network. Figure 12-7 shows an example of a DeviceNet network. The DeviceNet standard is based on the CAN chip. The CAN chip is a smart communications chip. It is a receiver/transmitter and has powerful diagnostic capabilities.

Rockwell Automation used the CAN chip when it developed DeviceNet in 1993. Any manufacturer can participate in the Open DeviceNet Vendor Association Inc. (ODVA, www.odva.org). DeviceNet is an open network standard. The standard is not proprietary. It is open to any manufacturer. ODVA is an independent organization that manages the DeviceNet specification. DeviceNet is a broadcast-based communications protocol.

By using devices that adhere to the standard, you can interchange devices from different manufacturers. DeviceNet allows devices from various manufacturers to be interchanged and makes interconnectivity of more complex devices possible.

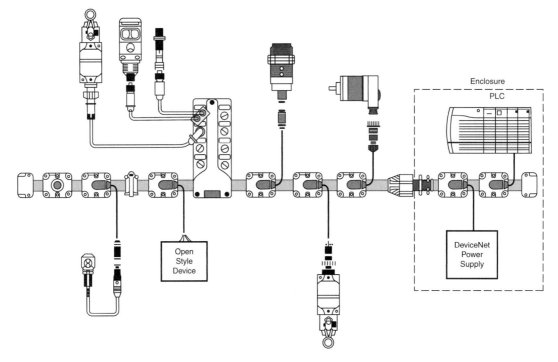

Figure 12-7 A DeviceNet network.

The user can choose master-slave, multimaster, peer-to-peer, or a combination configuration depending on device capability and application requirements. Higher-priority data gets the right-of-way. A DeviceNET network may have up to 64 node addresses. Each node can support many I/Os.

DeviceNet supports strobed, polled, cyclic, change-of-state, and application-triggered data communications.

Strobed

In strobed communications the DeviceNet scanner periodically strobes all devices for their status.

Polled

In polled communications the scanner polls individual devices for their status.

Change of State

Change of state means that a device reports only when the data changes. To be sure the scanner knows that the node is still alive and active, DeviceNet provides an adjustable, background heartbeat rate. Devices send data whenever their data changes or the heartbeat timer expires. This keeps the connection alive and lets the scanner know that the node is still alive and active.

Cyclic Transmission

Cyclic transmission can be used to reduce unnecessary traffic on a network. Devices can be set up to report their data on a regular basis that is sufficient to monitor their change rather than reporting too often and adding to the traffic on the network.

Figure 12-8 shows a simple example of a DeviceNet system. Note that there is only one cable running to the PLC. Note also the trunk line that all devices plug into. The PLC has a scanning module that acts as the master of the network.

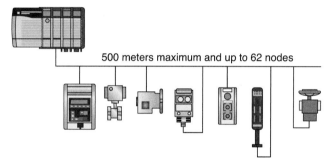

Figure 12-8 A DeviceNet system. Note that there is only one cable running to the scanner for all of the I/Os.

DeviceNet Components

Figure 12-9 shows the typical topology for a DeviceNet cable system. The trunk line is the backbone of the system. It is the one cable that all devices attach to. In DeviceNet terminology a device that has an address is called a node. There are two types of wire, thick and thin. There is also flat cable, which is shown in Figure 12-7. Thick wire is usually used for the trunk line. The wire that connects devices to the trunk lines is called a drop line. Drop lines usually utilize thin cable. The most common way in which drop lines are connected to trunk lines is through a tap, or T. The use of terminal blocks is also acceptable and common.

Wiring

Figure 12-9 shows a simple, generic network. Note the trunk line. Thick cable is capable of handing 8 amperes [although National Electric Code (NEC) only allows 4 amperes]. Thin cable is normally used for drop lines. Thin cable can handle up to 3 amperes of power. Both thick and thin cable have power and communication lines. This is important because many devices can be powered directly from the DeviceNet thick or thin cable. Two wires supply 24 volts to the network. Two other wires are used for communication. You should also notice in Figure 12-9 that the ends of the trunk line must be terminated with 121-ohm resistors. Terminating resistors should never be used on drop lines.

The DeviceNet cable system uses a trunk/drop line topology.

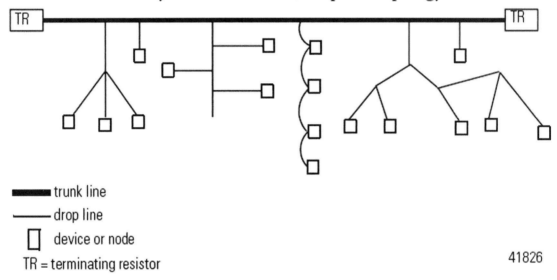

Figure 12-9 DeviceNet topology. (Courtesy of Rockwell Automation, Inc.)

DeviceNet is quite flexible in wiring topology. Figure 12-9 shows several possible topologies for wiring. Devices are shown as squares. The left-most example shows a tree-type structure. Next there is a single node attached by a drop line to the trunk line. Then we see a bus topology with three devices attached through a drop line to the trunk line. Next we see a daisy-chain topology. Daisy chain means that a communication line comes into one device and then another leaves the device to connect to the next device in line, and so on. This called daisy chaining. Then there is a more complex tree topology. The last is a single device attached by a drop line to the trunk line.

There is a maximum allowable drop line length of 20 feet. There is also a maximum cumulative drop line length. The maximum length varies by the network speed. The slower the speed, the longer the line length allowed. The faster the network, the shorter the cumulative line length that is allowed. The maximum drop line length is based on the cumulative drop line length. Figure 12-10 shows the maximum cumulative line length for the three network speeds allowed in a DeviceNet system. Note that the higher the speed that is used, the lower the maximum line length.

Network Speed	Maximum Cumulative Drop Line Length
125k bits/s	512 feet
250k bits/s	256 feet
500k bits/s	128 feet

Figure 12-10 Allowable line lengths for DeviceNet network speeds.

The most important device in a DeviceNet network is the scanner. The scanner acts as an interface between the PLC CPU and the inputs and outputs in the DeviceNet system. When the programmer configures a DeviceNet system he/she creates a scanlist in the scanner. The scanlist identifies which devices are included in the system and must be scanned by the scanner. The scanner reads inputs and writes to outputs in the system. The scanner is also to monitor the devices for faults. The scanner can download configuration data to each device. The scanner may also be equipped with a readout to help with troubleshooting the network. A Rockwell Automation scanner will flash a number that represents the error code and the number of the node that has the problem along with other diagnostic information.

Flex I/O uses one node address to connect many I/O devices. The I/O that is connected to the modules does not have to be DeviceNet capable. Any digital or analog I/O can be connected to the modules. Flex I/O modules are available for multiple inputs or outputs in digital or analog. Up to eight modules can be plugged together. This means that up to 128 discrete devices or 64 analog channels can be connected to a DeviceNet network using a FLEX I/O system. All of these devices can be connected and only use one node address. The modules are then attached to a DeviceNet communications module that is connected to the network. Flex I/O can also be used to connect non-DeviceNet devices to a DeviceNet network. Flex I/O and a DeviceNet bus is also useful to connect a PLC to devices that may be concentrated in one area of a machine a long way from the controller.

Communications Flow in a DeviceNet Network
The scanner module coordinates and controls all of the communications in a DeviceNet system. The data that a scanner receives or transmits is stored in its memory. The CPU (processor) of the CLX can utilize this information from the scanner. The CPU can receive or transmit information in two modes: I/O and explicit. I/O messaging is used for time-critical, control-oriented data exchange. I/O messaging makes the communications very transparent for the user. I/O is simply addressed as it would be in any card after the addressing is set up in the scanner.

Explicit messaging enables the CPU to transmit and receive between 1 and 26 words. Explicit messaging is typically used to transmit and receive information for a device that requires or produces more data than a simple input or output. An example that might utilize explicit messaging would be a drive.

The logic in the PLC would determine speeds, acceleration/deceleration, and so on. The PLC would use explicit messaging to communicate this information to the scanner module.

The scanner then transmits the message to the drive as a block of information. The PLC can also use explicit messaging to request the scanner to get information from the drive. The DeviceNet scanner then requests the information from the drive and makes it available to the PLC.

The DeviceNet scanner handles all communications with devices (nodes) and makes the information available to the CPU. In a Rockwell Automation system, explicit messaging is initiated by an instruction in the PLC logic. Reads and writes from the CLX utilize the Message (MSG) instruction. These reads and writes are only performed when they are called by the PLC logic. The programmer should be careful to only make these calls when they are needed. They should not be called every scan as they consume significant network bandwidth.

Troubleshooting

One of the greatest strengths of DeviceNet is the troubleshooting information that is available on the alphanumeric display on the scanner module. The ControlLogix DeviceNet scanner module displays alphanumeric codes that provide diagnostic information. The alphanumeric display on the module flashes the codes at approximately 1-second intervals. Under normal circumstances, the display for RUN toggles between the node address of the scanner and the mode of the 1756-DNB scanner module. In the example below the scanner is node 1 and it is in the run mode.

A#01
RUN

If there is a problem, the display shows the node number of the problem node, then the error code. The display toggles through these elements until the error is corrected. An example is shown below. The scanner is node 1 and it is in the run mode. The next two lines show that node number 33 has an error code. Error 72 means that the node stopped communicating.

A#01
RUN
N#33
E#72

CONTROL-LEVEL COMMUNICATIONS

Communications at the control level take place between control-type devices and large racks of remote I/Os. Control devices include PLCs, robots, CNC controllers, and so on.

SERCOS

SERCOS stands for Serial Real-Time Communications System. It is a digital control bus that is used to connect motion controllers, drives, and I/Os for motion control applications such as numerically controlled machines. It is very widely used in motion control applications. It does not fit neatly into the device level or the controller level of communications. It kind of straddles both.

The SERCOS standard makes it possible to use devices from various manufacturers. The standard specifies format for parameters, data, commands, and feedback that are normally communicated between controllers and devices. The devices are typically motion related.

SERCOS I was introduced in 1987. It could communicate at speeds of 2 to 4 Mb/second. The second generation SERCOS II replaced SERCOS I in 1999. It can operate at speeds of 2, 4, 8, and 16 Mb/second. SERCOS II can connect up to 254 devices to a control device using a ring topology. A ring consists of one master and multiple slaves that are daisy chained on the ring. Fiber optics are used for the communications medium.

SERCOS III was introduced in 2005. SERCOS III uses the basic elements of previous SERCOS standards, thus maintaining backward compatibility. SERCOS III uses Industrial Ethernet to transmit data at speeds of 100 Mb/second. The standard combined the low cost and high bandwidth of Ethernet with the deterministic performance of SERCOS. Ethernet is normally not deterministic. SERCOS III adds a real-time, collision-free channel that runs in parallel with an optional non-real-time channel. The non-real-time channel is used in Ethernet environments. SERCOS III uses Industrial Ethernet physical media instead of fiber. SERCOS III retains the ring configuration and allows multiple rings.

SERCOS III can connect up to 510 devices. In SERCOS III the servoloops are normally closed in the drive, not the controller. This reduces the load on the controller and enables more devices to be connected.

SERCOS III allows hot plugging of devices in the ring or at the end of a line during operation. Computers or devices using standard Ethernet can be connected to unused version III ports to communicate with devices. This is very helpful for setup and troubleshooting activities. SERCOS III is not supported by Rockwell.

ControlNet

There is a need for standardization of control-level communications so that control equipment from various manufacturers can communicate. ControlNet is one of those standards. ControlNet is a high-speed, deterministic network developed by Allen-Bradley for the transmission of critical automation and control information. ControlNet was originally developed by Allen-Bradley in 1995. It was proprietary but was then released to the general public under IEC-61158.

ControlNet is deterministic. It has very high throughput, 5 Mb/second for I/Os, PLC interlocking, peer-to-peer messaging, and programming. A ControlNet network can perform multiple functions. Multiple PLCs, human/machine interfaces, network access by a PC for programming, and troubleshooting from any node can all be performed on the network. The capability of ControlNet to perform all these tasks can reduce the need for multiple networks for integration.

A ControlNet network can be extended up to 1000 meters with two devices on it or up to 250 meters with 48 devices. If repeaters are used it can be extended it to 5 kilometers and handle 99 devices. If fiber-optic cable is used, the distance can be up to 30 kilometers.

ControlNet is compatible with Rockwell Automation PLCs, I/Os, and software. ControlNet supports bus, star, or tree topologies. It utilizes RG6-U cable, which is nearly identical to cable television cable but has four shields as opposed to cable TV cable's two shields. This means that cable and connectors are all easy to obtain and very reasonable in price. ControlNet also has a dual-media option (see Figure 12-11). This means that two separate cables can be installed to guard against failures such as cut cables, loose connectors, or noise. The figure also shows some of the types of computers and controllers that can be integrated as well as the wiring.

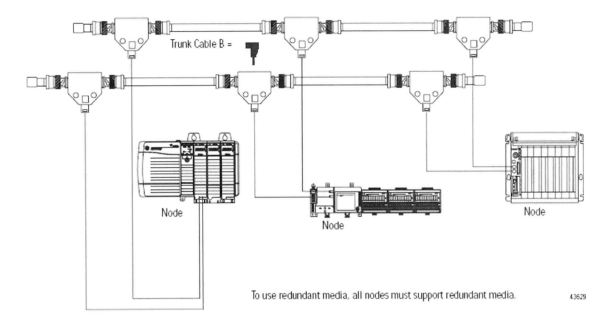

Figure 12-11 Redundant cabling. (Courtesy of Rockwell Automation, Inc.)

A device that can link different types of communications protocols so all can communicate with each other is called a gateway. Imagine a box that could do control or act as a communications gateway or do both simultaneously. ControlLogix (CL) is a modular platform for multiple types of control and communications. As a controller, the user can utilize a CL system for sequential, motion control, and process control in any combination. As a communications gateway, CL also enables multiple computers, PLCs, networks, and I/O communications to be integrated.

One of the largest advantages of ControlLogix is its capability to be used as a communications gateway to all of these various levels of communications and various network protocols. There are several communications modules that are available, including EthernetIP, ControlNet, DeviceNet, Data Highway Plus (DH+), remote I/O, Fieldbus, and serial communications modules.

Maximum speed of a ControlNet network is 5Mb/second, which is significantly slower than Ethernet. It is quite fast for a deterministic network, however. It achieves the fast throughput because it is optimized for control of I/O and is scheduled.

Every device is assured a turn to communicate on the network every 2 to 100 ms. The network update time (NUT) is user selectable from 2 to 100 ms. The NUT has three components: scheduled times, unscheduled time, and guardband. Each device on the network reserves scheduled time in advance when it is configured. If we add all of the device scheduled times, the sum equals the total schedule time for the NUT. Whatever time is leftover in the NUT is then used for unscheduled time transmission and guardband data.

Unscheduled time is not reserved in advance by devices. It is used as needed until time expires and the guardband is created. Figure 12-12 shows the NUT communications cycle.

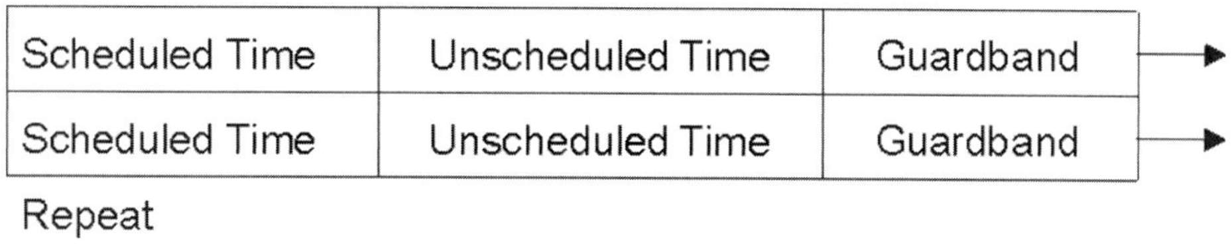

Scheduled Time	Unscheduled Time	Guardband
Scheduled Time	Unscheduled Time	Guardband

Repeat

Figure 12-12 The NUT cycle.

Within the scheduled time slot, each device is allowed to transmit data if it has the token. The first logical device receives the token; if it has data to send, it sends it and then passes the token to the next logical device. That device then sends data if it needs to and passes the token to the next logical device. This continues until the last logical device has received the token. That device then passes the token back to the originator of the token. At this point the scheduled time is over and the unscheduled time begins.

Similarly to scheduled time communications, the first logical device sends any necessary data and passes the token to the next logical device. This continues until the total allocated unscheduled time is reached. Then the guardband time begins.

In the guardband period, the device with the lowest address can send a maintenance message. This device is called the moderator and the message is called the moderator frame. The maintenance message typically includes

The synchronization of timers in each device
The NUT
The scheduled time
The unscheduled time
Miscellaneous maintenance data

In the event that there is no moderator frame in two consecutive NUTs, the device with the next-lowest address automatically becomes the new moderator. If a device with a lower address comes onto the network, that device will assume the role of moderator.

Communication can be peer to peer or master-slave. Multiple masters are allowed to exist on a ContolNet network. ControlNet is a token-ring system. Each device waits until it has the token before it talks. Then the token is passed to the next logical device. This ensures that each device waits no longer than the user selectable time to access the network. One of the largest advantages to using ControlNet is that it is deterministic. This can be crucial if the application is mission critical.

ControlNet uses a producer/consumer method. In a producer/consumer model, multiple devices can get the same data all at once.

The devices on a ControlNet network are arranged logically (not necessarily physically). Each device knows the address of the device to its logical left and logical right. If a device has the token, it can send data until it is done or the token time limit is reached. In either case the device then passes the token to the next logical device.

If we take the maximum time each device can hold the token, we can calculate how long it will take for the token to return to a specific device. This is deterministic. We can determine the worst-case scenario. If the application requires that updates must be done every X ms, it can be determined if that will occur.

If the token hasn't been passed in a given period of time, the network assumes that the device failed and the logical-next device automatically regenerates the token and takes over for the failed device.

ENTERPRISE-LEVEL COMMUNICATIONS

The enterpise level incorporates the office areas of an enterprise. The business systems, email, sales, design, and so on, all typically operate at this level. Ethernet is almost exclusively used at this level to create computer networks.

Ethernet

Most people think of Ethernet when they think of networking. Most home and office computer networks are Ethernet. Ethernet was developed by Xerox in the 1970s. The IEEE published the 802.3 standard in 1985, and it was adopted by the ISO as a worldwide networking standard. The 802.3 standard is normally known as Ethernet by most people.

On an Ethernet network only one device can talk at a time. Ethernet is a decentralized network. No particular device has control over the wire.

Ethernet is a bus system. It is based on a bus that is shared by all devices on the network. Typically unshielded twisted pair wires are used. 10/100BaseT is the most common type of wire that is used.

Ethernet is not deterministic. It cannot guarantee that data will get from one device to another within a certain period of time. Ethernet relies on the Carrier Sense Multiple Access/ Collision Detection (CSMA/CD) access method.

CSMA/CD is a set of rules determining how devices on a network respond when two devices attempt to use the network simultaneously. If they do, it causes a data collision. Devices on an Ethernet network use CSMA/CD to physically monitor the traffic on the line at participating stations. If no transmission is taking place at the time, a device can transmit. If two devices try to transmit simultaneously, a collision occurs, which is detected by all participating devices. The devices that collided wait a random amount of time and try to transmit again. If another collision occurs, the time intervals from which the random waiting time is selected are increased. Networks using CSMA/CD do not have deterministic transmission characteristics.

The best safeguard is to underload an Ethernet network. Ethernet transmission speeds are very high compared to other networks, and if the load is kept low, the delays may be acceptable for an industrial network. If the application is mission critical, however, a deterministic, token-passing network may be required.

Ethernet is becoming more popular in control- and even device-level networking. It is the most prevalent personal computer networking standard and is based on CSMA/CD access methods and bus structure.

Ethernet is popular because it is widely used for computer networking and it is relatively inexpensive to implement. Most controllers can now be purchased with Ethernet capability, and most PLCs have Ethernet communication modules available. Figure 12-13 shows an example of a modular I/O block that has Ethernet capability. It can be easily configured as ControlLogix I/O and can be controlled over a regular Ethernet network. The hardware shown in the figure has digital and analog I/O. Industrial Ethernet is the name given to the use of the Ethernet protocol in an industrial environment for automation and machine control.

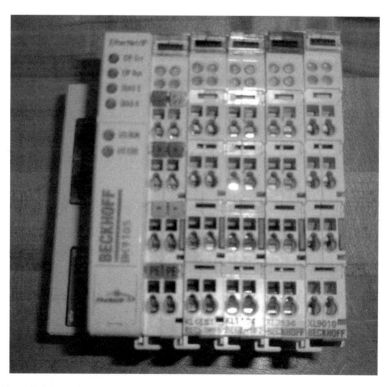

Figure 12-13 I/O block that uses Ethernet for communications to a CLX controller.

Ethernet for Control Networks
Ethernet is nondeterministic. This has often been used as a case against using it for control applications. Determinism enables users to accurately predict the worst-case data transmission. Improvements in Ethernet technology have improved Ethernet's determinism, repeatability, and performance. Switches can be used to break up communications networks into single devices or small groups of devices.

Using switches in place of hubs and running full duplex completely eliminates the chances of collisions on the wire. Only two devices are connected to any wire, and with full duplex they can both send and receive simultaneously. The switch will store and forward packets it receives to the other port, eliminating collisions within the switch (a symptom hubs suffered).

The universal acceptance of Ethernet TCP/IP has made it a popular choice for many users. Ethernet has a wide variety of compatible products and components that are available at low cost. The use of Ethernet for control applications will probably continue to grow rapidly. In fact, EthernetIP is now used nearly as often as ControlNet for controlling I/O, AC drives, and so on.

QUESTIONS

1. Name and describe the three levels of communications.
2. Define the term *asynchronous*.
3. Define the term *synchronous*.
4. Define half duplex.
5. Define full duplex.
6. What is daisy chaining?
7. What is token passing?
8. What is DH+, and what is it used for?
9. What is DeviceNet used for?
10. How many nodes can a DeviceNet network have?
11. What is a scanner, and what does it do?
12. What is a scanlist?
13. What is SERCOS?
14. What is ControlNet used for?
15. Which of the following can a ControlLogix system be used for?
    a. A stand-alone controller
    b. A process controller
    c. A motion controller

  d. A communications gateway
  e. All of the above.
  f. a, c, and d

16. Define the acronym *NUT*.
17. What is CSMA/CD?
18. What does the term *deterministic* mean, and why is it important in industrial communications?

CHAPTER 13

Motion and Velocity Control

OBJECTIVES

On completion of this chapter the reader will be able to:

- Explain the typical components in a motion control system.
- Explain the typical inputs to a drive.
- Explain terms such as *ball screw, lead, pitch, resolution, linear interpolation, circular interpolation,* and so on.
- Explain how motion control is accomplished in ControlLogix.

INTRODUCTION

Motion control is very common in automated systems. Figure 13-1 shows an example of an XY servosystem. This is a two-axis application. The X axis is the horizontal axis of motion. The Y axis is the vertical axis of motion.

CONTROLLOGIX CONTROLLER

A ControlLogix controller is used to coordinate the motion of the two axes in the system shown in Figure 13-1. The system has the capability to move each axis independently or to coordinate the two axes to move in a linear or even a circular path. Position and velocity are both controlled. The XY table in this application is just one component in the system; there are additional components and I/Os. In many applications control would be accomplished by having a PLC control everything but the motion aspect of the cell. A separate controller and drives, probably from a different manufacturer, would control the motion. The motion controller and the PLC would typically utilize digital I/O to handshake. The motion controller would utilize a manufacturer's proprietary software and language to develop the motion program. The technician would have to be able to program and integrate the PLC and the separate motion controller. A ControlLogix system can do it all.

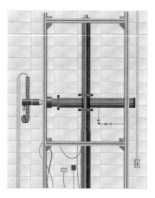

Figure 13-1 An XY (two-axes) motion control system.

SINGLE-AXIS LOOP

Let's examine an axis of motion. A single-axis of motion typically consists of a ball screw, a motor, an encoder, over-travel switches, a home switch, a motor drive, and a controller.

The ball screw will have a lead value for the spiral threads. The lead value determines how far the nut will move in one revolution. Figure 13-2 shows an axis of motion and the motor and encoder, ball screw, home and limit switches, and table that moves on the ball screw.

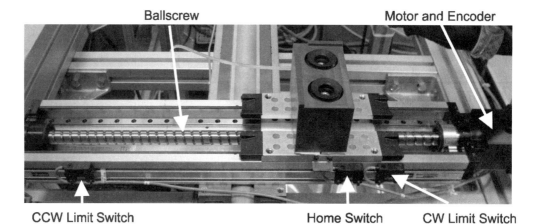

Figure 13-2 A single axis of motion showing over-travel limit switches and the home switch.

Ball Screw

A ball screw is used to translate rotational motion to linear motion. The threaded shaft on a ball screw (see Figure 13-3) provides a spiral raceway for ball bearings. A ball screw is essentially a low-friction high-precision screw. Ball screws are manufactured to very close tolerances and are used in applications that require high precision. The ball assembly acts like the nut in a regular screw system. The threaded shaft is the screw. The nut has a mechanism to recirculate the ball bearings, thus reducing friction and wear.

Backlash (slop) is effectively eliminated by a preload that is applied to the ball bearings by the nut assembly. Low friction in ball screws yields high efficiency. A ball screw is typically 90 percent efficient. An Acme lead screw of the same size would be about 50% efficient due to the higher friction. The higher efficiency of ball screws enables smaller motors to be used.

Note the distance between the spiral threads on the ball screw (see Figure 13-3). We call the distance between two threads the pitch. We call the distance the nut on a screw advances in one revolution the lead. The pitch is equal to the lead. The screw in Figure 13-3 has a pitch of 0.375 inch. This means that the lead is also 0.375. The nut would advance 0.375 inch if the screw is rotated one revolution. If the screw were rotated 180 degrees, the nut would move 0.375/2, or 0.1875, inch. The lead is used when the resolution of an axis of motion is calculated.

Figure 13-3 Ball screw. The pitch is the distance between two adjoining threads.

Figure 13-4 is an illustration of a single, servo-controlled axis. Note the PLC is the overall controller for the system. The PLC will send signals to the drive to control the motion. The drive will control how much and how fast the motor turns. The encoder will provide feedback that the drive will use for position and velocity information. For example, if the PLC commanded the drive to move the axis to the right 10 inches at a velocity of 5 inches per minute the drive will monitor the encoder counts to make sure how far it is moving and also how fast it is moving. The position and the velocity are closed loops, meaning that the drive monitors the feedback from the encoder and automatically adjusts so that the axis moves to the correct position and also at the correct velocity. Note the home switch and the limit switches in the figure, which are inputs to the drive.

The axis has a positive (CW) and a negative (CCW) hard over-travel input. Hard in this case means a mechanical/electric connection. This is a real-world switch to monitor the axis and make sure it does not go past the limit. The over-travel inputs are used to make sure the drive cannot turn too far in either direction. Study Figure 13-4. If the table goes too far to the left, it contacts a CCW normally closed limit switch. This opens the switch, and the input to the CCW over-travel switch becomes false. This inhibits the drive from moving any more to the left and prevents damage to the drive or mechanical components. It will allow positive motion to move to the right and move off the limit. The limit sensors are normally closed and provide fail-safe protection. If a sensor fails or a wire is cut, the drive will have a low on that over-travel input and the drive will be inhibited from moving in that direction. In CLX the programmer can configure how the drive will react to an over-travel condition.

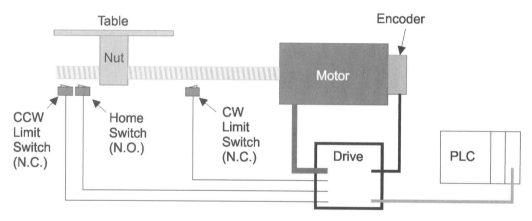

Figure 13-4 A single-axis motion control system.

Homing

A home switch is used to calibrate the position of each axis to a known reference. Typically, the home switch is located along the axis of movement between the CCW limit switch and a CW limit switch. A home switch is especially important in many systems because the controller will lose position information during a power cycle or reinitialization. The programmer can choose how home is established in a CLX system and can utilize the home switch to establish the home position. If this method and a motion axis home (MAH) instruction are used, the axis will move until the home switch is encountered and this position will be established as home. A more accurate method is to use the switch and the index pulse of the encoder. This is the switch/marker method. The index pulse is a part of the encoder that produces a signal once every revolution. It probably takes many revolutions of the motor to move the table through the whole range of motion. The index pulses once every revolution. When the switch/marker method is used and a MAH instruction is executed, the axis will move until it finds the home switch. This gets it close to home and within one revolution of perfect home. When the drive sees the home switch, it reverses direction and moves slowly until it sees the index pulse from the encoder. This is used to establish perfect home. The programmer can use this as actual home or offset the position to meet the needs of the application.

Resolution of a System

The combination of the lead of the screw and the encoder counts per revolution determine the resolution of the axis. Figure 13-5 shows an example. In this example the lead of the ball screw is 0.375 inch. For every revolution of the screw, the nut will move 0.375 inch. In this example there are 2000 encoder counts per revolution. So one revolution will provide 2000 counts and the nut will move 0.375 inch. If we divide the 0.375 inch by the 2000 counts, we find that the resolution is equal to 0.0001875 inch per count.

Lead of the ball screw = .375 inch
Encoder counts per revolution = 2000
.375/2000 = 0.0001875 inch resolution

1 inch / 0.0001875 = 5333 counts per inch

Figure 13-5 How resolution is determined.

Drives also have an enable input. This input must be true (high) for the drive to be enabled and operate. The drive must be enabled in order for any motion commands to execute. This is accomplished by using an output from the PLC to the enable input of the drive.

INCREMENTAL AND ABSOLUTE POSITIONING

Absolute position means actual position in a Cartesian coordinate system. For example, in Figure 13-6 point A's absolute position would be X2, Y2.5. If the machine's axis were presently at point D and we wanted it to move to point B in absolute mode, we would just specify the actual location of X1, Y2.

An incremental move involves a distance to move and the direction to move in. For example, if we wanted to program a move from point D to point C, it would be X1.5, Y0. In other words X has to move to the right (positive) 1.5 inches and Y does not change position to move from point D to point C.

If we wanted to move from point A to point D in incremental mode, it would be X2, Y-1.5. In other words, the X axis would have to move to the left (negative direction)

2 inches and the Y axis would have to move down (negative direction) 1.5 inches. Remember that incremental is direction and distance. Absolute is the actual position.

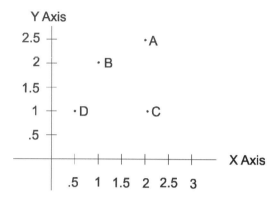

Figure 13-6 Incremental and absolute.

Understanding Interpolation

If only one linear axis is moved, there is no interpolation involved. If we have two axes and want to move both axes from the current location at the same time to a new position and we want the move to be a straight line between the start and end position, we need to interpolate. For example, in a move from point 1 to point 2 in Figure 13-7 to achieve a perfectly straight line to get to the programmed endpoint, the control would have to perfectly synchronize the X and Y axes moves and speeds. This would be called linear interpolation.

In a linear interpolation move, the control calculates a series of extremely small, single-axis steplike moves for each axis, which keep the move as close to the programmed linear path as possible. The step size is equal to the machine's resolution, usually 0.0001 inch, or 0.001 millimeter. With the accuracy of motion controllers, it will create an almost perfectly straight-line motion between the motions of the two axes.

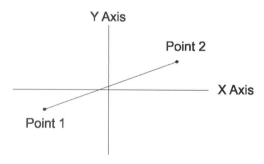

Figure 13-7 Linear interpolation.

Circular Interpolation

Many industrial applications require that the machine be able to move in circular paths. This requires circular interpolation. ControlLogix is also capable of circular interpolation.

Consider Figure 13-8. A circular move from point 1 to point 2 would involve two axes of motion. The X-axis motor must start out at a fairly high speed while the Y-axis motor is barely moving. As the move progresses, the Y increases in speed in relation to the X until the arc is half done. At this point the Y axis is moving at a slightly higher speed than the X axis. As the move progresses, the X axis is continually slowing down in relation to the Y axis until point 2 is reached. Circular motion is classified as clockwise or counterclockwise.

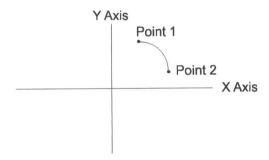

Figure 13-8 Circular interpolation.

A Motion Coordinated Circular Move (MCCM) can make a circular move using up to three axes. A MCCM circular move instruction is specified as either absolute or incremental. The actual speed of the circular move is either at a commanded speed or at a percentage of maximum speed. The MCCM instruction requires data that will enable it to interpolate (calculate) the path. There are four different methods to provide the information the instruction needs to interpolate the path.

The first method of programming a circular path is to specify the center of the circle and the endpoint of the move. In Figure 13-9 the endpoint of the circular path is X2.0, Y2.5. The center of the arc is at X2.0, Y1.0.

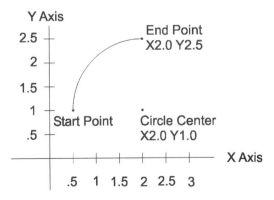

Figure 13-9 The circle center method of programming a circular move.

The next method involves specifying the endpoint of the circular move and a Via point. A Via point is a point somewhere on the circular path between the start point and the endpoint. Figure 13-10 shows an example of a Via point.

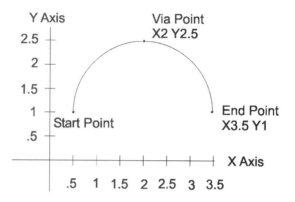

Figure 13-10 The Via point method of programming a circular move.

The next method is to specify the endpoint and the radius of the arc. In Figure 13-11 the endpoint would be X3.5, Y1 and the radius would be 1.5. This is enough information for the circular move instruction to make the move.

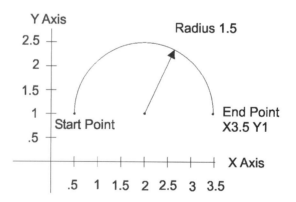

Figure 13-11 The radius method of programming a circular move.

The last method is by specifying the endpoint and the incremental location of the center of the arc from the start point (see Figure 13-12). The endpoint in this example is X3.5, Y1. The other information that must be specified is the incremental distance and direction of the center of the arc from the start point. The center of the arc is 1.5 inches to the right of the start point in the X direction. The Y location of the center of the arc in relation to the start point is the same. There is no change up or down (Y direction) between the start point and the center of the arc. So for this example the X value would be X1.5 (it would be positive 1.5 because the center is to the right of the start point). The Y value would be 0, because there is no change in the Y direction between the start and center points.

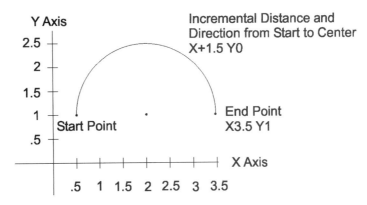

Figure 13-12 The incremental method of programming a circular move.

CONTROLLOGIX MOTION CONTROL

Figure 13-13 shows how a typical motion control system would be implemented in a CL system. Modules are chosen and installed in the chassis. RSLogix5000 software is used to configure each axis of motion in your project. The motion application is written in logic in the ControlLogix project. The application is then downloaded to the controller and can be run.

There is an international open standard for multiaxis motion synchronized motion control. It is called the Serial Real-Time Communication System (SERCOS) and is designed to be a protocol over a fiber-optic medium. Modules that employ the SERCOS standard are also available. Figure 13-4 shows an example of a SERCOS system. The SERCOS interface uses a ring topology with one master and multiple slaves (axes). The fiber-optic ring begins and ends at the master.

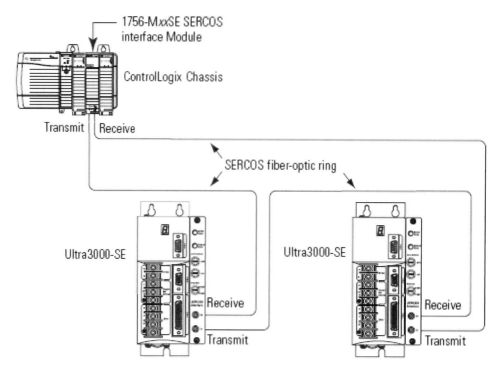

Figure 13-13 A typical ControlLogix Servosystem. (Courtesy of Rockwell Automation, Inc.)

SERCOS

The interfacing of the control device (PLC) and the drives that control motion is especially important in the precise coordination and control of axes of motion. The SERCOS interface is a global standard for the communication between industrial controls, motion devices (drives), and I/O devices. It is classified as standard IEC 61491 and EN 61491. The SERCOS standard is designed to provide hard real-time, high-performance communications between motion controllers and digital servodrives.

SERCOS-I was released in 1991. The transmission medium is optical fiber on a ring topology. The data rates that are supported are 2 and 4 Mb/second. Cyclic update rates are as low as 62.5 microseconds. SERCOS-I supports a Service Channel that allows asynchronous communication with slave devices for less time-critical data.

SERCOS-II was introduced in 1999. It increased data rates to 2, 4, 8 and 16 Mb/second.

Figure 13-14 shows a drive for a CLX SERCOS servosystem. Each drive in a SERCOS system is assigned a node address. The node addresses are set on the right side of the drive. There are two switches that are used to set the node address. The drive also has the motor connections, limit, home, and other input connections. The drive also has two fiber-optic connections for the SERCOS communications to the SERCOS servocard in the CLX chassis.

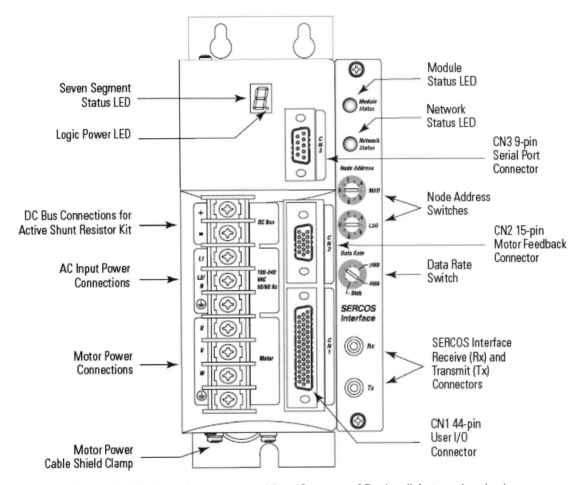

Figure 13-14 Typical servo motor drive. (Courtesy of Rockwell Automation, Inc.)

The pin numbers for the inputs on the drive are found in the drive manual. The drive's enable input is connected to a CLX output. The CLX output must be turned on to enable the drive input.

Sequence for Starting a Drive Application

Note that you must configure the project for the axis of motion and the parameters for each axis of motion. There is an example of configuring a two-axis CLX servosystem in Appendix E. The rest of the chapter will assume that a motion application has been correctly configured in a CLX project.

There is a correct sequence for starting a drive application. You must first enable the drive. An output from the CLX system is used to turn the drive enable (EN) input on. The drive EN input must be high for the drive to control an axis. The logic in the program is used to turn the drive EN on when the drive needs to be enabled.

Next a motion servo enable (MSO) command must be executed for each drive to close the servoloop so that the drive controls the position and velocity. An MSO is an instruction that is used in the program to close the servoloop for an axis. After the drive enable is on and the MSO instruction has been executed, the drive is ready to execute any motion commands it receives.

Next the drive should be homed with a home command. This establishes a known position (home position) for the axis. Note that once the drive EN has been turned on, the programmer can utilize Motion Direct commands, without logic, to test the axis. Note that the MSO and Motion Axis Home (MAH) can be executed with Motion Direct commands.

Motion Direct Commands

You must enable the drive to use motion direct commands. This can be done easily in a simple program. Construct a very simple ladder diagram that turns on the outputs that are connected to the EN input for each drive. This is covered in Appendix E.

Once the drives are enabled, you can try the motion direct commands. Motion direct commands can be used to test your axes before you write the actual logic. You must be online with the CLX in order for any of these commands to work. Right-click the axis icon for the axis you want to test, then click on Motion Direct Commands. The Motion Direct Commands window will appear (see Figure 13-15). The first command that must be executed is an MSO. First click on MSO and click on the execute button. The MSO instruction closes the servoloop for that axis. This must be done or the drive will not execute any commands. Once the MSO is executed, the drive will close the servoloop.

The axis is ready to be homed. When an axis is first started, it does not know where it is. The axis needs to be homed to establish its position. The motion direct command MAH will use the parameters that were set up above in the homing properties. Appendix E covers the configuration of these parameters. Click on MAH and then on the Execute button.

Once the axis is homed, other commands may be tried. A Motion Axis Jog (MAJ) instruction can be used to jog an axis. A MAJ instruction will make the axis move (jog) at a specified speed and direction. Note that you choose the speed and direction before you execute the MAJ command. A Motion Axis Stop (MAS) command must be used to stop the axis. When the MAJ command is executed in incremental mode, the drive will continue to move until it receives a MAS command. Be careful not to exceed the limits on your axis.

The table in Figure 13-16 shows some of the Motion Direct–Motion State commands that are available. Note that they are also available as motion instructions for programming logic. Motion State commands can be used to enable the drive, close the servoloop, reset faults, and so on.

CHAPTER 13—MOTION AND VELOCITY CONTROL   **327**

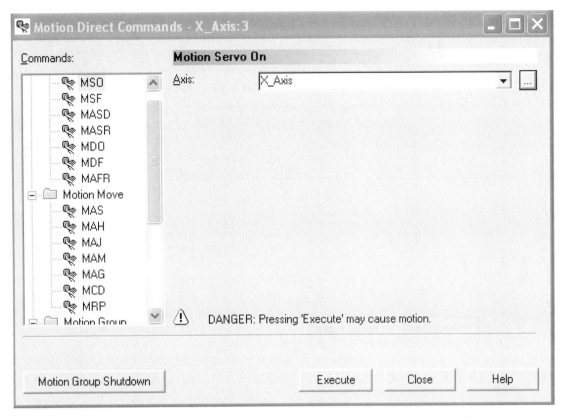

Figure 13-15   Motion Direct Commands screen.

Command	Name of Command	Description
MSO	Motion Servo On	Used to close the servoloop. The servoloop must be closed for the drive to execute commands.
MSF	Motion Servo Off	Used to deactivate the drive output for the specified axis and to deactivate the axis's servoloop.
MASD	Motion Axis Shutdown	Used to force a specified axis into the Shutdown state.
MASR	Motion Axis Shutdown Reset	Used to transition an axis from an existing Shutdown state to an Axis Ready state.
MDO	Motion Direct Drive On	Used to activate the module's Drive Enable.
MDF	Motion Direct Drive Off	Used to deactivate the module's Drive Enable.
MAFR	Motion Axis Fault Reset	Used to reset faults in an axis.

Figure 13-16   Motion Direct–Motion State commands. Note this is only a partial list of available commands.

The table in Figure 13-17 shows some of the Motion Direct–Motion Move commands that are available. Note that they are also available as motion instructions for programming logic. Motion Move instructions can be used to make the axis stop motion, home the axis, jog the axis, make an absolute or incremental move, and so on.

Command	Name of Command	Description
MAS	Motion Axis Stop	Used to make an axis stop motion.
MAH	Motion Axis Home	Used to make an axis perform a home routine to establish home position.
MAJ	Motion Axis Jog	Used to jog an axis plus or minus direction. Once active it will continue to move until a MAS command is used.
MAM	Motion Axis Move	Used to move an axis a commanded distance (incremental) or to a commanded position (absolute).

Figure 13-17 Motion Direct–Motion Move commands. Note this is only a partial list of available commands.

Motion Direct commands are a great way to test your axes and make sure they have been properly configured.

Programming Logic for Motion

The first step in your motion program should be to enable the drive. The logic should turn on a PLC output that is wired to the enable for each drive to enable the drive. This is covered in Appendix E. The drive enable must remain high for the drive to operate. After a drive is enabled, an MSO instruction is used to close the servoloop on each axis. This can be a momentary instruction; the MSO instruction does not have to remain high.

MSO Instruction

Step 1. Enable each drive with an output from an output module.
Step 2. Close the servoloop for each drive with a momentary MSO instruction.

Motion Commands

At this point motion instructions can be used to move and control the axes. When using any of the motion instructions in a program, you need to create a tag for the instruction. The data type of the tag will be motion instruction. Motion instructions need a motion-type tag to store configuration and operation data. Figure 13-18 shows an example of an MSO function block. Note that the axis must be chosen (Y_Axis in this example). Each motion function block must also have a tag. In this example the tag was named Y_MSO_Tag. It is a good idea to use tag names that will help to remember what they are used for.

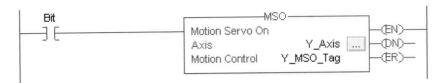

Figure 13-18 Note that in motion commands the axis must be chosen (Y_Axis in this example). A tag of type motion must be created for each instruction.

These motion tags have a variety of tag members available that may be useful in your programs. Figure 13-19 shows a MAH instruction.

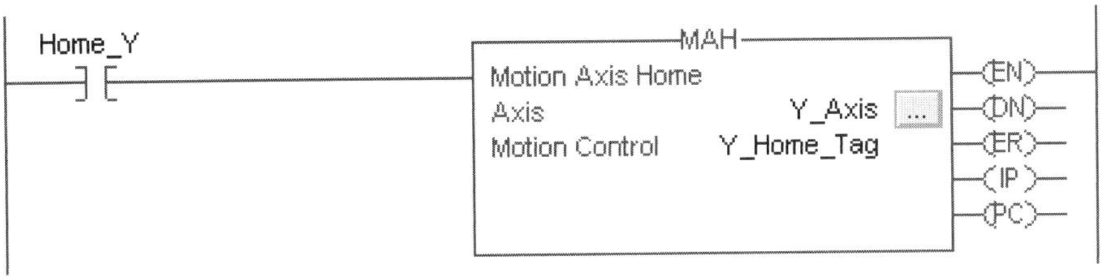

Figure 13-19 A MAH instruction.

Note the EN, DN, ER, IP, and PC to the right of the instruction. They are tag members for the tag named X_Home_Tag. These can be useful in your ladder program. Figure 13-20 shows the tag member bits and the conditions they represent. After the homing of an axis has been completed, other move instructions can be used.

Tag Member	Name and Function
EN	Enable. Set when the rung makes a false-to-true transition. Remains set until the servo message transaction is completed and the rung goes false.
DN	Done. Set when the axis's faults have been successfully cleared.
ER	Error. Indicates when the instruction detects an error, such as if an unconfigured axis was specified.
IP	In Process. Set by a positive rung transition. Cleared after the Motion Axis Stop is complete or by a shutdown command or a servo fault.
PC	Process Complete. Set after the stop operation has successfully completed.

Figure 13-20 MAH instruction bits

Jog Instructions

Jogging is accomplished with the MAJ instruction. A MAJ instruction is normally used with a MAS instruction (see Figure 13-19). The MAJ instruction can be used to jog an axis in the forward or reverse direction. Once the instruction is energized, it continues to jog even if the rung goes false. You must use an MAS instruction to stop the motion.

Study the example in Figure 13-21. The jog instruction is being used with an axis that was named Y_Axis. Note that Y_Axis was entered into the MAJ and the MAS instructions.

In this example a normally open (XIC) contact is used to control a MAJ instruction. A normally closed (XIO) contact with the same tag is used in the second rung to control the MAS instruction. The same tag was used for the contact in rung one and the contact in rung 2. This means that only one rung can be true. If the Jog_Bit is true, the MAJ executes. If it turns false, the MAS rung is true and the jog motion stops.

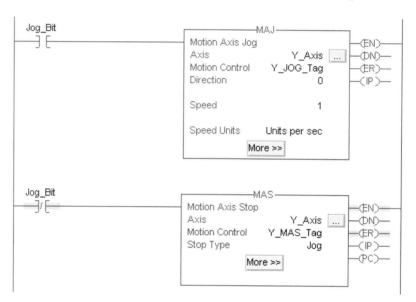

Figure 13-21 A simple ladder to control the jog of the Y_Axis.

Configuring a MAJ Instruction
There are several parameters that need to be configured in a MAJ instruction. Most of the parameters are similar in all motion instructions. The instruction Help file in RSLogix 5000 explains the parameters for each instruction.

Direction
After you have chosen the axis and assigned a tag name, you must enter a direction. The MAJ instruction can move in a positive or negative direction. Enter a 0 for forward or a 1 for reverse jogging. A tag name could also be entered so your logic could change the direction during operation.

Speed
Enter a number that represents the speed of the jog move in terms of units or enter a tag name for the speed. In the example in Figure 13-20, 2 was entered. In this

system the resolution was set up to be inches, so this would be a jog rate of 2 inches per second.

Speed Units
Enter a 0 if you want to use a percentage of maximum speed or a 1 if you want units per second. Units are dependent on how you set your resolution. The unit in this example is an inch.

There are a few more parameters to enter. Study Figure 13-22. Note the Less << button on the bottom of the MAJ instruction. If all of these parameters are not displayed there will be a More >> button. If they are not all visible click on the More >> button and enter the information. The parameters to be entered are shown in Figure 13-22.

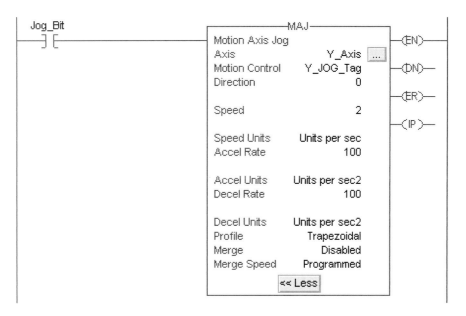

Figure 13-22 A MAJ instruction. Note the << Less button.

Acceleration and Deceleration
Next an acceleration parameter must be entered. A value is entered for the acceleration rate. The value 100 was entered in this example. Next the acceleration units must be configured. Enter a 0 to configure them to a percentage of maximum acceleration. Enter a 1 to enter in terms of units per second^2. In this example 1 was chosen and 100 was entered for a rate.

The next two are deceleration parameters. These are set just as the acceleration parameters in this application.

The next parameter is the move profile for acceleration and deceleration. If 0 is entered, it is a trapezoidal profile. If 1 is entered, it is an S-curve profile.

The trapezoidal profile is the most common. It also provides the fastest acceleration and deceleration times. Figure 13-23 shows an example of a trapezoidal profile.

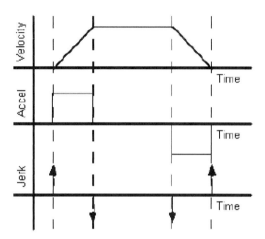

Figure 13-23 A trapazoidal profile. (Courtesy of Rockwell Automation, Inc.)

The S-curve profile is used for special circumstances. It is often used when there is an unusual stress on a mechanical system and the load on the system must be minimized. The S-curve profile is slower than trapezoidal because acceleration and deceleration values must be lower in the S-curve. Figure 13-24 shows an S-curve profile.

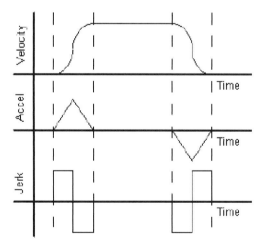

Figure 13-24 An S-curve profile. (Courtesy of Rockwell Automation, Inc.)

The next parameter to set is the merge parameter. If an axis is moving due to another command and the jog is activated, this choice determines whether the movement should be turned into a jog and also controls the speed of the merge between the current movement and the jog movement. If 0 is chosen, merge is disabled. If 1 is chosen, all movement would be turned into pure jog. Each of the choices is described below.

Merge Disabled

Merge disabled means that any single-axis motions that are currently executing are not affected by the activation of this instruction and results in superimposed motion on the affected axes. Any coordinated motion instructions for the same specified coordinate system runs to completion on the basis of its termination type.

Coordinated Motion

Any currently coordinated motion instructions that are executing are terminated. The motion that is active is blended into the current move at the speed that is specified in the merge speed parameter. Any single-axis motion instructions that are currently executing in the specified coordinate system will not be affected by this instruction being activated and will result in superimposed motion on the affected axes. Coordinated motion instructions that are pending are cancelled.

All Motion

Single-axis motion instructions that are currently executing in the specified coordinate system and any coordinated motion instructions that are currently executing are terminated. The prior motions are merged into the current move at the speed specified in the merge speed parameter. Pending coordinate move instructions are cancelled.

Lastly the merge speed is set. If you entered a 1 for Merge, then this field will determine the merge speed. If you choose At Current Speed as the type of Merge, the speed of the jog is automatically set to the current actual speed of the axis. In this case, any specified speed value or tag variable associated with the MAJ instruction is ignored. If you chose At Programmed Speed as the Merge Type, the speed of the jog is set to the entered speed value or tag variable. If this speed is different from the current speed of the axis, the axis is accelerated or decelerated as specified to the new speed.

Motion Axis Move (MAM) Instruction

The MAM instruction is one of the more useful move instructions. It is used to move an axis in the absolute or incremental mode. To use a MAM command, you first select the axis and create a tag for the command.

Figure 13-25 shows an example of a MAM instruction. The first parameter to be entered is the axis that is to be moved (X_Axis in this example). Next you must enter a tag name for a tag to be used by this instruction. X_Axis_MAM_Tag was entered. This must be a motion-type tag. The MAM instruction will store parameters and information related to the instruction in this tag. Next the type of move must be chosen. If 0 is entered, the move will be absolute. A 1 would make this an incremental move.

Incremental means that a distance and direction command in this instruction would move the axis that amount and direction from its current position. For example, a position/distance of 5.5 would make the axis move 5.5 inches from its current position in a positive direction. So an incremental move is really a distance and direction to move. If the instruction is executed and then reexecuted, the axis would move again, because it is an incremental move.

Absolute means that the instruction would move to a commanded position. For example, if the position/distance value was 5.5, the axis would move to actual position 5.5 from wherever it currently is. If the instruction is executed and then reexecuted, the axis would not move again, because it an absolute move and the axis would be at the absolute commanded position after the instruction was executed the first time.

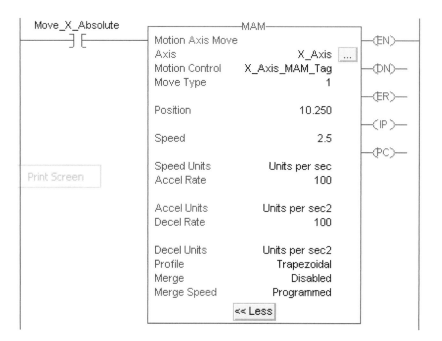

Figure 13-25 A MAM instruction.

The position that is entered is the absolute position or the incremental direction and distance if incremental is chosen. The rest of the entries are the same type as were entered for the jog command.

Note that tags could have been used for all of the parameters in the instruction. This would allow values to be changed in run mode. For example, a tag could be used to change from absolute mode to incremental mode. The speed could be varied by the program if a tag were used.

Motion-Coordinated Linear Motion (MCLM) Instruction

Use the MCLM instruction to make a multidimensional linear coordinated move for the specified axes within a Cartesian coordinate system. Imagine a two-axis system, like an XY table. If we would like to make an angular move, both axes would have to be precisely coordinated to make a smooth linear path. Figure 13-26 shows an example of a linear, two-axes interpolated move. Figure 13-27 shows an example of a MCLM instruction. You can define the new position as either absolute or incremental.

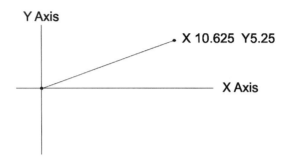

Figure 13-26 A linear interpolated move.

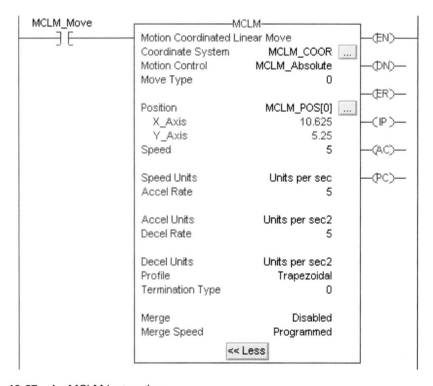

Figure 13-27 An MCLM instruction.

The first entry in an MCLM instruction will be the Coordinate System. Type a name in and then right-click on it. Choose Coordinate System for the type of the tag. Then click on the ellipsis to the right of your Coordinate System tag name. The screen shown in Figure 13-28 will appear.

The Coordinate System Properties General entry screen is used to choose the motion group and choose the axes that will be involved in the motion. Note the Dimension entry 2 was chosen in this example. This means there are two axes involved in this coordinate system.

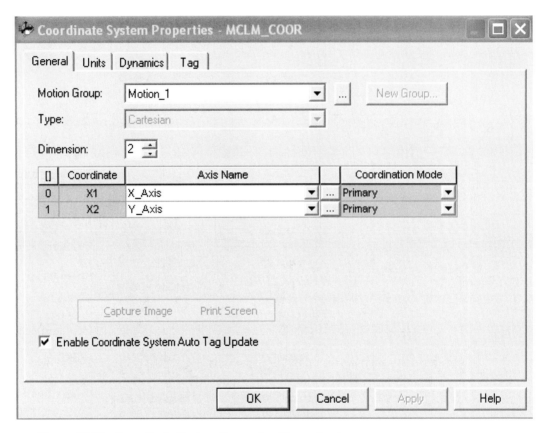

Figure 13-28   Coordinate System Properties General entry screen.

The Coordinate System Properties Units entry screen is used to enter conversion ratios (see Figure 13-29). In this example there is no conversion so 1/1 was chosen. This would be used if one axis motion needed to be scaled to the other axis motion.

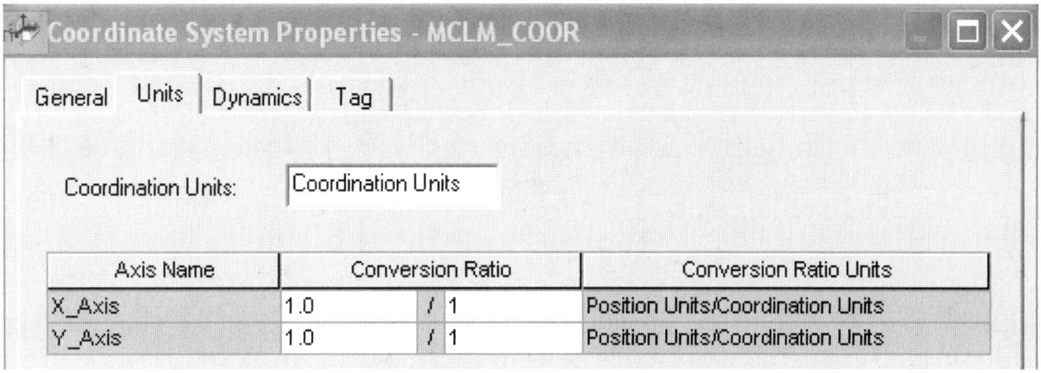

CHAPTER 13—MOTION AND VELOCITY CONTROL

Figure 13-29 Coordinate System Properties Units entry screen.

The Coordinate System Properties Dynamics entry screen shown in Figure 13-30 is used to enter a maximum speed, acceleration, and deceleration, as well as positional tolerances for moves.

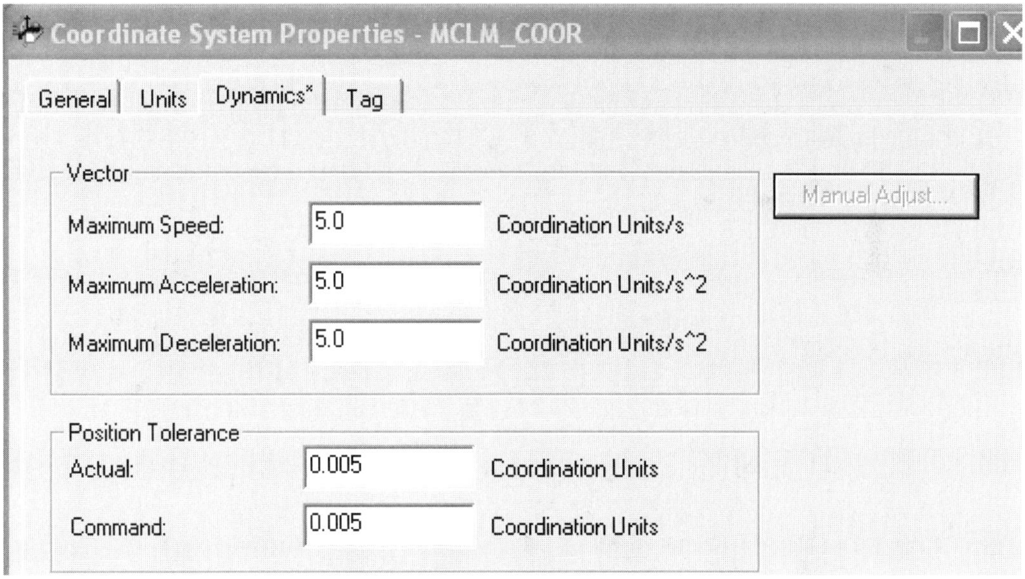

Figure 13-30 Coordinate System Properties Dynamics entry screen.

Figure 13-31 shows the Target Position Entry screen. The endpoint of the move is entered here.

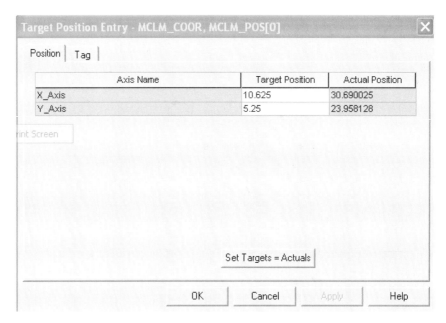

Figure 13-31 Target Position Entry screen.

Next a tag name must be entered in the Motion Control entry. This tag will be used by the instruction.

Move Type is chosen next. The choices are absolute or incremental. Enter 0 for absolute or 1 for incremental.

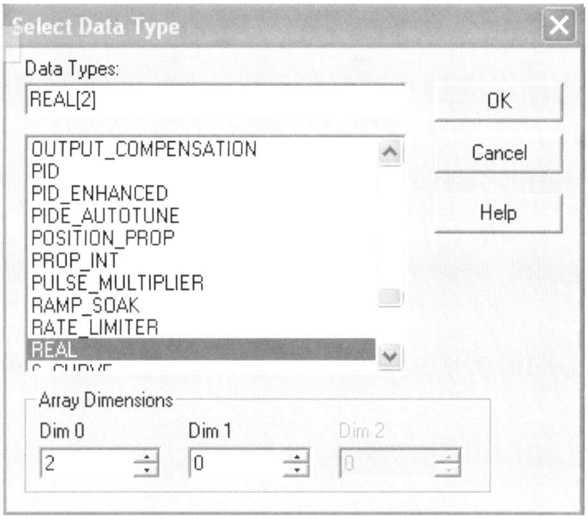

Figure 13-32 Declaring the data type for the tag array that will hold the X and Y position to move to.

Figure 13-33 shows the end position for the instruction in the tag editor. The end position was entered in Figure 13-31.

MCLM_POS	{...}
MCLM_POS[0]	10.625
MCLM_POS[1]	5.25

Figure 13-33 The array tag that holds the X (10.625) and the Y (5.25) position values for the move.

The rest of the instruction entries are similar to the previously covered motion instructions.

CIRCULAR INTERPOLATION

An example of a circular interpolated move is shown in Figure 13-34. In this example the first move was a linear move to X10, Y10. The next move was a circular path. The endpoint for the circular move is (X15, Y15). The center of the circle is (X15, Y10).

A Motion Coordinated Circular Move (MCCM) instruction can be used to make the circular portion of the move in Figure 13-34. The MCCM instruction is shown in Figure 13-35. The entries to the instruction are similar to the MCLM instruction. A coordinate system must be created and configured. The coordinate system is created just as it was for the MCLM instruction. The Coordinate System for this MCCM example was called XY.

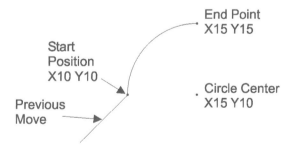

Figure 13-34 A circular move.

Next a tag name must be entered for the control tag for the instruction. The instruction uses this tag to store parameters and other information it needs. In this example the tag was named MCCM_Tag. It is a motion-type tag.

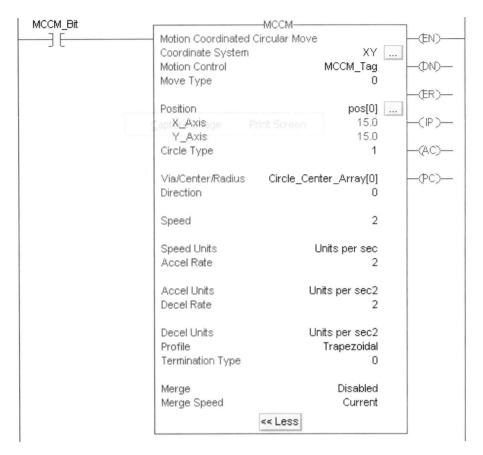

Figure 13-35 An MCCM instruction.

To program a circular move, you must specify the endpoint of the move. The endpoint is called the Position. A tag was created of type real. It was named pos[0]. It is an array tag that holds two values: the X position and the Y position of the endpoint (see Figure 13-36). The Position for this instruction is X15, Y15.

POS	{...}
POS[0]	15.0
POS[1]	15.0

Figure 13-36 The array tag that is used for the endpoint of the move.

There are three methods for specifying the needed position information: radius, center, and incremental center. A 1 was entered in the Circle Type parameter (center method). The Circle Center programming method was chosen for this MCCM instruction so the center of the circle was entered as 15 (X_Axis position) for the first element in the array and 10 for the second (Y_Axis position). Note that your values must be correct. If your positions are not accurate, the controller will not be able to interpolate the moves and a fault will occur.

Next the direction parameter is entered. A 0 is for clockwise and 1 is for counterclockwise; 2 is for clockwise full circle, and 3 is for counterclockwise full circle.

The rest of the parameters that must be entered are just like the other motion instructions that have already been covered.

Once you have mastered these few instructions, there are many other motion instructions that can be used.

USE OF CONTROLLOGIX TO CONTROL ROBOTS

One of the new trends in motion control is to utilize the ControlLogix controller to control a robot. There are a few companies that now offer the robot arm, minus the controller. They have designed their robot to work with ControlLogix hardware and software. There are several advantages to this. Programmers only need to know CLX well. They do not need to learn another system and language for a robot controller. One CLX controller can control the whole system.

The most complicated part of robotics is translating several axes of generally rotary motion into Cartesian moves. Robot kinematics involves the mathematical transformations that are done by a controller to calculate how to move multiple axes of motion to accomplish Cartesian motion. In other words, how can multiple axes of motion be coordinated to make a point-to-point move, a straight-line move, or a circular move to a specific point in space? It is very complex. Robot kinematics involve forward kinematics and inverse kinematics. In forward kinematics, the length of each link and the angle of each joint is given and we have to calculate the position of any point in the work volume of the robot. In inverse kinematics, the length of each link and position of the point in work volume is given and we have to calculate the angle of each joint.

Kinematics has been integrated into the ControlLogix software and controller. This significantly reduces the time and costs for integrating, programming, and maintaining a robot in a system. The robot can be programmed in simple Cartesian coordinates because the controller handles the kinematic transformations. RSLogix 5000 software allows control of two- and three-axis, articulated, independent/dependent, SCARA, H-bot and delta geometry robots.

A library of Add-On instructions is available to simplify robot program development and integration.

When developing a new application, we must enter the parameters for each axis of the robot into the configuration screens. The CLX controller uses this information to do the transformations.

Having one CLX controller control the whole application enables the system developer to easily integrate and synchronize a robot's motion with other parts of the application. It makes it much easier to integrate conveyor tracking and vision systems into an application.

QUESTIONS

1. List at least three reasons why ball screws are used in motion systems.
2. What is the function of an over travel switch?
3. Describe how homing is accomplished in a system.
4. A system has a ball screw with a pitch of 0.200 inch. There are 4000 encoder counts per revolution of the ball screw. Calculate the resolution for this system.
5. What is absolute positioning?
6. What is incremental positioning?
7. Describe linear interpolation.
8. What position information needs to be entered for a circular interpolated move?
9. What are motion direct commands, and why are they useful?
10. Describe what needs to be done in a program to enable a drive and close the servo-loop so that movement can take place.

CHAPTER 14

Risk Assessment and Safety

OBJECTIVES

On completion of this chapter the reader should be able to:

- Explain the purpose of risk assessment and risk reduction.
- Describe some of the standards that apply to risk reduction.
- Perform a risk assessment.
- Explain risk reduction strategies.

THE IMPORTANCE AND COST OF SAFETY

Safety may be the most important part of the manufacturing process. Injuries often happen when a system is stressed. Downtime and production stress tend to make operators or maintenance personnel do things they might not ordinarily do to keep production high or to get a system running. Safety problems can also arise from poor maintenance procedures or poor design that tends to create safety problems. Operators may feel pressure to keep production up. Maintenance personnel may feel pressure to repair machines as quickly as possible to get production going. A maintenance person may try quick fixes or work-arounds to get a machine running. One might see emergency stops or safety interlocks (light curtains, limit switches, proximity sensors) overridden to temporarily fix a problem. Quick fixes and work-arounds may get the machine running but can cause immediate and future safety hazards.

An American Society of Safety Engineers study calculated the ratio of indirect to direct costs of an industrial accident as high as 8:1. We usually just think of the direct costs of an accident. We think about the lost time or the cost of repairing the machine. But the indirect costs can be 8 times greater than these. This means that the costs associated with accidents are huge.

RISK ASSESSMENT FUNDAMENTALS

Machine safety is essentially a two-step process: risk assessment and risk reduction. All machines and systems have risks associated with them. When a new machine is purchased, one might assume the machine has been designed to be as safe as possible. But risk assessments should be performed on new equipment also. A company might also purchase machines or attachments that are integrated into/on other machines or equipment. This can create new risks, so a risk assessment should be performed. Some companies design and build their own equipment. Risk assessment must be performed. If accidents or close calls occur, a risk assessment should be carried out. These risk assessments are done to identify potential safety hazards.

There can be many hazards present on industrial equipment. Risks include the operator being exposed to hazards such as cutting, crushing, shearing, clamping, trapping, perforating, puncturing, risking shock, noise, heat, slipping and falling, and so on. Where necessary, additional protective measures must be implemented to protect operators and other individuals from these and other types of hazards.

Note that in the United States, the responsibility for protecting personnel falls on the end user because of Occupational and Safety Health Administration (OSHA) guidelines. Only in the last few years have consensus standards begun to explicitly require risk assessments by machine builders. Such standards include ANSI PMMI B155.1-2006, ANSI RIA 15.06-1999, and ANSI B11-2008.

Risk Assessment

To aid the machine manufacturer with the task of risk assessment, there are standards that define and describe the process of risk assessment. A risk assessment is a sequence of logical steps that permits the systematic analysis and evaluation of risks. The machine is designed and built taking into account the results of the risk assessment.

Risk reduction follows a risk assessment, when necessary, by applying suitable protective measures. New risks shall not result from the application of protective measures. The repetition of the entire process, risk assessment and risk reduction, may be necessary to eliminate hazards as far as possible and to reach residual risk that is tolerable.

ANSI B11.TR3-2000

The American National Standards Institute (ANSI) developed ANSI B11.TR3-2000, Risk Assessment and Risk Reduction--A Guide to Estimate, Evaluate and Reduce Risks Associated with Machine Tools. ANSI B11.TR3-2000 calls for risk assessment and reduction of risk through a systematic review. Many hazards are missed during machine design because task identification and risk assessment are not part of the process. ANSI B11.TR3 came out in late 2000. The ANSI B11.TR3-2000 standard puts the responsibility on machine manufacturers and users. It brings U.S. standards in line with European/International standards.

There is no industry or government requirement that mandates that the content of ANSI B11.TR3 be followed. The current OSHA regulations provide no methodology to address machine hazards and risks. Risk assessment is, however, already part of robotic standards (ANSI/RIA-1999, "Industrial Robots and Robot Systems").
Task-based risk assessment and reduction involves the following:

1. Determine the machine limits.
2. Identify all machine tasks.
3. Identify hazards associated with each task.
4. Rate the severity.
5. Rate the probability for each hazard.
6. Determine the level of risk.
7. Eliminate the hazard or reduce its severity.
8. Determine the type of safeguarding and performance level of the system necessary to achieve the desired risk level.

Tolerable risk is the term used to refer to a level of residual risk for a given hazard after applying risk reduction measures. ANSI B11.TR3-2000 defines tolerable risk as: risk that is accepted for a given task and hazard combination. Lately more focus is being directed at integrating safety into all phases from womb to tomb of a machine life cycle.

Acceptable risk is a newer term that is beginning to appear in updated standards. It more clearly represents the implied intent of evaluation and mitigation. The assumption is that risk can never be totally eliminated from a hazard but that every risk should be evaluated for risk reductions and mitigated to the smallest amount possible. Therefore, more current standards are defining acceptable risk as the level at which further risk reduction will not result in significant reduction in risk or that additional expenditure of resources will not result in significant advances toward increased safety.

ANSI B11.TR3-2000 makes it clear that although zero risk does not exist and cannot be attained, a "good faith approach" to risk assessment/risk reduction should achieve a tolerable risk level. This level may not be the same for every company. Risk is often defined with three major elements: frequency, probability, and severity. In other words how often someone is exposed to the hazard, how likely they are to become injured, and how bad the injury is likely to be. The idea is to reduce or eliminate exposure to all recognized hazards.

In a risk assessment following the ANSI B11.TR3-2000, standard ratings are chosen for severity, frequency, and probability. The first is severity. Severity is a measure of how bad the potential injury might be. The severity of harm or injury is rated in terms of minor, serious, major, or fatal. Figure 14-1 shows the ratings. A minor rating would get 1 point, while a fatal rating would get 10 points.

Ratings

Severity Rating	Effects
10	Fatal
6	Major – normally irreversible, permanent disability, e.g., loss of sight, amputation
3	Serious – normally reversible, cuts, broken bones, burns
1	Minor bruising, cuts, first aid care

Figure 14-1 Severity rating table.

The second factor to be rated is the frequency of exposure to the hazard. Frequency is rated in terms of seldom, occasional, or frequent. Figure 14-2 shows the ratings for the three potential frequencies of exposure.

Frequency Rating	Frequency
4	Frequent – several times per day
2	Occasional – daily
1	Seldom – weekly or less

Figure 14-2 Frequency rating table.

The third rating factor deals with the probability of the injury occurring. The rating factors are unlikely, possible, probable, or certain. Figure 14-3 shows the ratings.

Probability Rating	Probability
6	Certain
4	Probable
2	Possible
1	Unlikely

Figure 14-3 Probability rating table.

The severity, frequency, and probability ratings are then added and other factors are considered. If more than one person is exposed to the hazard, the number of people is multiplied by the severity factor. If a person spends more than 15 minutes per access in the danger zone without lockout/tagout protection, 1 point is added to the frequency factor. If the operator is unskilled or untrained, 2 points are added to the total. The total is then compared to a ratings scale. Figure 14-4 shows an example. In this example the ratings were added and the total was found to be 9. A 9 on the chart represents a medium risk.

CHAPTER 14—RISK ASSESSMENT AND SAFETY   **347**

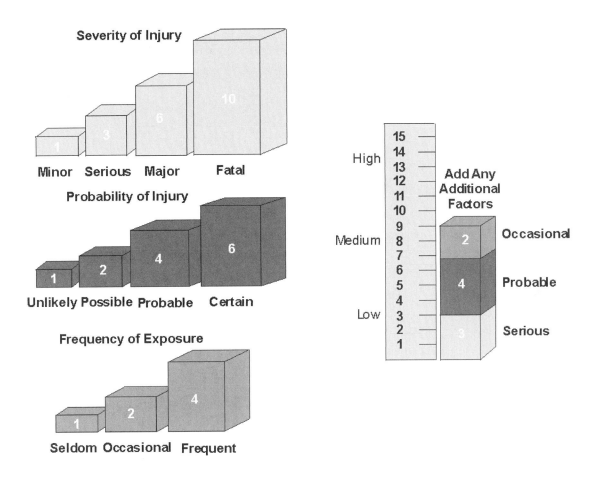

ANSI B11.TR3-2000 Risk Assessment/Risk Reduction

Figure 14-4 Rating risk.

Risk Estimation Example

Imagine a metal stamping machine. Figure 14-5 shows a stamping machine with an operator removing a completed part. Depending on how the machine is designed, an operator could have very different potential injuries. On some machines there may be a danger of the operator losing an arm if an arm is in the way of the machine's moving parts in a cycle. On another machine the most danger might be getting their finger trapped during a cycle.

Figure 14-5 A metal stamping machine.

One machine may have more than one access point, each having different potential injuries associated with them. Different measures for safety control could be used for each. For example one area of potential contact might be eliminated by putting a non-removable guard in place. Another area might require safety devices and interlocks such as a light curtain.

Let's rate the risk for the stamping machine. The possibility of losing an arm would be irreversible and would be rated a 6 for severity (see Figure 14-1). The frequency rating would have to be a 4 because the operator is exposed to the risk frequently (see Figure 14-2). A probability rating of 2 was given because the injury was rated possible (see Figure 14-3). The total would be 12 (6 + 4 + 2). The rating scale in Figure 14-4 would show that a 12 would be a high-risk level.

ANSI B11.TR3 Safeguarding

ANSI B11.TR3 describes a hierarchy of four levels of safeguarding that can be applied, depending on the level of risk reduction needed for that machine.

Hierarchy of Hazard Controls

- Eliminate, by design.
- Control access to exposures by safeguarding.
- Provide other safety measures, like awareness barriers, signals, and so on.
- Institute administrative controls, procedures, and personal protective equipment (PPE).

EUROPEAN SAFETY STANDARDS, EN ISO 12100-1

EN ISO 12100 is a European safety standard that covers the "Safety of Machinery – Basic Concepts, General Principles for Design." Part 1 of the standard covers "Basic Terminology, Methodology," and Part 2 covers "Technical Principles."

There are three main clauses in Part 1 in addition to explaining the scope of the standard and normative references.

Clause 3 of EN ISO 12100-1 covers terms and definitions.

Clause 4, "Hazards to be taken into account when designing machinery," is a checklist of hazards, including mechanical, electrical, and thermal hazards, hazards generated by noise or vibration, and so on.

Clause 5 of EN ISO 12100-1 is the "Strategy for risk reduction." Subclause 5.1.3 lays out the basic steps to be taken in a risk assessment.

- Specify the limits and the intended use of the machine.
- Identify the hazards and associated hazardous situations.
- Estimate the risk for each identified hazard and hazardous situation.
- Evaluate the risk and take decisions about the need for risk reduction.
- Eliminate the hazard or reduce the risk associated with the hazard by protective measures.

Another standard, ISO 14121-1:2007, provides guidance on the information that is required to perform a risk assessment. Procedures are described for identifying hazards and estimating and evaluating risk. The standard also provides guidance on decisions relating to the safety of machinery and on the type of documentation required to verify the risk assessment.

EN ISO 14121-1:2007, "Safety of machinery–Risk assessment," Part 1: "Principles," also provides a more detailed list of hazards and hazardous situations. ISO 14121-1:2007 establishes general principles intended to be used to meet the risk reduction objectives in ISO 12100-1:2003, Clause 5.

Figure 14-6 shows a flowchart for performing a risk assessment and risk reduction for a machine or system, based on safety standard EN ISO 14121. Note that it is an iterative process.

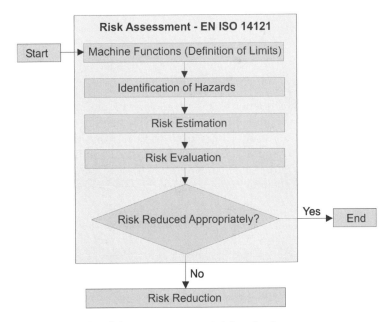

Figure 14-6 A flowchart for risk assessment and risk reduction.

RISK ASSESSMENT

A team approach is recommended for risk assessment studies. Operators, maintenance personnel, engineering, and other appropriate personnel should be involved in risk assessment studies. Each will have a different perspective on the machine and will be able to provide insight on potentially hazardous conditions. The assessment process needs to cover the entire life cycle of the machine – design to discard.

Machine Limits

The process of performing a risk assessment begins with studying the machine or system. This is often called defining the limits of the machine. It has to do with studying the functions and operation of the machine. Following are some considerations that should be taken into account in machine limits:

- The intended use of the machine, production rates, cycle times, materials used, and so on
- Range of movement of the machine, physical boundaries, the expected place of use, and so on
- Noise, temperature, humidity, and so on
- Maintenance and wear of tools, fluids involved, and so on
- Other machines, auxiliary equipment, energy sources, and so on
- Malfunctions and faults that are to be expected
- The correct use of the machine and also the unintentional actions of the operator or the reasonably foreseeable misuse of the machine or system

Once the machine limits of the machine or system have been defined, the next step in the process is to identify the various tasks and associated hazards associated with operating the machine.

Determining the Necessary Safety Level

This entails a risk estimation analysis for each task and its associated hazard(s). There are a variety of standards and approaches that have been developed for risk estimation.

Risk estimation methods are used to determine the probability of occurrence of harm, the exposure to the hazard, the ability to avoid injury, and the seriousness of the potential injury. Figure 14-7 shows a simple illustration of risk estimation. Risk is equal to the extent of the potential injury multiplied by the probability of its occurrence.

Risk Estimation

Figure 14-7 Risk estimation.

Safety standards were created to help designers in the definition of categories for safety-related control. The following categories are often used to assess risk:

- The possible severity of injury (S)
- The frequency or duration of exposure to the hazard (F)
- The possibility of preventing the hazard (P)

The diagrams shown in Figures 14-8 and 14-9 show one method for classifying risk in accordance with safety standards EN 954-1. EN 954-1 was used until 11/29/2009. It has been superseded by EN ISO 13849-1.

Let's consider the stamping machine example again. Figure 14-8 shows the severity, frequency, and possibility of avoidance factors for EN 954-1. The severity of potential injury is considered first. In our example, an operator has to load and unload parts into a stamping machine. The operator could lose a hand if it were in the machine during the operation. This is a serious injury–normally irreversible, fatal, or requiring more than first aid–so S2 would be chosen for the severity level (see Figure 14-8).

Next we need to consider the frequency. The operator has to make several pieces per hour. This would be considered frequent exposure so F2 would be chosen for the frequency factor (see Figure 14-8).

The last thing to be considered is the possibility of avoiding the hazard. The moving ram that punches the metal on the machine is very fast so it is unlikely the operator can move out of the way. Thus a factor of P2 is chosen (see Figure 14-8).

S Severity of the Injury	
S1	Slight Injury: normally reversible, requires only first aid
S2	Serious injury: normally irreversible, fatal or requiring more than first aid

F Frequency (or duration of the exposure to the hazard)	
F1	Infrequent: Typical exposure to hazard is less than once per hour.
F2	Frequent: Typical exposure to hazard is more than once per hour.

P Possibility of avoiding the hazard	
P1	Likely: Can move out of the way or sufficient warning/response time or the robot speed is less than 250 mm/second.
P2	Not likely: Cannot move out of the way or inadequate response time or the robot speed is greater than 250 mm/second.

Figure 14-8 Risk classification based on ANSI R15.06-1999.

When these factors are applied to the decision tree in Figure 14-9, we see that this would require a category 4 solution. This applies to the EN 954-1 standard's method. In EN 954-1 categories are used to describe the level of safety requirements that should be followed.

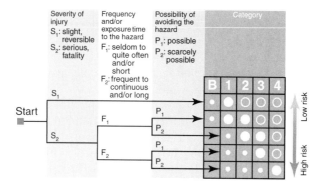

Figure 14-9 The EN 954-1 method of risk estimation and safety category selection. This is also called the risk graph.

EN 954-1 sets five safety system classifications depending on the level of assessed risk.

Category B is the first. Category B has no special requirements for safety. It requires that designers follow good design practices and choose components that are consistent with the parameters of the application. Category B forms the base requirements for the rest of the categories.

Category 1 is concerned with the prevention of faults. The principles of category 1 can be complied with by the use of suitable design principles, materials, and components.

Categories 2, 3, and 4 require that faults that cannot be prevented must be detected and appropriate actions taken. Category 2, 3, and 4 systems monitor and check safety critical functions. Redundancy is one of the more common ways of monitoring. Critical safety functions are duplicated, monitored, and compared to assure they are operating correctly.

Categories 2 and 3 allow a single fault in the safety circuit to lead to the loss of the safety function.

Category 3 improves on category 2 and requires that a single fault not lead to the loss of the safety function. Category 3 does not, however, require that all faults be detected. An accumulation of undetected faults that could lead to the loss of safety function is permitted. Appropriate steps must be taken to minimize the possibility of occurrence.

Category 4 requires that there be no loss of safety function with an accumulation of multiple faults.

Figure 14-10 explains the five categories in more detail.

Category	General Safety Requirements	Safety System Behavior
B	The safety system is designed to meet the operational requirements and to withstand expected external influences. Category B requirements are usually satisfied by selecting components that are compatible with the application conditions, e.g., load, temperature, voltage, etc.	A single fault or failure of a component in the safety system can lead to the loss of the safety function.
1	Category 1 safety systems must meet the requirements of safety category B and must also employ well-tried principles and components. Well-tried principles include Reduction of probability of faults, for example, overrated selected components, overdimensioned components for structural integrity, etc. Avoidance of certain faults, such as short circuits Detection of faults early, ground fault protection, for example Assurance of mode of the fault, for example, ensuring an open circuit when it is vital that the power be interrupted or unsafe conditions arise Limitation of the consequences of a fault	A single fault or failure of a component in the safety system can lead to the loss of the safety function. Well-tried safety principles and components are used to increase the reliability of the safety system.
2	Category 2 safety systems must meet the requirements of safety category B. Additionally, the machine shall be prevented from starting if a fault is detected upon the application of machine power, or upon periodic checking during operation. The requirement suggests the use of a safety controller with a startup test. Single-channel operation is permitted provided that the input devices are tested for proper operation on a regular basis. Input devices would include emergency-stop (E-stop) pushbuttons, machine guard interlocks, etc.	A single fault or failure of a component in the safety system can lead to the loss of the safety function. Periodic checking may detect faults and permit timely maintenance of the safety system.
3	Category 3 safety systems must meet the requirements of safety category B. In addition the safety system must be designed so that a single fault will not result in the loss of the safety function. Where practical, the single fault will be detected. This requires redundancy in the safety monitoring device and the use of dual-channel monitoring of input and output devices such as E-stop pushbuttons, machine guard interlocks, safety relays, etc.	A single fault or component failure in the safety system will not lead to the loss of the safety function and, where possible, will be detected. Some but not all faults will be detected. An accumulation of undetected faults can lead to the loss of the safety function.
4	Category 4 safety systems must meet the requirements of safety category B. In addition the safety system must be designed so that a single fault will not result in the loss of the safety function and will be detected at or before the next demand on the safety system. If this is not possible, then the accumulation of multiple faults must not lead to the loss of the safety function. This requires redundancy in the safety monitoring device and the use of dual-channel monitoring of input and output devices such as E-stop pushbuttons, machine guard interlocks, safety relays, etc. Here the application, technology used, and system architecture will determine the number of allowable faults.	A single fault or component failure in the safety system will not lead to the loss of the safety function, and it will be detected in time to prevent the loss of the safety function. If detection of the fault is not possible, then an accumulation of faults will not lead to the loss of the safety function. Faults will be detected in time to prevent the loss of safety functions.

Figure 14-10 Safety requirements and system behavior.

The EN ISO 13849-1 standard has succeeded the EN 954-1 standard. EN ISO 1389-1 also uses a risk graph to determine the necessary safety level. The parameters **S**, **F**, and **P** are used to determine the magnitude of the risk the same as they were in EN 954-1. The difference is that the result of the analysis is a required performance level (PLr). The performance level is defined in five discrete steps: a, b, c, d, and e (see Figure 14-11).

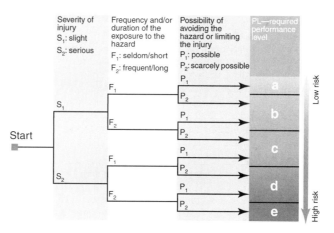

Figure 14-11 A risk graph based on EN ISO 13849-1.

IEC/EN 62061

The IEC/EN 62061 standard is another widely used safety standard. IEC/EN 62061 is named the "Safety of Machinery – Functional Safety of Safety Related Electrical, Electronic and Programmable Electronic Control Systems." The functional requirements used to determine a safety level with this standard's method include details like frequency of operation, required response time, operating modes, duty cycles, operating environment, and fault reaction functions. The safety integrity requirements are expressed in levels called safety integrity levels (SILs). The SILs are the approximate equivalent of categories in EN 854 and performance levels in EN ISO 13849-1. Figure 14-12 shows an approximate comparison between the safety levels of the three methods. Note that the SILs and categories and performance levels may be compared but are not equivalent. For example, 62061 and 13849 can be applied to electronic components but 62061 has no means of defining SIL ratings for hydraulic and pneumatic components.

Figure 14-12 An approximate comparison of EN ISO 13849-1, EN954-1, and EN 62061 categories.

RISK REDUCTION

The goal of risk reduction, according to safety standards, is to reduce risk to personnel to a tolerable level. The responsibility for the definition of a tolerable level of residual risk rests on the decision of the owner of the machine.

If the risk evaluation shows that measures are necessary to reduce the risk, the three-step method shall be used. The machine manufacturer shall apply the following principles during the selection of the measures and in the following order:

- The first step is a safe design. The design should eliminate or minimize residual risks as far as possible in the machine or system.
- The second step is technical protective measures. Necessary protective measures must be used against risks that cannot be eliminated by design.
- The third step is the provision of information to users on residual risks.

General principles on the process of risk reduction are covered in: EN ISO 12100, 12. Figure 14-13 shows a three-step risk reduction method as a decision tree. Note the first step would be to ask if the safe design adequately reduced the risk. If yes, we would need to make sure that no new hazards resulted from the design. If safe design did not adequately reduce the risk, we would need to implement protective measures such as a light curtain or other protective measures. Then we would need to decide if the protective measures adequately reduced the risk. If the answer were yes, we would need to make sure that no new hazards were created by the additional protective measures. Lastly, we would look at providing information to users about any residual risk. If information to users adequately reduced the risk, we would just need to make sure that no new risk was created. If user information did not properly reduce the risk, the risk reduction process would have to be repeated.

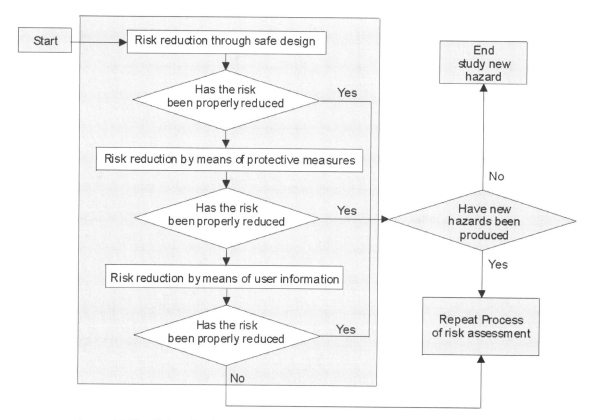

Figure 14-13 Risk reduction method.

Risk reduction strategies are shown in more detail in the table in Figure 14-14. The table shows five strategies and examples of how each might be used to reduce risk.

Risk Reduction Strategy	Examples
Changes in machine design	■ Reduced or eliminated human interaction ■ Eliminated pinch points ■ Automated material handling
Engineering controls (safeguarding technology)	■ Mechanical hard stops ■ Barriers ■ Interlocks ■ Presence-sensing devices ■ Two-hand controls
Awareness	■ Lights, beepers, horns ■ Computer warnings ■ Signs, labels ■ Restricted space painted on the floor
Training and procedures	■ Safety procedures ■ Equipment safety inspections ■ Training ■ Lockout/tagout
Personal protective equipment	■ Safety glasses ■ Ear plugs ■ Face shields ■ Gloves

Figure 14-14 Risk reduction strategies.

SUMMARY

There are standards that define and describe the process of risk assessment. A risk assessment is a sequence of logical steps that permit the systematic analysis and evaluation of risks. The machine shall be designed and built taking into account the results of the risk assessment.

Risk reduction follows a risk assessment, when necessary, by applying suitable protective measures. New risks shall not result from the application of protective measures. The repetition of the entire process, risk assessment and risk reduction, may be necessary to eliminate hazards as far as possible and to sufficiently reduce the risks identified.

QUESTIONS

1. *True or False:* OSHA guidelines place the responsibility for machine safety on the end user.
2. Describe the process of risk assessment according to ANSI B11.TR3-2000.
3. What is the difference between tolerable risk and acceptable risk?

4. A machine task was studied and found to have a risk of injury associated with it. The injury was evaluated for its potential severity. It was decided that it could result in cuts or even a broken arm. The operator was exposed to the hazard about three times a week. It was determined that the probability rating for being injured was possible. What would the rating be using ANSI B11.TR3-2000?
5. ANSI B11.TR3-2000 specifies four levels of safeguarding that can be applied, depending on the level of risk reduction. What are they?
6. What are machine limits?
7. Describe the S, F, and P factors that are used in a basic risk assessment.
8. EN 954 uses categories to describe the necessary safety level. What does EN ISO 13849-1 use?
9. What does SIL stand for?
10. Describe the three-step method for risk reduction.
11. Risk assessment and risk reduction is an iterative process. Explain.

CHAPTER 15

Safety Devices for Risk Reduction

OBJECTIVES

On completion of this chapter the reader should be able to:

- Describe machine guarding.
- Describe the operation of a safety relay.
- Explain terms such as *safety relay, ESPE, force guided,* and so on.
- Describe the use of safety switches, light curtains, and laser scanners for safeguarding.
- Explain the function of a safety controller.

INTRODUCTION

Safety is becoming an ever-increasing part of the technician's job responsibilities. It is imperative that the technician understand the role and importance of safety in the workplace as well as the devices that can help reduce the risks inherent in automated equipment. This chapter focuses on the safety technology that is available to reduce risk in the workplace.

STANDARDS

ANSI B11 2008

One of the standards that defines safety for machines is the ANSI B11 2008, "General Safety Requirements Common to ANSI B11 Machines." The ANSI B11 2008 safety standard applies to new, modified, or rebuilt power-driven machines, not portable by hand and used to shape or form metal or other materials by cutting, impact, pressure, electric or other processing techniques or a combination of these processes. The table in Figure 15-1 shows the subsections for ANSI B11 and the machines they apply to. These are very useful when developing risk reduction strategies for particular types of equipment.

ANSI B11.1	Safety Requirements for Mechanical Power Presses
ANSI B11.2	Safety Requirements for Construction, Care, and Use of Hydraulic Power Presses
ANSI B11.3	Safety Requirements for Power Press Brakes
ANSI B11.4	Safety Requirements for Shears
ANSI B11.5	Safety Requirements for Construction, Care, and Use of Ironworkers
ANSI B11.6	Safety Requirements for Manual Turning Machines with or without Auto Control
ANSI B11.7	Safety Requirements for Construction, Care, and Use of Cold Headers and Cold Formers
ANSI B11.8	Safety Requirements for Manual Milling, Drilling, and Boring Machines
ANSI B11.9	Safety Requirements for the Construction, Care, and Use of Grinding Machines
ANSI B11.10	Safety Requirements for Metal Sawing Machines
ANSI B11.11	Safety Requirements for Gear and Spline Cutting Machines
ANSI B11.12	Safety Requirements for Roll Forming and Roll Bending Machines
ANSI B11.13	Safety Requirements for Single- and Multiple-Spindle Automatic Bar, and Chucking Machines
ANSI B11.14	Coil Slitting Machines - Safety Requirements for Construction, Care, and Use
ANSI B11.15	Safety Requirements for Pipe, Tube, and Shape Bending Machines
ANSI B11.16	Safety Requirements for Powder/Metal Compacting Presses
ANSI B11.17	Safety Requirements for Horizontal Hydraulic Extrusion Presses
ANSI B11.18	Safety Requirements for Machines Processing or Slitting Coiled or Non-coiled Metal
ANSI B11.19	Performance Criteria for Safeguarding
ANSI B11.20	Safety Requirements for Integrated Manufacturing Systems
ANSI B11.21	Safety Requirements for Machine Tools Using Lasers for Processing Materials
ANSI B11.22	Safety Requirements for Turning Centers and Automatic NC Turning Machines
ANSI B11.23	Safety Requirements for Machining Centers and Automatic NC Milling, Drilling, and Boring Machines
ANSI B11.24	Safety Requirements for Transfer Machines
ANSI B15.1	Safety Standards for Mechanical Power Transmission Apparatus
ANSI/ISO 12100–1	Safety of Machinery–Basic Concepts, General Principles for Design Part 1
ANSI/ISO 12100–2	Safety of Machinery–Basic Concepts, General Principles for Design Part 2

Figure 15-1 Safety standards for machines.

Electrosensitive Protective Equipment (ESPE) Standards

Light curtains, laser scanners, photosensors, and other devices have been used for many years in machinery-related safety applications. These devices and others are referred to as electrosensitive protective equipment (ESPE). Requirements governing the use of ESPE have been specified in a number of different national regulations and standards including the new international standard IEC 61496. ESPE can be used for a variety of safety functions.

They can be used to prevent an operator's fingers, hands, or arms from entering a hazardous part of a machine, to scan the path of an automated vehicle, or to encircle and safeguard an area around an industrial robot. In each of these applications, the ESPE will produce an output signal(s) when a person or an object comes within the detection zone; the dangerous movement of the machine can then be stopped or reversed. Such equipment has long been subject to national regulations and standards, but over the past several years it has become clear that an international framework of safety requirements and a recognized terminology are also badly needed.

The IEC established standards for electrosensitive protective devices. IEC 61496 "Safety of Machinery—Electrosensitive Protective Equipment" is accepted as the default standard for ESPE. The standard includes light curtains, laser scanners, and so on. Underwriters Laboratories (UL) adopted the IEC 61496 standard into their own UL standards.

SAFETY CONSIDERATIONS

There are three main safety considerations when a new machine is designed or if existing machinery is upgraded.

- The design of safety should be done with maintenance in mind.
- Interlocking principles and devices must be considered.
- Safety controls and the use of safety PLCs must be considered to prevent the machine from injuring personnel, the operator from damaging the machine, or the machine from damaging itself.

SAFEGUARDING METHODS

There are many ways to safeguard machines. Things that must be considered include the layout of the work area; the type of operation; the workpiece size, material, weight, and so on; the method of handling; and the production requirements. These and other considerations will help determine the appropriate safeguarding methods for the machine.

In general, power transmission components are most effectively protected by fixed guards that enclose the danger areas. There are several possible safeguarding methods for hazards located at machine access points where moving parts perform work on stock. When selecting a safeguarding method, you must always choose the most effective and practical one that is available.

Guarding

According to definitions found in ANSI B11 safety standards for metalworking machine tools, a guard is used to prevent a person from reaching over, under, around, or through the machine, avoiding both intentional and unintentional access to hazardous machine areas. A shield only prevents unintentional contact with hazardous machine areas. Awareness barriers are better than a yellow warning line on the floor because they take an intentional effort to get beyond. Warning barriers include things like a railing, a chain, or a cable suspended at waist height.

Guards are mechanical barriers that can be used to prevent access to hazardous areas. There are four general types of mechanical guards.

Fixed Guards

Fixed guards are designed to be permanently mounted on the machine. Fixed guards are typically constructed of sheet metal, bars, screen, plastic, or other material that is durable enough to stand up to impacts they may experience. Fixed guards are usually preferred over other types of guards because of their low cost, simplicity, and effectiveness. A properly designed fixed guard can provide the most protection. Maintenance is usually minimal for fixed guards. Fixed guards can usually be designed and made in the plant.

There are a few disadvantages to fixed guards. Fixed guards can interfere with visibility.

Machine adjustment and repair may require that the guard be removed, thus requiring additional protection for maintenance personnel.

Adjustable Guards

Adjustable guards provide a barrier that can be adjusted to meet different product requirements. Adjustable guards can be adjusted to admit varying sizes of stock, and they can be designed to meet the needs of a variety of applications. Disadvantages of adjustable guards include the guard may be made ineffective by the operator, it may affect visibility of the machine, it may be possible for hands to enter the hazardous area, it may not provide complete protection at all times, and it may require frequent adjustment and maintenance.

Self-Adjusting Guards

Self-adjusting guards move to adjust the stock size and still protect the operator. The safety guard on a table saw moves up and down as a piece of wood is cut to adjust for the thickness of the wood and allow the wood to go through and at the same time prevent the operator from getting her or his fingers or hand in the blade. The openings of self-adjusting guards are determined by the movement of the stock. On a table saw, as the operator moves the wood into the hazard area, the guard is pushed up, providing an opening which is only large enough to admit the wood. After the wood has been pushed through the saw blade (cut), the guard returns to the rest position. A self-adjusting guard protects the operator by placing a barrier between the hazard area and the operator.

Advantages of self-adjusting guards include they provide a barrier that moves according to the size of the stock entering the hazardous area and they may be commercially available off the shelf. Disadvantages of self-adjusting guards include they do not always provide maximum protection, they may interfere with visibility, and they may require frequent maintenance and adjustment.

Interlocked Guards
The machinery used in discrete manufacturing usually has many moving parts and mechanisms. Moving parts can injure personnel. Protective measures must be implemented to protect the worker and the machine from the moving devices. A safety design team should consist of operators, personnel responsible for safety, maintenance personnel, and engineering personnel. Simplest is not always best. It may be a simple to use a limit switch to detect whether a guard door is open. But standard limit switches are easy to cheat and also prone to failure. An operator can actuate the limit switch with his or her hand or duct tape and override the safety. New safety hardware can reduce the possibility of operators cheating the safety features of a system.

An interlocked guard has sensors or switches that link the guard into the safety circuit. If an interlocked guard is opened or removed, the tripping mechanism is activated or power is automatically shut off or disengaged, and the machine cannot run or be started until the guard is back in place. Interlocked guards can provide the maximum protection, and they allow access to a machine without the removal of fixed guards. This can save time. The interlock should require the machine to be stopped before the worker can reach into the hazardous area.

There are three categories of stop functions.

- Category 0 is an uncontrolled stop by immediately removing power to the machine actuators.
- Category 1 is a controlled stop with power to the machine actuators available to achieve the stop and then removed when the stop is achieved.
- Category 2 is a controlled stop with power left available to the machine actuators.

An interlocked guard may use mechanical, hydraulic, pneumatic, or electric power or any combination for the interlock. Replacing the guard should not automatically restart the machine. Movable guards should be interlocked to prevent hazards. This should make sense; if the operator can move the guard, she or he might forget or leave it open on purpose and the hazard would not be prevented.

Switches for Guard Interlocking
Safety limit switches used for guard interlocking must meet special requirements. The requirements are listed in standards EN 60 204-1, EN 1088, and EN 60947-5-1, "Control Circuit Devices and Switching Elements: Electromechanical Control Circuit Devices."

The placement and design of safety switch interlocks must protect them from inadvertent operation, damage, and changes in position. The switch and the control cam (key) must be secured by its shape, not by force.

Switches used for safety interlocking must be protected by their actuation method, or their integration in the control must be such that they cannot be easily bypassed or jury-rigged. Safety interlock switches must have normally closed contacts for fail safing.

Safety interlock switches must be mounted so that they are not used as a mechanical stop for the guard. It must be possible to check the safety interlock switches for correct operation. If possible, safety interlock switches must be easily accessible for inspection. Safety interlock switches must be mounted such that they are protected against damage due to external effects.

Interlocked Guard Design
Guards that are frequently opened or are removed or opened for setup must be interlocked with the hazardous movement. It must not be possible to override the interlock with simple means. In other words, it should not be easy for an operator to use wire, a magnet, some other tool, or a piece of metal to override the interlock switch.

Ergonomic Considerations for Guards
Ergonomics are also important when designing guards. Guards should not hinder employees doing their job. Guards should not interfere or slow down activities such as setup, maintenance, and other similar ones any more than necessary. Employees will not accept guards that unduly interfere with them or slow them down.

Guards that are only removed or opened for maintenance work or not removed or opened very often (if they are not interlocked to prevent dangerous movement) must be fastened to the machine such that they can only be removed with tools.

Guards that are opened for setup or frequently opened must be interlocked with the dangerous movement. This means that after opening or removing the guard, dangerous movements come to a stop in a safe amount of time.

In terms of opening guards, frequently means guards are opened at least once per shift. Locks must be used if hazards are to be expected when the guard is opened. In other words, locks must be used if it takes a long time for the machine to safely stop after a guard is removed.

Protection against reaching through the barrier must also be considered. The permissible mesh size (opening) for chain link fence is dependent on the distance between the fence and the hazardous point. The larger the openings in the fence, the farther the fence must be from the hazardous point (EN 294). In other words, the safe distance between the fence and the hazard is much different if a person cannot get a hand through the fence opening than for a fence that an arm can fit through.

Figure 15-2 shows two examples of the use of a guard (barrier). The one on the left is unacceptable because it allows fingers to reach the hazardous area. In this example if the opening in the barrier remained this large, the barrier would have to be located farther away from the hazardous area. The example on the right is acceptable because the openings in the barrier are too small to allow a finger to reach the hazardous area.

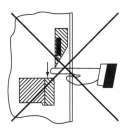 

EN 294 not observed   EN 294 observed

Figure 15-2 On the left the barrier opening allows a finger to get to the hazardous area. On the right the openings in the barrier are too small to allow a finger to reach the hazardous area.

The following standards establish requirements for guards: EN 953, is named "Safety of Machinery Guards" is a standard that establishes general requirements for the design and construction of fixed and movable guards.

Safeguarding by Location

To safeguard a machine by location, the dangerous moving part of a machine must be located such that the hazardous areas are not accessible or do not present a hazard to personnel during normal operation. Safeguarding by location can be done by locating the hazardous parts of the machine away from operator workstations or other areas where personnel may be present. Walls or fences can be used to restrict access to machines. Safeguarding by location is also accomplished by having the dangerous parts of the machine located high enough to be out of the normal reach of any worker.

The placement of the operator's control station is another way to provide safeguarding by location. If the operator's control station is positioned properly, the operator cannot be in the hazardous area while operating the control.

SAFETY DEVICES

Safety has become more complex as manufacturing systems have become more automated. Until relatively recently machines tended to be small and were run by one operator who had control and a view of the entire machine. The operator could stop the machine with one E-stop button. Now, operators may have to tend several locations around larger manufacturing systems. Several E-stop pushbuttons and other devices might be required to improve safety. E-stops are limited in protective capability. They require intentional human action.

Other methods may also be used to protect operators. There are safety devices to stop the machine if a hand or any part of the body enters the danger area. There are restraint-type devices to withdraw the operator's hands from the danger area during operation.

Restraint Devices

The restraint, or holdout, device uses cables or straps that are attached to the operator's hands. The cables or straps must be adjusted to let the operator's hands travel within a predetermined safe area. The restraints do not allow the operator to move her or his hands into the danger area. If parts need to be placed into the danger area, tools are used to place the workpiece. Note that these only protect the operator; no one else is protected.

Pullback Devices

Pullback devices consist of a set of cables that are attached to the operator's hands, wrists, or arms. The cables actually pull the operator's hands away from the danger area during the dangerous time in the cycle if the operator has them too close. This type of device is typically used on machines with stroking action, like a punch press. When the machine is between cycles, the operator is allowed access. When the slide/ram starts its descent, a mechanical linkage automatically pulls the cables to keep the operator's hands away from the point of operation. Note that these only protect the operator; no one else is protected.

Safety Trip Controls

Safety trip controls are used to deactivate a machine in an emergency situation. One type of safety trip control is the safety edge or safety bumper. These are long safety edges that can be used to detect pressure. A pressure-sensitive safety edge, when pressed, will deactivate the machine. If the operator or anyone trips, loses balance, or is drawn toward the machine, pressure to the safety edge will stop the operation. Safety edges must be positioned so that the machine will be stopped before a part of the operator's body can reach the hazardous area.

Safety trip wires are slightly different. Safety trip-wire cables are located around the perimeter of a work area or near a hazardous area. The operator must be able to reach the safety trip cable with either hand. The trip wire is normally attached to a safety switch that is connected to the safety circuit. If the switch is triggered, the machine is shut down.

Many devices have been developed to increase the safety of machines and systems. Some of the devices include light curtains, safety mats, safety gate switches, laser scanners, key switches, two-hand switches, and so on.

INTERLOCKING

Power Interlocking

In power interlocking the power source of the hazard is directly interrupted by the opening of a guard. The mechanical movement of a guard door is interlocked with the direct switching of the power to the hazard. A trapped key system is one method of accomplishing this. This is very difficult for an operator to override without the key.

Control Interlocking

In control interlocking the power source of a hazard is interrupted by switching the circuit that controls the power-switching device.

E-STOPS

E-stop pushbuttons must be red. The background behind an E-stop should be yellow. Once an E-stop pushbutton is pressed, it must lock into position before generating the E-Stop command. This is intended to prevent the potential manipulation of the actuators. Figure 15-3 shows a variety of safety switches.

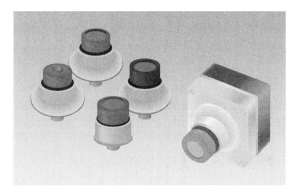

Figure 15-3 E-stop switches.

When should E-stop pushbuttons be used?
The machinery directive states that E-stop pushbuttons must always be present with only two exceptions.

- Handheld or hand-guided machines
- Machines in which an E-stop device would not reduce the risk, either because it would not reduce the stopping time or because it would not enable special measures that might be necessary

European standard 60204-1 and NFPA 79 distinguish between the types of stop categories. An E-stop must have priority over all other functions.
Where should E-stop pushbuttons be located?

A pushbutton must be present at each operator station that can be used to start or control a dangerous movement. It is also desirable to install E-stop devices on access points such as operator and safety gates or in areas where there is poor visibility.

ISO EN138490-1 and -2 standards require that, when a group of machines are working together in a coordinated manner, all E-stop control devices shall be able to signal an E-stop condition to all parts of the system. In some applications it may be necessary to link the E-stop circuits of stand-alone machines. On integrated machines such as transfer lines, this requirement is specified in the machinery directive. This means that upstream or downstream sections need to be incorporated into the main machine's E-stop circuit.

Series Connection of E-Stops

Safety requirements can be implemented cost effectively if individual E-stop pushbuttons are connected in series. In this method all E-stops in a line are electrically connected in series and operate with a common monitoring device. This can meet the E-stop pushbutton requirements up to Category 4 of EN 954-1. The E-stop is not a primary protective device like a safety gate. It can be assumed that several E-stop buttons will not be operated simultaneously. The frequency with which the function is requested can also be considered low.

Illuminated E-stops should be considered because it is easier to immediately identify which E-stop has been operated and isolate the problem. Illuminated pushbuttons can prevent injuries in dark areas by making the E-stop easier to find.

TWO-HAND SWITCHES

Two-hand control requires that the operator apply his or her hands almost simultaneously to two switches. Some also require constant, concurrent pressure by the operator to activate the machine. Two-hand trip devices initiate a machine cycle. Two-hand control devices must be pressed throughout the machine cycle. With the two-hand switch, the operator's hands are required to be on control buttons at a safe location away from the danger area while the machine completes the dangerous part of its cycle.

Figure 15-4 shows a two-hand switch that could used on a press-type machine. Both switches have to be closed to make the press move. This is meant to ensure that the operator's hand cannot be in the press when the press moves. Originally these would have been regular switches connected in series. If both switches were down, the press could move. Operators soon discovered that it might be easier or quicker not to use two hands. They would use tape and lock one switch on. Then one hand could hit the second switch and make the press move. This might be quicker, require less effort, and so on. But the operator has overridden the safety measure. It would be possible to have one hand in the danger area while the other hand hit the second switch.

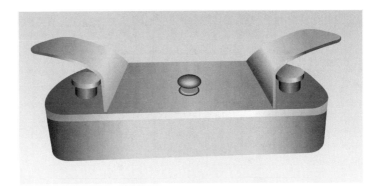

Figure 15-4 A two-hand safety switch.

New safety technology was designed to eliminate this possibility. There are two contacts under each of the switches in the two-hand switch shown in Figure 15-4. Positively guided (also called force-guided) contacts in each switch are mechanically interlocked such that two contacts on the relays will not contradict each other, even in the event that the relay welds. This two-hand switch is designed to be connected to a safety relay. The safety relay performs several tasks. The first is that it will not allow the operator to tape one side down and still have the switch work. The safety relay ensures that both switches must open before they close again or the two safety outputs of the relay will not be on. The two outputs from the safety relay are used to disable or enable the system operation. The safety relay also makes sure it receives signals from both switches within 500 ms of each other. The operator must hit both switches at almost the same time for the safety relay to accept the inputs and turn on the safety outputs.

GATES AND GATE SWITCHES

A gate is very similar to a movable guard that protects the operator at the point of operation before the machine cycle can be started. It may be necessary for the operator to enter a cell and set a part up for an operation. A gate is included to allow entry when the machine is not operating to allow setup. The gate is interlocked with the safety system to prevent the cell from operating when it is opened. The operator must then be outside of the cell with the gate closed to reset the system to operate. Gates also prevent other personnel from entering danger areas. Figure 15-5 shows a gate safety switch.

Figure 15-6 is a table that shows some safeguarding methods as well as their advantages and disadvantages.

Figure 15-5 A gate interlocking safety switch.

Method	Action	Advantages	Disadvantages
Photoelectric	Machine will not start when the light field has been interrupted. Immediate machine braking is activated.	Simple to use. Protects everyone. No adjustment is required.	Limited to machines that can be stopped at any point in the machine cycle.
Safety trip controls: Pressure-sensitive bar Safety trip wire Safety trip rod	Stops the machine when tripped.	Simple.	Must be manually activated. May be difficult to activate controls because of their location. Only protects the operator (may protect others with proper positioning). May require special fixtures to hold work. May require a machine brake.
Two-hand control	Concurrent use of both hands is required, thus preventing the operator or the operator's hands from entering the hazardous area.	The two-hand control can be mounted in a safe location from the hazardous area. Can be set up so that the operator's hands are free to pick up a new part after the first half of a cycle is complete to increase efficiency.	Only protects the operator. Requires a partial-cycle machine with a brake. Some two-handed controls can be made unsafe by holding with an arm or blocking, thereby permitting one-hand operation.
Gate	Provides a barrier between the hazardous area and the operator and other personnel.	A gate can prevent entry or reaching into the hazardous area. Protects everyone.	Should be inspected regularly. May interfere with operator's ability to see into the cell.

Figure 15-6 Safety methods.

German automotive engineers developed safety standards in the 1980s that defined the need for redundancy and monitoring in hazardous areas. The standard was named VDE 0113 and became the basis for the European safety standards. Europe was a world leader in the development of safety standards.

SAFETY RELAY

German engineers invented the safety relay. Regular electromechanical relays are not sufficient for use in safety systems because they are mechanical and will eventually fail. They can easily fail in an unsafe state. An electromechanical relay may fail without the operator being aware of it. If the relay is being used in a safety application, it could be very dangerous for personnel.

Solid-state relays are also not suitable because they depend on power transistors for their output. Transistors can fail in an open or closed state. If a solid-state relay were to fail in the on state, an E-stop would not be able to stop the machine. Pilz GmbH and Company developed the safety relay in 1987. The safety relay they developed was a multipole relay. It had a combination of three relays that had positive-guided contacts.

Positive guiding means that the mechanical linkage (actuator) works directly on the switching elements (contacts). The contacts do not rely on spring action to open or close. Positively guided relays have contacts that are mechanically interlocked such that two contacts on the relays will not contradict each other, even in the event that the relay welds.

The first safety relays had two sets of normally open contacts and one set of normally closed contacts. When the two normally open contacts are closed to energize the circuit, the normally open set of contacts is forced open. If either of the two sets of normally open contacts were to weld, the normally closed set of contacts would be forced to stay open. This would prevent the system from being restarted, because the normally closed contacts are used to restart the circuit. The failed relay would have to be replaced before the machine could be restarted.

Figure 15-7 shows some potential safety input devices for one type of safety relay. On the left of the drawing there is a hand above a normally open switch. This would be used for a reset switch, which is used to reset the module after a stop condition occurs. The safety protection devices that could be used are shown around the module. Note that they are shown as OR. Only one may be used. On the upper left of the diagram there is a light curtain. A light curtain has two outputs for safety. These can be connected as safety inputs to the safety relay. If the beam is broken between the transmitter and receiver of the light curtain, the light curtain outputs turn off and the safety module turns off the two safety relay outputs from the module. For this example, let's assume the two outputs from the safety relay are connected to a robot's E-stop input circuit. If one or both of the inputs to the robot become false, the robot will stop and prevent motion. Two are used for safety. Assume now that the obstruction from the light curtain has been removed. The two outputs from the light curtain will be true. The safety module will sense that they are true, but will not change the state of the safety outputs so the robot is still in an E-stop condition. The safety relay will not energize the safety outputs until the manual reset switch is closed momentarily. This is also for safety. Imagine if a person walks through a light curtain into a robot cell. The outputs from the light curtain go false, the safety relay turns the safety outputs off, and the robot goes into an E-stop condition. The person is now in the cell and the light is not blocked so the light curtain outputs are now true. The safety relay does not turn on the safety outputs until it sees the reset input. This protects the operator and anyone else who wanders into the hazardous area. The reset switch must be located outside of the danger area and in a position where a person can see the entire danger area so that he or she does not reset the safety relay circuit until it is safe to do so. This is an example of perimeter guarding.

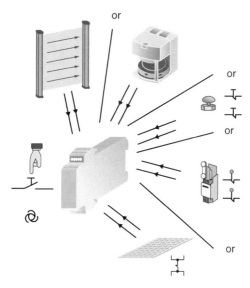

Figure 15-7 A safety relay with some possible safety input devices.

Consider a few of the other safety input devices that could be used with this relay. The next one in the diagram is a laser scanner. A laser scanner is programmable. The user can program a protective plane, which can be contoured to fit the area to be guarded, into the laser scanner. The user can also program a warning area that can be used to warn personnel that they are getting too close to a dangerous area. Warning areas warn people not to go further. The second area that can be programmed is the alarm area. If a person enters this area, the laser's two safety outputs are turned off. The outputs are inputs to a safety relay so the safety relay's two output relays are deenergized.

The next safety device shown is an E-stop switch. Note that it has two contacts. It would provide two signals to the safety relay. The next device is a mechanical limit switch with two outputs. The last device shown is a safety mat. Note that it is shown as a normally open switch. A person must be standing on this mat to have the two outputs on. Safety relays also are able to check when the two safety inputs come true.

Study the functional wiring diagram of the safety relay in Figures 15-8 and 15-9. Pins A1 (+ 24 VDC) and A2 (0 VDC) are used to provide 24 VDC to power the relay. Pins S21 (−) and pins S11 (+ 24) and S33 (+ 24) provide DC voltage for control. Pins S34 and S35 can be used for a reset circuit (automatic or manual depending on how they are wired). The safety outputs from the module are shown as K1 and K2. The relays K1 and K2 are not wired independently. Look closely and you see that K1 and K2 control a series contact in each of the output circuits 13-14 and 23-24. Each output circuit has redundant contacts.

Relays K1 and K2 are not wired independently. K1 and K2 control a series contact in each of the output circuits 13-14 and 23-24. Each output circuit has redundant contacts. The safety module also provides another set of output contacts (pins 31 and 32) that can be used, but they are not safety outputs.

Figure 15-8 A functional wiring drawing of one type of safety relay.

Inputs from safety devices such as light curtains or safety switches are connected as inputs to pins S12 and S31. The wiring controls how the module controls the outputs of the module. This safety relay can be used in single- and dual-channel applications by using different wiring configurations. It can also be wired for manual reset or for automatic reset.

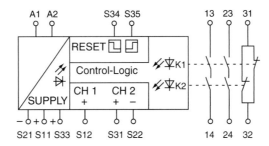

Pin	Function
A1	Voltage supply (+24)
A2	Voltage supply (0 V)
S11/S33	+ 24 V (Control voltage)
S21	0 V (Control voltage)
S33 - S35	Automatic reset
S33 - S34	Manual reset
S12	Input circuit 1 (K1)
S31	Input circuit 2 (K2)
13 - 14	Output contacts 1 (Safe)
23 - 24	Output contacts 2 (Safe)
31 - 32	Signal circuit (Non-safe)

Figure 15-9 Safety Relay Terminals. A diagram of a safety relay module is shown on the left. Note the two rows of four terminals on the top and bottom of the module. Also note the numbering of the terminals. K1 and K2 are status LEDs for the state of the output relay contacts. The pins and their functions are shown in the right two columns. (Courtesy of SICK)

Let's consider an application that has a single-channel gate switch. Figure 15-10 shows an example of the safety relay and a circuit to accomplish a safety circuit for a single-channel gate switch. It is shown as a normally closed switch in the diagram. This is done to help fail-safe the circuit. If the switch fails or a wire is cut, the safety relay would normally shut the system down because the input would go false. It does not perfectly protect against failure.

The number one cause of system faults is wiring failure. What happens if there is a short circuit to +24 volts between the switch and the safety relay? The relay would think everything is fine even if the gate were opened. So although using a normally closed switch helped reduce the possibility of a failure, that would not be safe. It did not eliminate the possibility of failure because it is only single channel and the switch could short-circuit to +24 volts.

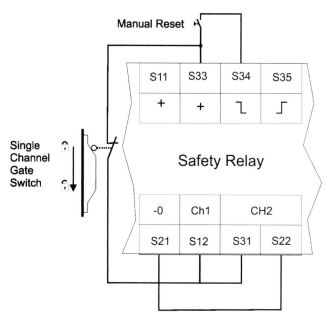

Figure 15-10 A single-channel gate switch connected to a safety relay. Note the manual reset.

The next example will use a dual-channel gate switch (see Figure 15-11). Note that the gate switch has two sets of contacts. The safety relay will monitor both outputs to assure a failure cannot occur without being sensed. This diagram also has a manual reset and monitors two additional contacts in the reset circuit. The K_a and K_b contacts shown in the reset circuit are external device monitoring (EDM) contacts from contactor coils being driven by the outputs of the safety relay. A failure of the driven contactor prevents the safety relay from resetting. This type of safety configuration should be acceptable up to a level 4 classification.

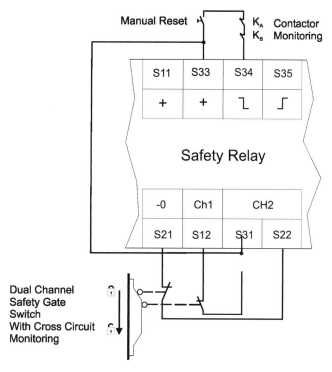

Figure 15-11 A dual-channel gate switch connected to a safety relay. Note the manual reset and the two contacts that are being monitored in the reset circuit.

THE USE OF OPTOELECTRONIC DEVICES FOR SAFEGUARDING

Optoelectronic safety devices are often a good choice when the risks cannot be eliminated through design. Optoelectronic safety devices can be used to reduce the risk to a tolerable level. When an operator is exposed to a hazard in the operation of a machine, a protective device should be employed to prevent the operator from exposure to dangerous machine movement.

Light Curtains

Light curtains can be used to either detect or prevent an operator entering or reaching into a hazardous area on a machine. Optoelectronic safety devices have advantages over mechanical safeguarding devices such as fixed or movable guards and two-hand control switches. Opening and closing a mechanical guard takes time. A light curtain requires much less time than a mechanical movable guard. A light curtain requires much less access time than any other method. The operator saves time in loading the machine; this makes the machine more productive.

A light curtain has two components: a sender and a receiver (see Figure 15-12). The sender has several emitters that emit infrared light beams. The receiver has many receivers to sense the emitted light from the sender. When the emitted beams reach the receivers, the light curtain energizes its two safety outputs.

Interruption of any beam in the light curtain deenergizes the two safety outputs from the light curtain. The safety outputs from the light curtain can be used as inputs to a control circuit to stop any hazardous motion or prevent the machine from initiating a start sequence.

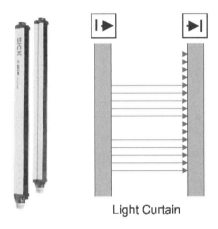

Figure 15-12 On the left is a programmable light curtain. On the right is an illustration of a light curtain showing the transmitter, light beams, and receiver. (Courtesy of SICK)

Blanking

Blanking is a term to describe allowing certain beams of a light curtain to be blocked without turning off the light curtain outputs. For example, we might need to allow two beams to be blocked by incoming material. This is called *fixed blanking*. The light curtain could be programmed to allow these two particular beams to be blocked. The rest of the beams would still be used to prevent hands or arms from entering the hazardous area. Not all light curtains have blanking capability. Another term is *floating blanking*. Some light curtains can be programmed to allow a certain number of beams to be blocked. For example, we may need four beams to be blocked to allow the operator to insert a new part, but the area where the part enters may shift. Floating blanking would allow four beams to be blocked in any programmed area of the light curtain and still not shut the machine down.

Muting

A light curtain with muting capability enables the safety (protection) function to be temporarily disabled in an application. The disabling of the safety function for a limited period of time is called muting.

There are generally two types of applications where muting is required. The first is to permit materials for production to enter a cell while at the same time preventing personnel from entering. An example would be to allow a pallet to enter the hazardous area, but prevent personnel from entering the area. The second is to permit personnel to access the hazardous area during the nonhazardous portion of the production cycle. An example would be removing a workpiece or setting up the next workpiece after a cycle. When used properly, muting can help speed up industrial processes and still protect personnel from hazardous situations. Machine control systems that initiate muting must be control reliable. Figure 15-13 shows the use of pattern recognition to detect the difference between a pallet and a person.

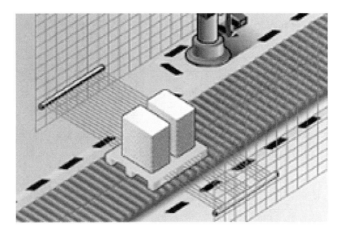

Figure 15-13 A palletizer, which uses pattern recognition to detect the difference between a pallet and a person. This is a less conventional way of doing muting than the more common muting sensors and light curtain combination. (Courtesy of SICK)

Figure 15-14 shows a light curtain attached to a safety relay. The light curtain has two safety outputs. If nothing is blocking the light, the two outputs will be true. Two are used for redundancy. These outputs also contain circuitry to detect short circuits between the outputs, to the power supply, and to DC common. If a single fault occurs, the outputs turn off and deenergize the safety relay.

Note that there are many types of safety relays that are available for different types of uses. This safety relay is appropriate for taking two inputs. The first output from the light curtain is connected to S12. The second output from the light curtain is connected to S31. Note that S21 and S22 are tied together for this application. Note the series circuit on top the relay. It is connected to pins S33 and S34. There is a manual reset in this example. If something breaks the light beam in the light curtain, the light curtain outputs will become false. The relay will sense the inputs are false and turn off the safety contacts from the safety relay. If the obstruction in the light curtain is removed, the outputs of the light curtain will become true, but it will not change the state of the safety relay output contacts. The outputs from the safety relay will not change until the manual

Description	LED Color	Function
Supply Voltage	Green	Shows voltage present
K1	Green	Shows safety output relay K1 energized
K2	Green	Shows safety output relay K2 energized

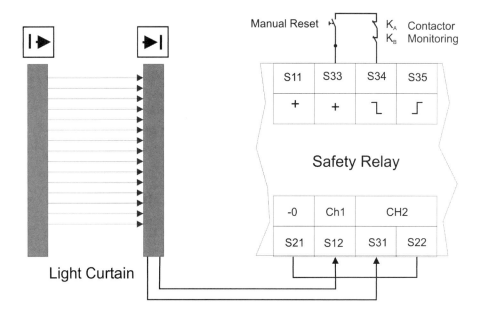

Figure 15-14   A light curtain connected to a safety relay.

reset switch is pushed. Note that we could add additional contacts (K_A and K_B) in the manual reset circuit. These contacts come from contactors driven by the safety relay outputs and must be used to achieve control reliability.

Input Monitoring

When the input circuits are energized the safety relay monitors their synchronization. The output circuits will only close if input 2 closes by no later than 70 ms after input 1. If input 2 closes before input 1, the synchronization will not be affected, and the output circuits will close. This monitoring only takes place in this relay if this relay is wired for automatic reset.

Laser Scanners

A laser scanner is an optical sensor that can scan an area with infrared laser beams and detect entry into the area.

The principle behind a laser scanner is time-of-flight measurement. A laser scanner emits very short light pulses which are sent out and reflected back when they hit an object. The laser scanner measures the time it takes the light pulses to return and determines distances from the object by the time.

Scanners are programmable. Computer software is used to program how the scanner will operate. Scanners can be programmed to safeguard areas of any shape, as well as multiple zones. Many scanners can have a warning zone also. This would be an area larger than the protective zone. It would be used to warn personnel that they are getting close to the area where the laser scanner would shut the system down. This area is also programmable. Scanners can be programmed to ignore fixed objects in hazard areas such as posts or fixtures. Scanners can be mounted in different orientations, horizontally, vertically, or at an angle depending on the needs of the application.

The application on the left in Figure 15-15 uses a laser scanner on an automatically guided vehicle. The laser scanner scans for obstructions and slows down if an object is detected in the protective area. On the right a laser scanner is mounted vertically to scan the entry point to a robot work cell. Muting would be employed in this example to differentiate between a product and a person.

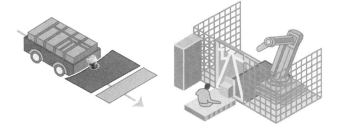

Figure 15-15 Two laser scanner applications.

Figure 15-16 shows a wiring diagram for a dual-channel safety switch on a door.

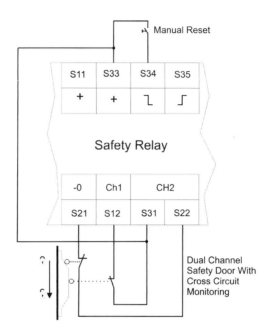

Figure 15-16 Dualchannel safety door switch connected to a safety relay. Note the manual reset.

A safety relay for two-hand control is shown in Figure 15-17. S1 and S2 are the two switches. Note that each switch has two contacts, one normally open and one normally closed. The normally closed contact of buttons S1 and S2 must have opened before the normally open contact closed.

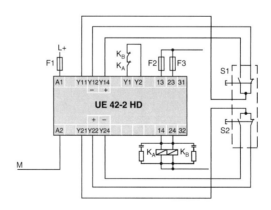

Figure 15-17 A safety relay for two-hand control and its wiring.

CONTROL AND INTERFACE REQUIREMENTS

When considering the use of a safeguarding device, you must also consider the control of signals related to the safeguarding devices. The system, in other words, the combination of the protective device, machine control, and main stop elements, must satisfy regulatory requirements.

An example would be the use of a non-safety-rated (regular) PLC to control actuators that cause hazardous motion. Even if safety relevant signals were used as inputs into the PLC, this would not generally satisfy regulatory requirements. A regular PLC cannot meet the requirements. A safety-rated PLC or certified safety controller should be used.

Until relatively recently, safety-relevant control has been implemented using positively guided relays (safety relays). In other words, safety circuits were hardwired. Standards have been developed to determine the suitability of programmable devices in safety applications. IEC 61508, "Functional Safety of Electrical/Electronic/Programmable Electronic Safety-Related Systems," is one of the more notable.

IEC 61508 establishes a method of determining the probability of failure of devices. IEC 61508 has become the basis for determining the suitability of programmable devices in safety control applications based on the associated SIL.

This enables users to utilize programmable safety devices and still comply with the regulations and standards that apply to their application. Industry standards now provide support for utilizing devices that have been approved by a nationally recognized testing laboratory for use in safety relevant applications.

IEC 61508 has dramatically changed the safety-relevant control possibilities. Safety controllers and safety field-bus networks are rapidly gaining in use and popularity.

Another important standard is ISO EN 13849. This standard is the successor to EN 954 and is useful in assessing control safety systems that employ pneumatic and hydraulic actuators. IEC 61508 and its subset IEC62061 only address electronic components.

CONTROL RELIABILITY

OSHA 1910.211 defines the term *control reliability* as a control system designed and constructed so that a failure within the system does not prevent normal stopping action from being applied when required, but does prevent initiation of a successive cycle until the failure is corrected. The failure must be detectable by means of a simple test or indicated by the control system.

ANSI defines control reliability in Standard B11.192003 (3.14) as "the capability of the machine control system, the protective device, other control components and related interfacing to achieve a safe state in the event of a failure within their safety related functions".

Control reliability can be attained by the use of, but not be limited to, one or both of the following:

The use of two or more dissimilar components, modules, devices, or systems, with the proper operation of each being verified (monitored) by the other(s) to ensure the performance of the safety function(s)

The use of two or more identical components, modules, devices or systems, with the proper operation of each being verified (monitored) by the other(s) to ensure the performance of the safety function(s)

Control reliability must be taken into account in the development of safety-related systems. Control reliable circuits should be hardware-based and include monitoring at the system level.

ANSI/RIA R15.06-1999 (4.5.4) provides a practical guide to implementing control reliability by requiring the following:

1. The monitoring generates a stop signal if a failure is detected. A warning is provided if a hazard remains after cessation of motion.
2. Following detection of a failure, a safe state is maintained until the fault is cleared.
3. Failures with a common cause (e.g., overvoltage) are taken into account if the probability is high that such a failure may occur.
4. The single failure is detected at time of failure. If not practical, the failure is detected at the next demand on the safety function.

SAFETY CONTROLLER

Safety controllers are programmable. They can take multiple inputs from various types of safety devices. Figure 15-18 shows a programmable safety controller. The controller can take multiple inputs and then has outputs that can be used to control the inputs to a safety relay. The controller is programmable from a computer. Note that the safety controller is not a PLC; it is a programmable safety device that can take multiple inputs and generate safety outputs. The diagram on the right of Figure 15-18 shows potential inputs, the software logic modules for processing and the potential outputs.

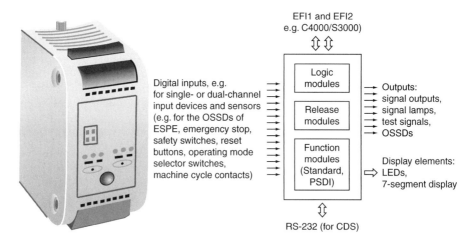

Figure 15-18 A Sick-brand safety controller is shown on the left. The diagram on the right shows a functional diagram of the safety controller.

A Safety Controller Example

Figure 15-19 shows a picture of a robotic work cell that utilizes several safety devices. In this system several safety devices are used as inputs to a programmable safety controller. The controller will take the inputs and generate output signals to be used as inputs to a safety relay. The two contacts from the safety relay will be used as inputs to a robot

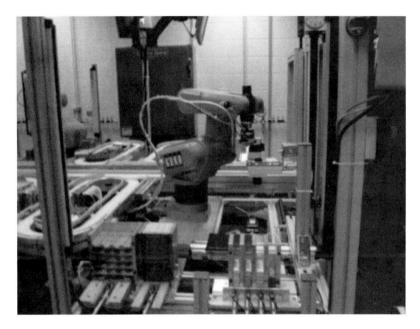

Figure 15-19 A robotic work cell.

safety circuit. The devices in this system that can be used to create an E-stop condition will be a laser scanner, a light curtain, a robot gripper crash switch, and E-stop. A reset switch will also used as an input to the safety controller to reset the system from an E-stop. Figure 15-20 shows a picture of the safety controller and relay in the control cabinet. The software that is used to program the safety controller also helps the technician plan the wiring. This example uses Sick safety devices and controllers as well as their programming software. Figure 15-21 shows a screen capture of the software. The basic procedure is that the user adds the devices to the application. The software shows all of the potential connection points for each device and the user chooses one set of connections. When all input and output devices are shown, the software then shows the pin numbers for all connections. The user also chooses logic and connection blocks for how the controller should respond to the inputs. In this example you see that an AND logic block was chosen. The AND logic is used to make sure all four inputs are true. If any of the inputs are deenergized or fail, the logic will turn off the two safety outputs from the controller. The two safety outputs from the controller are connected to the safety relay. If either or both outputs from the safety controller go false, the safety relay turns its two outputs off. Those outputs in this application immediately stop and prevent further motion of the robot.

Figure 15-20 A safety controller on the left and a safety relay on the right.

The left of the diagram shows the inputs to the safety controller. The topmost input is an E-stop switch. Right under the E-stop input is a reset switch. Then there is a crash switch for the robot gripper. If the robot gripper runs into something, the switch will open. The next input is a two-contact gate switch. The next input device is two inputs from a light curtain. The last input device is a laser scanner with two inputs to the controller. The outputs to the robot are shown on the right of the figure. Out1 provides two outputs: O2.0 and O2.1. These must both be true for the robot to be able to move in auto mode. If any of the input devices goes false, the outputs go false and the robot is stopped or prevented from moving until the system is reset.

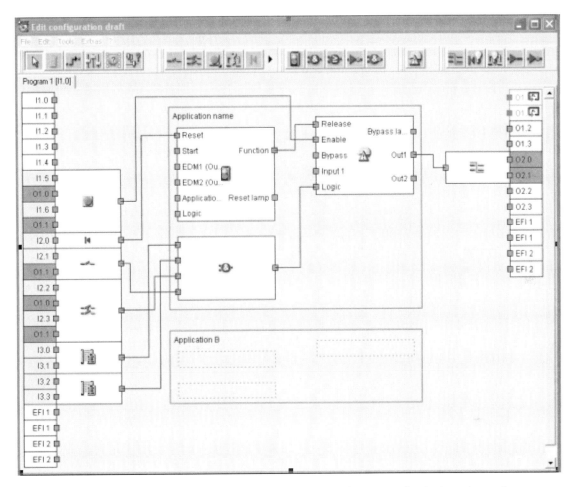

Figure 15-21 Screen capture of the logic for the safety controller in the robot cell.

SAFETY PLCS

The concept of the safety PLC is not really new. The safety PLC concept traces its history to the 1970s. Originally when PLCs were used for safety, standard PLCs were used in pairs. One was the primary PLC and the other one was the redundant PLC. If the primary PLC failed, the redundant PLC could safely shut the system down. Dual PLCs was an expensive method for implementing safety.

The concept of a safety PLC incorporates redundancy into a single PLC by having multiple processors that execute the same logic, checking each other and only writing the outputs on agreement. Although a safety PLC is more expensive than a regular PLC, it is substantially cheaper than using two PLCs. Rockwell Automation has a safety PLC that they call the GuardPLC.

QUESTIONS

1. Describe the three main safety considerations when a new machine is designed or if existing machinery is upgraded.
2. With respect to guarding, what is interlocking?
3. What is the difference between a guard, a shield, and an awareness barrier?
4. What is the difference between a safety relay and a regular relay?
5. Describe the typical outputs from a light curtain.
6. What is muting?
7. What is blanking?
8. What two zones does a typical laser scanner have?
9. Explain the term *control reliability*.
10. Name two differences between a safety controller and a safety relay?
11. What is a safety PLC?

CHAPTER 16

Installation and Troubleshooting

Automated systems have become ever more prevalent. They have also become more reliable. In a way, this reliability makes maintenance more difficult for technicians. If you do not have to repair a system for a period of time, you tend to forget things about the system. This makes effective troubleshooting even more important.

Safety has also become a greater concern for operators and for maintenance personnel. Increasing use of safety systems such as safety relays and safety controllers has also added to the complexity and importance of proper planning and installation. The increasing complexity and cost of automation has also made maintenance and troubleshooting more important. It is very costly for a system to be down. Technicians must be able to quickly find and fix problems.

OBJECTIVES

On completion of this chapter, the student will be able to:

- Explain such terms as noise, snubbing, suppression, and single-point ground.
- Explain correct installation techniques and considerations.
- Explain proper grounding techniques.
- Explain a typical troubleshooting process.

INSTALLATION

Installation of an automated system is one of the most crucial phases of any project. It can be a frustrating, exciting, and rewarding time. A project usually involves a tight time schedule and a looming deadline.

Documentation

The proper and efficient installation and wiring of a system is dependent on the quality of the associated documentation. Documentation should include the following:

A system description
A diagram of the entire system
All programs including cross-referenced memory usage and clearly labeled I/Os
A clear wiring diagram
A list of peripheral devices and their manuals
A system manual showing start-up and shutdown procedures

Manuals should also make provision for notes concerning maintenance. When the system requires maintenance or repair, notes should be kept to facilitate future troubleshooting and repair.

Fusing and Wiring

Proper fusing within the system is important. Devices must be fused to protect the device, the wiring and personnel. The system must also be wired so that the devices are protected from overcurrent situations. There are various types of enclosures available to protect the PLC.

Enclosures

The devices in the control cabinet must be protected from the environment, from coolant, chips, and other contaminants in the air. PLCs are mounted in protective enclosures. There are various types of enclosures available to protect the PLC, and there are rating systems for the protective enclosures. An enclosure is chosen on the basis of how severe the application environment is. Enclosures typically protect the control devices from airborne contamination. Metal enclosures can also help protect control devices from electrical noise.

The International Electrotechnical Commission (IEC) and the National Electrical Manufacturers Association (NEMA) both have ratings for enclosures. The IEC system uses an Ingress Protection (IP) rating. Figure 16-1 has a table that shows the IP ratings. The IP system has two numbers associated with the rating for an enclosure. The first number represents the protection against ingress by a solid object. The second number represents the protection against ingress by liquids. An IP 54 enclosure, for example, would prevent an object larger than dust from entering the cabinet and would also protect against splashing water from all sides.

Level	First Numeral: Protection against Ingress by Solid Objects	Second Numeral: Protection against Ingress by Liquids
0	Nonprotected	Nonprotected
1	Protected against objects greater than 50 mm	Protected against dripping water
2	Protected against objects greater than 12 mm	Protected against dripping water when tilted up to 15 degrees
3	Protected against objects greater than 2.5 mm	Protected against spraying water up to 60 degrees from both sides
4	Protected against objects greater than 1.0 mm	Protected against splashing water from all sides
5	Dust protected	Protected against water jets
6	Dust tight	Protected against heavy seas
7		Protected against immersion
8		Protected against submersion
9K		Protected against high-temperature, high-pressure spray

Figure 16-1 IP enclosure ratings for nonhazardous locations.

NEMA's standard for enclosures is shown in Figure 16-2. The bottom row of the table shows EIC's IP equivalent enclosure.

Heat is generated by devices. One must make sure that the PLC and other devices to be mounted in the cabinet can perform at the temperatures required. The temperature in an enclosure is often higher than the temperature in the atmosphere around the enclosure because of the heat generated by the devices in the enclosure. Fans can be used in the enclosure to increase circulation and to reduce hot spots. Fans should have filters to clean incoming air so that contaminants are not introduced to the cabinet and components by the cooling air.

Adequate space must be provided around devices in an enclosure. This will allow air to flow around the devices and enable them to remain cool enough. Hardware installation manuals for each device will show minimum clearance dimensions for each device.

Clearance distances around devices should be equal to or greater than those shown in installation documentation. Figure 16-3 shows an example of mounting documentation for a ControlLogix controller. Check the specifications for all of the devices to be installed in the cabinet. In most cases devices will be able to operate in the temperatures that will be experienced without additional cabinet cooling. Observing proper clearances around devices is normally sufficient for heat dissipation.

Protection against	1 Indoor	2 Indoor	3 Outdoor	3S Outdoor	4 Indoor or Outdoor	4X Indoor or Outdoor	6 Indoor or Outdoor	6P Indoor or Outdoor	12 Indoor	13 Indoor
Accidental bodily contact	X	X	X	X	X	X	X	X	X	X
Falling dirt	X	X	X	X	X	X	X	X	X	X
Dust, lint, fibers (non-volatile)	–	–	X	X	X	X	X	X	X	X
Windblown dust	–	–	X	X	X	X	X	X	–	–
Falling liquid, light splash	–	X	X	X	X	X	X	X	X	X
Hosedown and heavy splash	–	–	–	–	X	X	X	X	–	–
Rain, snow, sleet	–	–	X	X	X	X	X	X	–	–
Ice buildup	–	–	–	X	–	–	–	–	–	–
Oil or coolant seepage	–	–	–	–	–	–	–	–	X	X
Oil or coolant spray and splash	–	–	–	–	–	–	–	–	–	X
Occasional submersion	–	–	–	–	–	–	X	X	–	–
Prolonged submersion	–	–	–	–	–	–	–	X	–	–
Corrosive agents	–	–	–	–	–	X	–	X	–	–
IEC equivalent	IP10	IP11	IP54	IP54	IP56	IP56	IP67	IP67	IP52	IP54

Figure 16-2 NEMA enclosure ratings for nonhazardous locations.

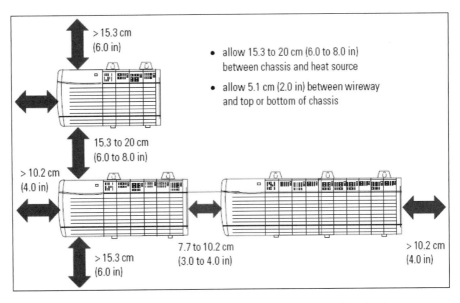

Figure 16-3 Mounting clearances. (Courtesy of Rockwell Automation, Inc.)

If you are adding devices to a cabinet, you must be very careful not to allow metal chips or components such as screws to fall into devices or other components. Metal in the wrong place can cause equipment to short circuit. It could also cause later intermittent or permanent problems. Proper wiring of a system involves choosing the appropriate devices and fuses (see Figure 16-4). Normally, three-phase power (typically 480 V) will be used in manufacturing. This will typically be the electric supply for the control cabinet.

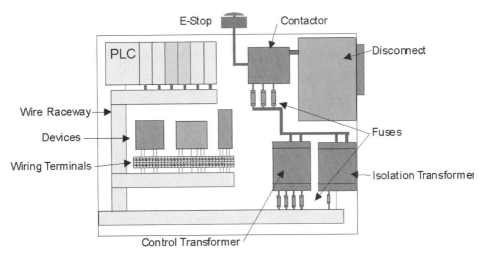

Figure 16-4 A block diagram of a typical PLC control cabinet.

Disconnects

Three-phase power is brought into the enclosure by a mechanical disconnect. This disconnect is turned on or off by the use of a lever on the outside of the enclosure. This disconnect should be equipped with a lockout. This means that the technician should be able to put a lock on the lever to prevent anyone from accidentally applying power while the system is being worked on. The three-phase power must be fused. Normally, a fusible disconnect is used. This means that the mechanical disconnect has fusing built in. The fusing is to make sure that too much current cannot be drawn.

The main power disconnect should be placed so that it is easily accessible for operators and technicians. Disconnects must ensure that power can be turned off before the enclosure is opened. NFPA 79 provides information and guidelines on disconnects.

Master Control Relay

The three-phase power is then connected to a contactor. The contactor acts as a master control relay (MCR) for the system. The MCR is used to turn power off to devices in case of an emergency. The contactor is attached to hardwired emergency circuits in the system. If someone hits an emergency stop switch, the contactor drops out and cuts the power to devices that might be dangerous. The MCR must be able to inhibit all machine motion by removing power to the machine I/O devices when the MCR relay is deenergized.

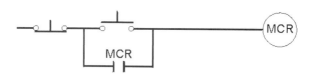

Figure 16-5 An MCR circuit.

If a DC power supply is used, the MCR should interrupt power on the load side rather than on the AC supply side. This provides quicker shutdown of power. The DC power supply should be powered directly from the fused secondary of the transformer. DC power from the power supply is then connected to the devices through a set of MCR contacts.

Emergency stop switches can be placed at multiple locations. The locations should be chosen to provide maximum safety for anyone in the area. They are wired in series (unless safety relay technology is utilized) so that if any one of the switches is activated, the MCR is deenergized. This removes power from all input and output circuits. These circuits must never be altered or bypassed to defeat their function. Severe injury or damage could occur if they are altered or bypassed. These switches are designed to fail in a safe mode (fail-safe). If there is a failure in the switch, they should open the master control circuit to disconnect power. It is possible for a switch to short out and not offer any protection. The newer technology safety switches and relays can provide the required protection. Switches should be tested periodically. Their testing should assure that they will still stop all machine motion if activated.

Figure 16-6 shows an example of a machine that cuts stock to length. There is a guard door safety switch that is used to shut the system down if the guard is opened. There is also an E-stop switch. The figure shows the wiring diagram. Note that the E-stop or the guard-limit switch can shut the MCR off. Note that the MCR contacts are then used to shut power to the PLC and the output module off.

Transformers

Three-phase power coming into the enclosure must then be converted to single-phase for the control logic. Power lines from the disconnect are fused and then connected to transformers. In the example shown in Figure 16-4 there are two transformers: a control transformer and an isolation transformer. The isolation transformer is used to clean the power for the PLC. Isolation transformers are normally used when there is high-frequency conducted noise. Isolation transformers also step down the line voltage.

The control transformer is used to supply the correct voltage to other devices in the enclosure. The lines from the power supplies are fused to protect the devices they will supply. These individual device circuits should be individually fused to match their current draw. DC power is usually required also. An enclosure typically has a power supply to convert the AC to DC. Motor starters are typically mounted in separate enclosures. This protects the control logic from the noise these devices generate.

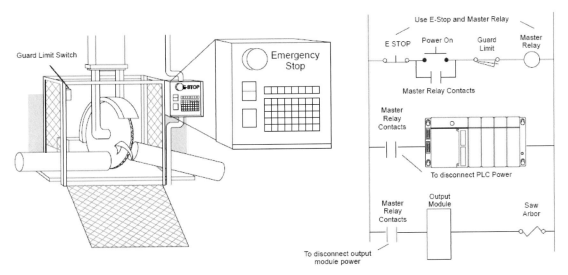

Figure 16-6 The use of a hardwired E-stop and MCR. (Courtesy Automationdirect.com)

CODES

NFPA 70 National Electrical Code

This code covers the installation of electric conductors, equipment, and raceways; signaling and communications conductors, equipment, and raceways; and optical fiber cables and raceways.

NFPA 79: Electrical Standard for Industrial Machinery

NFPA-79 is an electrical standard that was developed by the National Fire Protection Association. It was "intended to minimize the potential hazard of electrical shock and electrical fire hazards of industrial metalworking machine tools, woodworking machinery, plastics machinery and mass produced equipment, not portable by hand." This is the same association that is responsible for the National Electrical Code (NEC).

> **Scope:** "The standard shall apply to the electrical/electronic equipment, apparatus, or systems of industrial machines operating from a nominal voltage of 600 volts or less, and commencing at the point of connection of the supply to the electrical equipment to the machine."

WIRING

Wire Color

Within the control cabinet certain wiring conventions are typically used. Red wiring is normally used for ungrounded AC control conductors at less than the line voltage. Black wiring is used for ungrounded line, load, and control conductors at line voltage. Blue wire is used for ungrounded DC control conductors. Yellow wire is used to show that the voltage source is separately derived power (outside of the cabinet). Green wire (with or without one or more yellow stripes) is used to identify the equipment grounding conductor where it is insulated or covered.

Signal wiring is usually low voltage/low current and can be affected by being too close to high-voltage wiring. Signal wiring should be run separately from 120-volt wiring. When possible, signal wires should be run in a separate raceway or conduit. Some raceway is internally divided with a barrier to isolate signal wiring from higher voltage wiring.

General Wiring Suggestions

Do not run system and field wiring close to high-energy wiring. Use cable trays for wiring. Separate DC and AC wiring when it is possible. A good ground must exist for all components in a system (0.1 ohm or less). If long return lines to the power supply are needed, use separate wires for input and output modules. Separate return lines will minimize the voltage drop on the return lines of the input connections.

CHAPTER 16—INSTALLATION AND TROUBLESHOOTING   **395**

GROUNDING

Proper grounding is crucial for the safety and proper operation of a system. Grounding also helps limit the effects of electromagnetic interference (EMI).

The PLC and components are connected to the subpanel ground bus. See Figures 16-7 and 16-8. Ground connections should run from the PLC chassis and the power supply for each PLC expansion unit to the ground bus. The connection should exhibit very low resistance. Connect the subpanel ground bus to a single-point ground, such as a copper bus bar to a good earth ground reference. There must be low impedance between each device and the single-point termination. A rule of thumb would be less than 0.1-ohm DC resistance between the device and the single-point ground. This can be accomplished by removing the anodized finish and using copper lugs and star washers.

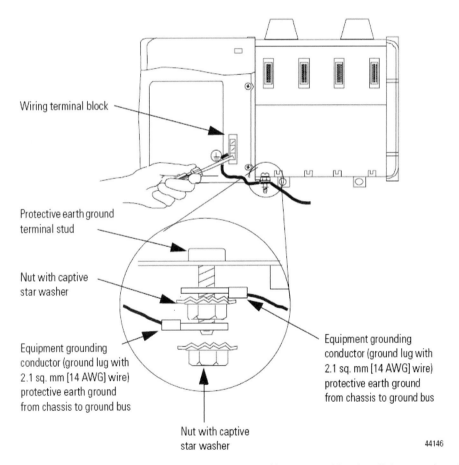

Figure 16-7 Proper grounding of CPU and chassis. (Courtesy of Rockwell Automation, Inc.)

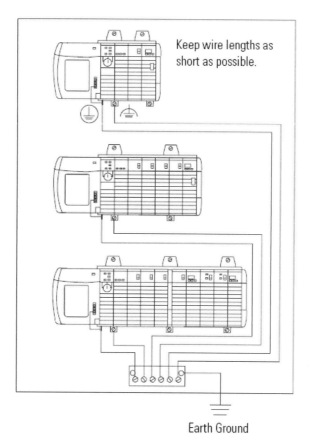

Figure 16-8 Proper grounding of ControlLogix controllers. (Courtesy of Rockwell Automation, Inc.)

PLC manufacturers provide details on installation in hardware installation manuals. The NEC is also an authoritative source for grounding requirements.

Grounding Guidelines

Refer back to Figures 16-7 and 16-8. Grounding braid and green wires should be terminated at both ends with copper eye lugs to provide good continuity. Lugs should be crimped and soldered. Copper No. 10 or 12 bolts should be used for those fasteners which are used to provide an electric connection to the single-point ground. This applies to device mounting bolts and braid termination bolts for subpanel and user-supplied single-point grounds. Tapped holes should be used rather than nuts and bolts. Note that a minimum number of threads are also required for a solid connection.

Paint, coatings, or corrosion must be removed from the areas of contact. Use external-toothed lock washers (star washers). This practice should be used for all terminations: lug to subpanel, device to lug, device to subpanel, subpanel to conduit, and so on.

Instructions for a functional ground (Courtesy Rockwell Automation Inc.)

- Use 2.54-cm (1-in.) thick copper braid or 8.3-mm^2 (8 AWG) copper wire to connect the equipment grounding conductor for each chassis, the enclosure, and a central ground bus mounted on the back panel.
- Use a steel enclosure to guard against EMI.
- Make sure the enclosure door viewing window is a laminated screen or a conductive optical substrate (to block EMI).

Rockwell instructions for an equipment protective earth ground

- Use 2.1-mm^2 (14 AWG) copper wire for the equipment grounding conductors.
- Install a bonding wire for electric contact between the door and the enclosure; do not rely on the hinge.

You must check the appropriate electrical codes and ordinances to assure compliance and safety.

Ground versus Neutral

Grounding is one of the least understood things about electric systems. Grounding helps ensure safety but also ensures that electric devices do not interfere with each other in the cell. The terms *neutral* and *ground* can cause confusion. Ground is an electric path that is designed to carry current when there is an insulation breakdown in a system. An example would be a technician dropping a tool into a cabinet and causing a breakdown when the tool contacts a voltage source and the cabinet or other metal. Some current will always flow through the ground path. This is usually caused by inductive or capacitive coupling between the current-carrying conductors and the ground path.

Neutral represents a reference point within an electric distribution system. These wires should be sized to handle the short-term faults that may occur. Neutral can be grounded: ground is not neutral.

ELECTRICAL NOISE

Electrical noise is unwanted electrical interference that can affect the operation of control equipment. Noise can cause intermittent or consistent minor problems or severe damage to equipment and people.

Many devices in manufacturing create electrical noise. Manufacturing devices that switch high voltage and current are the primary sources of noise. Motors, starters, welding equipment, and contactors that are used to turn devices on and off are some of the worst offenders. Electrical noise is not always continuous; it is often an intermittent problem. It can be very difficult to find intermittent sources of noise.

Power line disturbances, transmitted noise, or ground loops can cause electrical noise. Power line disturbances are generally caused by devices that have coils. Coil-type devices include relays, clutches/brakes, contactors, starters, solenoids, and so on. When a coil is switched off, it creates a line disturbance. Line filters can be used to deal with power line disturbances. Surge suppressors such as MOVs or an RC network installed across coils can limit noise from the coil.

Some devices create radio frequency noise. This is called transmitted noise. Transmitted noise generally occurs in high-current applications. Welders cause transmitted noise. When contacts that carry high current open, transmitted noise is generated. Control wiring that carries signals can be disrupted by this type of noise. Transmitted noise can leak into control cabinets. Holes in electrical enclosures for switches and wiring allow transmitted noise to enter the cabinet. Properly grounding the cabinet can help reduce the chance that noise will cause problems.

Another example of transmitted noise would be a sensor that is connected to a PLC input card. Transmitted noise can cause false signals on the sensor wiring. Twisted pair shielded wiring and connecting the shield to ground can help alleviate this problem.

Ground loops can also generate noise. These are the most difficult problems to find. Ground loops often cause intermittent problems. Ground loop noise occurs when multiple grounds exist. The farther the grounds are apart, the more likely there will be a problem. A potential can exist between the power supply earth and the remote earth. This can create unpredictable results, especially in communications.

Proper installation can avoid problems with noise. There are two main ways to deal with noise: isolation and suppression.

Noise Isolation

In noise isolation the device or devices that are generating the noise are physically separated from the control system. The enclosure also helps separate the control system from noise. Field wiring is often located in very noisy environments. This presents a problem, especially when low voltages are used such as in sensors. Control wiring should be done with shielded twisted pair wiring. The shielding should only be grounded at one end. The shield should be grounded at the single-point ground.

The effects of noise can be minimized by

> Using a suitable enclosure to house the PLC and components
> Proper grounding of all equipment
> Proper routing of all wiring
> Installing suppression devices on noise-generating devices

Noise Suppression

Noise suppression is aimed at the devices that generate noise. When the current to an inductive load is turned off, a very high voltage spike occurs. The manufacturing environment is filled with inductive devices. Inductive devices include motors, motor starters, relays, solenoids, and so on. Suppression is even more important when an inductive device is in series or parallel with hard contacts such as those found in pushbuttons and switches.

High-voltage spikes can cause problems for manufacturing devices and PLCs. The problems are often sporadic. Some PLC modules have circuitry to protect against inductive spikes. A suppression network can be installed to limit voltage spikes. Surge suppression is connected directly across load devices. This helps reduce arcing of output contacts. Excessive noise can reduce the life of relay contacts.

The top of Figure 16-9 shows examples of surge suppression for AC outputs. Surge suppression should be used on all coil-type outputs. Noise suppression can also be called snubbing. Snubbing can be used to suppress the arcing of mechanical contacts that occurs when inductive outputs are turned off. An RC, a varistor circuit, or a surge suppressor can be used across an inductive load to suppress noise.

The bottom of Figure 16-9 shows noise suppression for a DC output. A diode or a surge suppressor can be used for DC output devices.

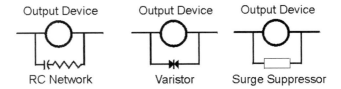

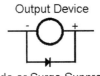

Figure 16-9 Surge suppression methods for AC and DC loads.

Figure 16-10 shows an example of a snubber installed across an output.

The components used for surge suppression must be sized to meet the characteristic of the inductive output device. The PLC installation manual will usually contain detailed information.

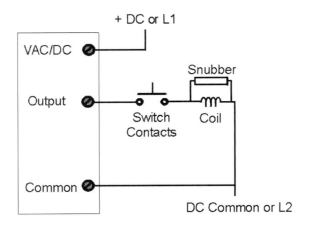

Figure 16-10 Snubbing.

PLC MAINTENANCE

PLCs are designed to be very reliable devices. Downtime is very expensive in manufacturing. Technicians must keep systems running. Downtime must be kept to a bare minimum. There are some tactics that can reduce downtime.

It is important to keep PLCs, modules, and devices clean and free from dust and debris. PLCs and control devices are typically mounted in electrical enclosures. Some electrical enclosures are cooled by a fan that is mounted in the wall of the enclosure. The fan is used to circulate fresh air through the cabinet and cool the devices in the enclosure. Dust, dirt, and other contaminants can accumulate on the devices. Dust and dirt can short out components and also cause heat problems.

Enclosure fans must have adequate filters. The filters must be cleaned regularly. Every system should have a preventive maintenance procedure that includes a check of the enclosure to make sure the inside is free of contamination. The procedure should include an inspection for loose wires or loose termination screws. Vibration in a system can loosen screws and cause intermittent or permanent problems.

Many control devices have battery backup. The procedure should specify battery replacement for devices that have batteries for memory backup. Lithium batteries are typically used. They typically have lifetimes of two or more years. Batteries always fail at the worst possible time. Their failure can cause a loss of memory that can cause unnecessary system downtime. This is very expensive and frustrating. One can imagine a technician frantically searching the plant for the backup copy of the program that was lost. Replacing the batteries once a year is a very inexpensive investment to avoid downtime.

Keeping the System Operational

Downtime is very expensive. Repairing boards or other components is usually not an option. The technician must be able to find and correct the problem in a minimum of time. This usually means replacing components. A rule of thumb is to inventory one spare for each ten devices used. This applies to CPUs, each type of I/O, and special-purpose modules. Parts that fail more often or take more time to get should be kept in stock. Spare parts should also be maintained for drives, motors, sensors, and other devices. In some cases it may be possible to return more expensive devices to the manufacturer for repair. A spare one assures that production can resume while the defective one is returned to the manufacturer for repair. A reasonable inventory of spare parts can dramatically reduce downtime, expense, and frustration.

TROUBLESHOOTING SYSTEM PROBLEMS

The first consideration in troubleshooting and maintaining systems is safety. When you encounter a problem, remember that less than one-third of all system failures will be due to the PLC. Most of the failures are due to input and output devices (up to 80 percent).

You must always be aware of the possible outcomes of changes you make. Many years ago there was a technician who isolated the problem in a machine to a defective part sensor. The technician crawled into the machine and bypassed the sensor. The system restarted with the technician in the machine. This does help explain the importance of following lockout/tagout procedures.

The use of the Web is indispensable to the technician. Manufacturers put their product manuals online. More importantly there are many websites devoted to helping solve PLC and automation problems. There are online forums where you can enter your questions for the problem you are experiencing. Users will respond with solutions in a very short time. Rockwell Automation also offers help on its website. The Web has become one of the technician's best friends.

People Skills in Troubleshooting

Involve the Operator

Was the system running when the problem happened?
Ask the operator what she or he thinks may have caused the problem.

Many technicians will assume that a symptom is the problem. Solving the symptom may get the system running again. But the real problem is bound to cause problems later.

An example of solving the symptom rather than the problem is taking an aspirin for a headache. If you have a headache, you may take an aspirin. The aspirin will help with the pain. But the pain is merely a symptom. You did not solve the cause for the headache. It may have been due to a hangover (in which case it will pass) or a brain tumor, stress, or many other causes.

The equivalent example in maintenance would be a blown fuse. The maintenance technician replaces the fuse and the system runs again. Replacing the fuse did not solve the cause of the problem.

The fuse blew because of excess current, but what caused the excess current? It could have been the result of a short circuit, a voltage spike, a short in the output device, or many other causes. If the technician doesn't find the cause before replacing the fuse, he or she did not solve the problem but only temporarily hid a symptom.

Logically Isolate the Probable Cause

Are you familiar with the game where you have to guess a number between 1 and 100? If you are wrong you are told if you are too high or too low and you guess again. This continues until you get it right. The best strategy is to split the problem in half. In other words guess 50 first. Then if you are too low, use 75. By continually splitting the problem in half, you solve the problem in the fewest possible steps.

Troubleshooting should involve similar thought. Too many times a technician may jump to too quick a conclusion when presented with a problem to troubleshoot. You should think the problem through and make a decision about what to test first. Each test should help divide the problem area in half.

Problems should be solved by replacing only the defective devices or components. Have you ever had your car fixed and found that the technician replaced several parts that you did not think you needed? The technician probably began replacing parts that were not defective before finding the root cause of the problem.

One of the most important things in troubleshooting is to establish a logical approach. A good troubleshooter uses an approach that enables her or him to logically and efficiently determine what is wrong.

Typically 80 percent of all PLC malfunctions can be traced to problems with I/O modules or field devices.

Troubleshooting Input and Output Problems

Typically the first task in I/O troubleshooting is to find out why the internal I/O states of the PLC do not agree with the external I/O states.

One of the best troubleshooting tools is the personal computer and the programming software. With the PLC online, you can monitor the I/O and check the program versus the real-world states.

Troubleshooting Discrete Input Modules

If you are troubleshooting an input, check the LED for the input first. If the input were true, the LED for that input should be on. It should also appear energized in the programming software if you are online with the PLC.

If the status LED on the module does not turn on, check to make sure there is input power. If input power is not present, determine and rectify the cause of the failure before proceeding. Remember that input power is normally not supplied by the module. There will probably be a power supply for the inputs.

PLC input modules normally do not supply their own power.

If there is proper input power, connect a meter across the input device, change the state of the input, and measure the voltage at the PLC input to see if it changes when the input device changes state. If you do not see the change of state, the input device or its wiring is most likely the fault.

If you do observe the correct voltage change, the input status LED on the module should change. This should also be observable in the programming software when online. If the status LEDs do not properly indicate the state of the input, the input module should be replaced.

Troubleshooting Analog Input Modules

Analog input modules are used to measure the actual value of a voltage or current.

Change the voltage or current level generated by the analog input device. For example, if it is a thermocouple, change the temperature at the thermocouple. Check the value of the input in the programming software. Make sure you are online. If you do not see a change in the address for that input, use a meter to see if the correct input level is present at the input module. If there is no input at the module input, check to make sure there is power to the device. If the device is properly powered, check and make sure there is output from the device. If not, replace the device. If there is output, but it is not at the correct level, try to calibrate the device. If the output from the device seems correct, check the wiring to the module. If the signal is correct at the input terminal but the numbers in the memory address are not correct, you may be able to calibrate the module to show the correct value. If not, replace the module.

Inputs can be checked and calibrated by using a calibration meter as the input source to the analog input. A calibration meter is a very accurate meter that can source various signals. They are calibrated at regular intervals to make sure they are accurate.

Troubleshooting Discrete Output Modules

The first step is to determine if the power for the output in question is present and to restore that power if it is not present. The power for PLC outputs is not typically supplied by the module. Locate the power supply for the outputs. Output modules are typically fused.

Faults in devices or field wiring can blow the module fuse. Most modules have a fuse-blown LED that shows which output or module has a blown fuse. Check the fuse-blown LED before proceeding. These fuses may be accessible from the front in some modules. Some modules have fuses that can be reset electronically with the programming software. In some modules the module may have to be removed and disassembled to replace fuses.

After the fuses have been checked and proper power has been verified, you can troubleshoot the digital outputs.

Use the computer and programming software to force the output on or off. Observe the LEDs on the module for a change. If the output status LEDs on the module do not agree with the forced condition, the output module should be changed.

If the output LEDs are observed to be reacting to the forced state but the problem still exists, measure the voltage across the output device to see that it's changing as the state of the output changes. This may seem a little strange, but if the output is on, you should read 0 volts between the common and the output pin on the module, assuming there is a load connected.

A common technician error is to disconnect the load and check the voltage at the output with respect to neutral when it is off. The meter will typically read full voltage (due to leakage). The technician assumes there is a bad module.

If the output is off, you should read full voltage between the module common and the output pin. Think this through; it is an important concept for understanding I/O and troubleshooting.

If the voltage is changing but the device is still not working, the problem is the output device or wiring.

If you find that the voltage is not changing, the problem is most likely in the field wiring. If this is the case, you can disconnect the actual output and connect a test load to the module.

If the test load operates correctly, the problem is in the field device or field wiring. A solenoid valve or a relay can be used for a test load.

Troubleshooting Analog Output Modules

Analog outputs are used to generate a variable voltage or current output. Usually there is no indication on the module to reflect the level of the output. First you must determine the resolution and range of the modules. For example, if it is a 14-bit module, you might expect the number 0 to represent 4 mA and 32,767 to represent 20 mA. These numbers can be looked up in the manual for the module.

Next you can enter test values into the PLC address associated with the output. It is good to test the minimum, half-scale, and maximum values while you measure the voltage or current generated at the output.

In the 14-bit current module example, a 0 in the appropriate PLC address should generate 4 mA at the output terminals, 16,383 should generate 12 mA, and 32,767 should generate 20 mA. If the field wiring or field device are in doubt, they can be temporarily disconnected and replaced by a test load. If the proper currents or voltages are not measured at the test load, the analog output module should be calibrated if possible or replaced. A properly sized resistor, typically between 250 and 1000 ohms, can be used as a test load in analog circuits.

Troubleshooting CPU problems

If the problem seems to be in the PLC CPU or system, one of the first things to look at is the PLC's power supply and ground. Visually inspect the power and ground wiring, looking for loose, corroded, or problems in connections. The ground can be electrically checked by measuring the voltage between the PLC ground terminal and a known ground. The AC and DC voltages should be 0 on a meter.

Test the power supply. If the PLC power supply has an AC power source, check the input voltage; make sure it is within the manufacturer's specified range.

Check the DC supplies for AC ripple. Use a digital meter set on a low AC range. The value should be well below the manufacturer's specifications for ripple. Excess ripple can have dramatic effects on the operation of the CPU and its memory.

The final power check is to check the batteries in the system. Batteries are often used to maintain processor memory during times when the PLC is not powered. Battery voltage should be within the specified values.

EMI or radio frequency interference (RFI) can also cause intermittent erratic operation. It is difficult when problems are intermittent, but try to correlate the erratic behavior with events such as ARC welding in the area, radio transmitters, the starting of a large motor, lightening, or other such events. Once a problem can be identified a solution can be implemented. Solutions to EMI and RFI problems typically involve improvements in power conditioning, shielding, and grounding.

If the battery or noise was eliminated as a problem, one should verify that the PLC program is correct. The program should be verified against a good copy or the program should be downloaded to the PLC. It is a good idea to verify the memory against a known good copy so that you know if there is a problem. Make sure you keep program backups and store them away from an EMI or RFI to prevent damage.

Review

Troubleshooting is a relatively straightforward commonsense process. The first step is to think. Many technicians are too quick to jump to improper, premature conclusions and waste time finding problems. The first step is to examine the problem logically. Think the problem through using common sense first. Troubleshooting is much like the game Twenty Questions. Every question should help isolate the problem. In fact, every question should eliminate about half of the potential causes.

Think logically.
Ask yourself questions to isolate the problem.
Test your theory.

Next use the resources you have available to check your theory. Often the error-checking present on the PLC modules is sufficient. The LEDs on PLC CPUs and modules can provide immediate feedback on what is happening. Many PLCs have LEDs to indicate blown fuses and many other problems. Check these indicators first.

Troubleshooting Example

The operator of a machine calls a technician and tells him or her that an output is not turning on when it should. There are many things that could cause this to happen.

The PLC output module or the output module that controls the output could be defective. A fuse could be blown. The output device could be faulty. One of the inputs that control the contact in the logic that controls the output could be defective. The logic in the PLC could be faulty. Logic can be written that performs perfectly the vast majority of the time but fails under certain conditions.

The technician should not immediately jump to a conclusion about what is causing the problem.

The I/O status LEDs provide the best and easiest source of information. If the status LED for that output is on, the problem is probably not the inputs to the PLC. In this case, the device is defective, the wiring is defective, or the PLC output is defective. This simple check eliminated half of the system from consideration.

A voltmeter can be used is to further isolate the problem. If the PLC output is off, a meter reading should read the full output voltage. If the output is used to supply 115 volts to a motor starter, the meter should read 115 volts between the output terminal and the module common. It may seem strange that you would read full voltage with the output off.

If the PLC output is on, the meter should read 0 volts. Remember that the output acts as a switch. If we measure the voltage across a closed switch, we should read 0 volts. If the switch is open, we should read the full voltage. If there is no voltage in either case, the power supply and wiring should be checked. If the power supply and wiring are okay, the device is faulty. There may also be fuses or overload protection present that can be the cause.

Now pretend that the output LED was not turning on. Turn your attention to the input side of the system. Study the logic to see which inputs may be involved in the rung that controls the output state. Assume there is only one sensor that controls the state. If the input status LED on the input module is on, the sensor would seem to be operational. If you can safely activate and deactivate the sensor, do so and watch the LED for the change. Next monitor the logic to see if the PLC really sees the input as true. The logic should be monitored in online mode to check the input. Monitor the logic to see if the contact is closing. If it is seeing the true input as false, the problem is probably a defective PLC module input. Figures 16-11 and 16-12 show troubleshooting for input and output modules.

Input LED	State of Device	Condition	Problem	Things to Check
Off	Off	The program operates as if the input is on.	Input may be forced on.	Check for forced I/O on the CPU, remove forces, and verify the wiring. Try an unused input or replace the module.
			An input circuit is bad.	Verify the wiring. Try a different input or replace the module.
	Off	The input device will not turn on.	The input device is bad.	Verify the operation of the device. Replace if defective.
On	On	The input device will not turn off.	The device is damaged or shorted.	Verify the operation of the device. Replace if defective.
		Program operates as if the input device is off.	An input is forced off.	Check the forced I/O on the CPU. Remove forces.
			An input circuit is bad.	Verify the wiring. Try a different input or replace the module.
	Off	Program operates as if the input is on or the input circuit will not turn off.	An input circuit is bad.	Verify the wiring. Try a different input or replace the module.
			An input device is bad or shorted.	Verify the operation of the device. Replace if defective.
			Leakage current of the input device exceeds the module's input specification.	Use a load resistor to bleed off excess current.

Figure 16-11 Chart for troubleshooting inputs.

CHAPTER 16—INSTALLATION AND TROUBLESHOOTING

Output LED	State of Device	Condition	Problem	Things to Check
Off	On	Output device is on, but program indicates it is off.	There is incorrect wiring.	Check the wiring. Disconnect and test the device.
			Output device is defective or shorted.	Verify that the device is OK, replace if necessary.
			The output circuit is defective.	Check the wiring. Move the device to an unused output or replace the module.
	Off	Program shows that the output is on, or the output circuit will not turn on.	The output circuit is damaged.	Use a force to turn the output on. If the output turns on, it is a program problem. If the output does not turn on, there is an output circuit problem. Try a different output or replace the module.
			This is a program problem.	Look to make sure outputs have not been duplicated in logic. If subroutines are being used, outputs are in their last state when not executing subroutines. Force the output on. If the output does not turn on, the output circuit is bad. Try a different output or replace the module. If the output does force on, check for a problem in the program.
			The output is forced off in the program.	Check the CPU to see if I/O has been forced. Remove forces.
On	On	The program shows the output is off, or the output will not turn off.	This is a program problem.	Look to make sure outputs have not been duplicated in logic. Outputs will remain in their last state when not executing subroutines. Force the output off. If the output does not turn off, the output circuit is bad. Try a different output or replace the module. If the output cannot be forced off, it is most likely a program logic problem.
			An output in the module is bad.	Force the output off. If the output turns off, there is a problem with the program. If not, there is an output circuit problem. Try a different output or replace the module.
			The output is forced on.	Check the CPU for forced I/O. Remove forces.
	Off	Output device will not turn on, but the program indicates it is on.	There is bad wiring or an open circuit.	Check the wiring and the connections to the common.
			The output in the module is bad.	Try a different output on the module or replace the module if necessary.
			Voltage across the load is low or absent.	Check the source voltage.
			Output device is not compatible with the module.	Check the specifications of the module - sink, source, etc.

Figure 16-12 Output troubleshooting chart.

POTENTIAL LOGIC PROBLEMS

Some PLCs allow the same output coil to be used on more than one rung. This means that there are multiple rungs with different conditions that control the same output coil. There are many technicians who were sure that they had a defective output module because when they monitored the ladder the output coil was on but the actual output LED was off. In many cases they had inadvertently programmed the same output coil twice with different rung conditions. The rung they monitored was true, but one farther down in the ladder was false, so the PLC deenergized the output.

Problems can be intermittent in logic. Timing can cause serious issues in poorly written logic. Devices often exchange signals to signify that an action has occurred. For example, a robot finishes a step and sends a digital signal to the PLC to tell it the step is done and the robot is ready for the next step in the sequence.

If the robot programmer just turns it on for one step in the program, the output may be on for less than a millisecond. The PLC may catch the signal every time, or it may miss it sporadically because of the length of the scan time of the PLC.

This is a difficult problem to troubleshoot, because it occurs intermittently. Handshaking is often used by programmers instead of relying on the PLC seeing a periodic input. Handshaking means that the devices work together to assure no signals are missed. If handshaking is used, the programmer would develop the robot program so that the robot's output would stay on until it has been acknowledged by the PLC. This means that the robot turns the output on and waits for an input from the PLC to assure that the PLC saw the output from the robot.

SUMMARY

The installation of automated systems must be carefully planned. Safety must be the prime concern. Risk analysis must be done to evaluate the safety of a system. People and devices must be protected though good design principles. Hardwired E-stop switching should be provided to drop all power to the system. Lockouts should be provided to assure safety while maintenance is being performed. Fusing must be done to protect individual devices and personnel. Electrical enclosures must be carefully selected to meet the needs of the application environment. Proper grounding procedures must be followed to ensure the safety of personnel and devices. Control and power wiring should be separated to reduce noise.

To be effective, troubleshooting should be logical. Think the problem through before acting. Ask questions that will help isolate the potential problems. Above all, apply safe work habits while working on systems.

QUESTIONS

1. What items should be included in system documentation?
2. Describe what each of the two numbers in an IP enclosure rating is used for.
3. What is a NEMA enclosure?

4. Describe what an IP 54 enclosure would be.
5. Describe a NEMA 4 enclosure.
6. What would be an equivalent enclosure in a IP rating for a NEMA 4 enclosure?
7. Describe how an enclosure is chosen.
8. What is a fusible disconnect?
9. What is a contactor?
10. What is the purpose of an isolation transformer?
11. What is the major cause of failure in systems?
12. Describe a logical process for troubleshooting.
13. Describe proper grounding techniques.
14. Describe at least three precautions that should be taken to help reduce the problem of noise in a control system.
15. Troubleshoot the following:

    An output device will not turn on. The output status LED on the module seems to be working fine. A voltmeter is placed across the PLC output with the output on; 0 volts is read. What is most likely wrong?
16. Troubleshoot the following:

    The output device is not turning on for some reason. The technician turns the output on and places a voltmeter over the PLC output; 115 volts is read. The output status LED is on but the device is not turning on. What is the most likely problem? What should the technician do?
17. Troubleshoot the following:

    An input on the module seems to be faulty. The technician notices that the input's status LED is never on. A voltmeter is placed across the PLC input; 0 volts is read. The technician removes and tests the sensor. (The LED on the sensor comes on when the sensor is activated.) The sensor seems fine. What is wrong?

CHAPTER 17

Lockout/Tagout

OBJECTIVES

On completion of this chapter the reader will be able to:

- Understand the sources of potentially harmful energy.
- Understand the purpose and importance of lockout/tagout.
- Identify sources of energy.
- Develop lockout/tagout procedures.

INTRODUCTION

An automated machine or system can have several dangerous sources of energy. There are many types of energy that can be dangerous for the operating personnel and especially for technicians performing maintenance or service on the equipment. Lockout/tagout was designed to reduce the risk for these personnel when maintaining and servicing equipment.

LOCKOUT/TAGOUT

On October 30, 1989, the Department of Labor released "The control of hazardous energy sources (lockout/tagout)" standard. It is the lockout/tagout standard numbered 29 CFR 1910.147. The standard was intended to reduce the number of deaths and injuries related to servicing and maintaining machines and equipment.

The lockout/tagout standard covers the servicing and maintenance of machines and equipment in which the unexpected startup or energization of the machines or equipment or the release of stored energy could cause injury to employees. The standard covers energy sources such as electric, mechanical, hydraulic, chemical, nuclear, and thermal. The standard establishes minimum standards for the control of these. Normal production operations, cords and plugs under exclusive control, and hot tap operations are not covered by the standard.

Normal production operations are excluded from the lockout/tagout restrictions. Normal production operation is the use of a machine or equipment to perform its intended production function. Any work performed to prepare a machine or equipment to perform its normal production operation is called setup.

If an employee is working on cord-and-plug-connected electric equipment for which exposure to unexpected energization or start-up of the equipment is controlled by the unplugging of the equipment from the energy source, and the plug is under the exclusive control of the employee performing the servicing or maintenance, this activity is also excluded from the requirements of the standard.

The standard does not apply to hot taps when they are performed on pressurized pipelines, provided that the employer can demonstrate that continuity of service is essential; shutdown of the system is impractical; documented procedures are followed; and special equipment is used that will provide proven, effective protection for employees.

Hot tap operations involve transmission and distribution systems for substances such as steam, gas, water, or petroleum products. A hot tap is a procedure used in maintenance that involves welding on a piece of equipment such as a pipeline or tank, under pressure, in order to install connections or devices.

The lockout/tagout standard defines an energy source as any source of electric, mechanical, hydraulic, pneumatic, chemical, thermal, or other energy. Machinery or equipment is considered to be energized if it is connected to an energy source or contains residual or stored energy. Stored energy can be found in pneumatic and hydraulic systems, capacitors, springs, and even gravity. Heavy objects have stored energy. If they fall, they can cause injury.

Servicing or maintenance includes activities such as installing, constructing, setting up, adjusting, inspecting, and modifying, as well as servicing or maintaining equipment. These activities include lubricatng, cleaning, or unjamming machines or equipment and making adjustments or tool changes where personnel may be exposed to the unexpected energization or start-up of the equipment or a release of hazardous energy.

Employers are required to establish (lockout/tagout) procedures and employee training to ensure that before any employee performs any servicing or maintenance on a machine or equipment where the unexpected energizing, start-up or release of stored energy could occur and cause injury, the machine or equipment is isolated and rendered inoperative. Most companies impose strong sanctions on employees who do not follow the procedures to the letter. Many companies terminate employees who repeatedly violate the procedures. The procedures are designed to keep personnel safe. Make sure you follow procedures. You should also be aware of the penalties of not following them. The greatest penalty could be severe injury or death.

Employers are also required to conduct periodic inspections of the procedures at least annually to ensure that the procedures and the requirements of the standard are being followed.

Only authorized employees may lockout machines or equipment. An authorized employee is defined as one who has been trained and has been given the authority to lock or tag out machines or equipment to perform servicing or maintenance on that machine or equipment.

An energy-isolating device is defined as a mechanical device that can physically prevent the transmission or release of energy. A disconnect on an electrical enclosure is a good example of an energy isolating device (see Figure 17-1). If the disconnect is off, it physically prevents the transmission of electric energy. Energy-isolating devices include disconnects, manual electric circuit breakers, manually operated switches by which the conductors of a circuit can be disconnected from all ungrounded supply conductors and no pole can be operated independently, line valves (see Figure 17-1), locks, and any similar device used to block or isolate energy. Pushbuttons, selector switches and other control-circuit-type devices are not considered to be energy-isolating devices.

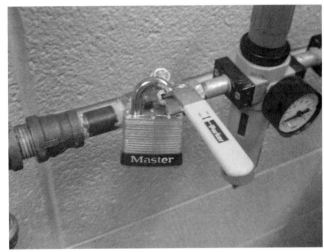

Figure 17-1 Electric disconnect on the left and a pneumatic disconnect on the right.

There has been a requirement since January 2, 1990, that when new machines or equipment is installed, energy-isolating devices for machines or equipment must be designed to accept a lockout device. An energy-isolating device is capable of being locked out if it has a hasp or other means of attachment to which or through which a lock can be affixed or if it has a built-in locking mechanism. Other energy-isolating devices are capable of being locked out if lockout can be achieved without the need to dismantle, rebuild, or replace the energy-isolating device or permanently alter its energy control capability.

Lockout

Lockout is defined as the placement of a lockout device on an energy-isolating device in accordance with an established procedure to ensure that the energy-isolating device can ensure that the equipment being controlled cannot be operated until the lockout device is removed.

A lockout device uses a positive means to hold an energy-isolating device in the safe position and prevent the energizing of a machine or equipment. A lock used for lockout may be a key or combination type.

Employee Notification

Affected employees must be notified by the employer or authorized employee of the application and removal of lockout devices or tagout devices. Notification shall be given before the controls are applied, and also after they are removed from the machine or equipment. Affected employees are defined as employees whose job requires them to operate or use a machine or equipment on which servicing or maintenance is being performed under lockout or tagout or whose job requires them to work in an area in which such servicing or maintenance is being performed.

Tagout

If an energy-isolating device is incapable of being locked out, the energy control program shall utilize a tagout system. Tagout is the placement of a tagout device on an energy-isolating device, in accordance with an established procedure, to indicate that the energy-isolating device and the equipment being controlled may not be operated until the tagout device is removed. A tagout device is a prominent warning device, such as a tag and a means of attachment, which can be securely fastened to an energy-isolating device in accordance with an established procedure, to indicate that the energy-isolating device and the equipment being controlled may not be operated until the tagout device is removed from each energy-isolating device by the employee who applied the device. Tagout shall be performed only by the authorized employees who are performing the servicing or maintenance.

When the authorized employee who applied the lockout or tagout device is not available to remove it, that device may be removed under the direction of the employer, provided that specific procedures and training for such removal have been developed, documented, and incorporated into the energy control program.

Tagout devices must be affixed in such a manner as to clearly indicate that the operation or movement of energy-isolating devices from the safe or off position is prohibited. Where tagout devices are used with energy-isolating devices designed with the capability of being locked, the tag attachment must be fastened at the same point at which the lock would have been attached. Where a tag cannot be affixed directly to the energy-isolating device, the tag must be located as close as safely possible to the device, in a position that will be immediately obvious to anyone attempting to operate the device.

Training

Training shall be provided by the employer to ensure that the purpose and function of the energy control program are understood by employees and that the knowledge and skills required for the safe application, usage, and removal of energy controls are required by employees. Training should include

- The recognition of hazardous energy sources
- The type and magnitude of the energy available in the workplace
- The methods and means necessary for energy isolation and control
- A thorough understanding of the purpose and use of the lockout/tagout procedures

All other employees whose work operations are or may be in an area where lockout/tagout procedures may be used shall be instructed about the procedure and about the prohibition against attempting to restart or reenergize machines or equipment which are locked out or tagged out.

When tagout procedures are used, employees must be taught about the following limitations of tags:

- Tags are really just warning devices and do not provide physical restraint on devices.
- When a tag is attached it is not to be removed without authorization of the authorized person responsible for it, and it is never to be bypassed, ignored, or otherwise defeated.
- Tags must be legible and understandable by all authorized employees, affected employees, and all other employees whose work operations are or may be in the area.
- Tags may create a false sense of security. Their meaning needs to be understood by all.

Retraining

Retraining is required for all authorized and affected employees whenever there is a change in their job assignments; a change in machines, equipment, or processes that present a new hazard; or a change in the energy control procedures.

Additional retraining must also be done when a periodic inspection reveals or whenever the employer has reason to believe that there are deviations from or inadequacies in the employee's knowledge or use of the energy control procedures.

The retraining shall reestablish employee proficiency and introduce new or revised control methods and procedures, as necessary.

The employer must certify that employee training has been accomplished and is being kept up to date. The training records must contain each employee's name and dates of training.

Lockout/Tagout Device Requirements

Lockout and tagout devices must be the only device(s) used for controlling energy and must not be used for other purposes.

Lockout/tagout devices must be durable. They must be capable of withstanding the environment to which they are exposed for the maximum period of time that exposure is expected.

Lockout and tagout devices must be standardized within the facility in at least one of the following criteria: color; shape; or size. Print and format must also be standardized for tagout devices. Tagout devices must be printed and made of suitable materials so that exposure to weather conditions or wet and damp locations will not cause the message on the tag to become illegible or cause the tag to deteriorate.

Tagout devices, including the attachment device, must be substantial enough to prevent inadvertent or accidental removal. Tagout devices must be attached with an attachment that is non-reusable. Tags must be attachable by hand, be self-locking, and also be non-releasable. They must have a minimum unlocking strength of at least 50 pounds. Lockout devices must be substantial enough to prevent removal without the use of excessive force or unusual techniques, such as with the use of bolt cutters or other metal-cutting tools.

An attachment device should have the general design and basic characteristics at least equivalent to a one-piece, all-environment-tolerant nylon cable tie.

Lockout devices and tagout devices must identify the employee who applied the devices.

Tagout devices must warn against hazardous conditions if the machine or equipment is energized and shall include a clear warning such as Do Not Start. Do Not Operate. Do Not Open. Do Not Close. Do Not Energize.

Application of Lockout/Tagout Procedures

The procedures for the application of lockout or tagout procedures shall cover the following and shall be done in the following sequence:

1. All affected employees must be notified that a lockout or tagout procedure is going to be used. All affected employees must understand the reason for the lockout. Before an authorized or affected employee turns off a machine or equipment, the authorized employee must understand the types and magnitudes of the energy, the hazards of the energy to be controlled, and the method or means to control the energy for the machine or equipment being maintained or serviced.
2. The equipment must be turned off or shut down using the procedures for the machine or equipment. An orderly shutdown shall be performed to avoid additional or increased hazards to employees as a result of the equipment stoppage.
3. All energy-isolating devices that are needed to control the energy to the equipment or machine shall be physically located and operated in such a manner as to isolate the equipment or machine from the energy sources.

4. Lockout or tagout devices must be affixed to each energy-isolating device by authorized employees. Lockout devices, where used, shall be affixed so that the energy-isolating devices are held in a safe or off position.
5. Stored energy must be dissipated or restrained by methods such as repositioning, blocking, bleeding down, and so on, after the application of lockout or tagout devices to energy-isolating devices. If there is a possibility of reaccumulation of stored energy to a hazardous level, verification of isolation shall be continued until the servicing or maintenance is completed or until the possibility of such accumulation of energy no longer exists.
6. Prior to starting service or maintenance work on machines or equipment that have been locked out or tagged out, the authorized employee shall verify that the machine or equipment has actually been isolated and deenergized. This is done by operating the pushbutton or other normal operating controls to make certain that the equipment will not operate. *Warning:* You must make sure that the operating controls are returned to the neutral or off position after the test.

The machine or equipment is now locked out or tagged out.

To Remove Lockout/Tagout Devices

Before lockout or tagout devices are removed and energy is restored to the machine or equipment, procedures must be followed and actions taken by the authorized employees to ensure that

The work area has been inspected to be sure that nonessential items, such as tools, have been removed from the work area and to ensure that machine or equipment components are operationally intact.

The work area has been checked to be sure that all personnel have been safely positioned or removed from the area.

Affected employees have been notified that the lockout or tagout devices have been removed.

Each lockout or tagout device must be removed from each energy-isolating device by the employee who applied the device. The only exception to this is that when the authorized employee who applied the lockout or tagout device is not available to remove it, that device may be removed under the direction of the employer, provided that specific procedures and training for such removal have been developed, documented, and incorporated into the employer's energy control program.

The employer must demonstrate that the specific procedure to be used if the authorized employee who applied the lockout/tagout is unavailable includes at least the following elements:

The employer must verify that the authorized employee who applied the device is not at the facility.

All reasonable efforts must be made to contact the authorized employee to inform him or her that his or her lockout or tagout device has been removed.

The authorized employee must be made aware of that his or her lockout/tagout device was removed before he or she resumes work at that facility.

Testing of Machines, Equipment, or Components

Situations may occur in which lockout or tagout devices must be temporarily removed from the energy-isolating device and the machine or equipment energized to test or position the machine, equipment, or component. If this situation arises, the following sequence of actions must be followed:

The machine or equipment must be cleared of tools and materials.
Employees must be removed from the machine or equipment area.
The lockout or tagout devices must be removed as specified in the procedure.
The machine or equipment must be energized and the testing or positioning proceeded with.
After testing, all systems must be deenergized and energy control measures reapplied in accordance with the standard to continue the servicing or maintenance.

Group Lockout/Tagout

When more than one person is involved in maintenance, the risk of injury is increased. When maintenance is performed by a group of people, they must use a procedure that protects them to the same degree that a personal lockout/tagout procedure designed for one person would.

The standard for lockout/tagout specifies requirements for group procedures. Primary responsibility is vested in an authorized employee for a set number of employees. These employees work under the protection of a group lockout or tagout device. This is typically a hasp (see Figure 17-2). The hasp allows several locks to be applied so that multiple people are protected. The group lockout device assures that no one individual can start up or energize the machine or equipment. All lockout or tagout devices must be removed to reenergize the machine or equipment. The authorized employee who is responsible for the group must ascertain the exposure status of individual group members with regard to the lockout or tagout of the machine or equipment. When more than one crew, type of maintenance personnel, department, and so on, is involved, overall job-associated lockout or tagout control responsibility is assigned to an authorized employee. This employee is designated to coordinate affected workforces and ensure continuity of protection. Each authorized employee working on the system must affix a personal lockout or tagout device to the group lockout device, group lockbox, hasp, or comparable mechanism when she or he begins work. The individual shall remove those devices when she or he stops working on the machine or equipment being serviced or maintained.

Figure 17-2 A lockout hasp.

Personnel or Shift Changes

Specific procedures shall be utilized during shift or personnel changes to ensure the continuity of lockout or tagout protection. This ensures the orderly transfer of lockout or tagout device protection between off-going and oncoming employees, to minimize exposure to hazards from the unexpected energization or start-up of the machine or equipment or from the release of stored energy.

LOCKOUT PROCEDURE FOR THE EXTRUDER

Purpose of the Procedure

This procedure shall be used to ensure that the extruder is stopped and isolated from all potentially hazardous energy sources and locked out before employees perform any servicing or maintenance where the unexpected energizing or start-up of the machine or equipment or the release of stored energy could cause injury.

Employee Compliance

All employees are required to comply with the restrictions and limitations imposed on them during the use of this lockout procedure. Authorized employees are required to perform the lockout in accordance with this procedure. All employees, on observing a machine or piece of equipment that is locked out to perform servicing or maintenance, shall not attempt to start, energize, or use that machine or equipment. Failure to follow this procedure exactly will result in the actions specified in the employee handbook.

Lockout Sequence for the Extruder

1. Notify all of the affected employees that servicing or maintenance is to be done on the extruder and that the machine must be shut down and locked out to perform the servicing or maintenance.
2. The extruder has pneumatic, electric, and heat energy sources.
3. If the machine or equipment is in operation, shut it down by the normal stopping procedure.
4. Deactivate the energy disconnects so that the extruder is isolated from the energy sources. There are two disconnects: the electrical disconnect on the electrical enclosure and the pneumatic disconnect on the air supply line to the extruder.
5. Lock out the two energy-isolating devices with your assigned locks according to the procedure.
6. The pneumatic disconnect released the air pressure in the system when you turned it off and applied your lock. Heat energy still exists. The protective guard is in place to protect you from exposure to the heat energy. If you have to remove the guard for service, you must wait one hour after disconnecting the energy and applying your locks before you remove the guard. Carefully check for residual heat before continuing.
7. You must ensure that the extruder is disconnected from the energy sources by first checking that no personnel are exposed. Next verify the isolation of the equipment by operating the operating controls to make certain the equipment will not operate. *Caution:* You must return the operating controls to neutral or off position after you verify the isolation of the equipment.
8. The extruder is now locked out.

RETURNING THE MACHINE OR EQUIPMENT TO SERVICE

When the servicing or maintenance is completed and the machine or equipment is ready to return to normal operating condition, the following steps shall be taken:

- Check the machine or equipment and the immediate area around the machine to ensure that nonessential items have been removed and that the components are operationally intact. Check the work area to ensure that all employees have been safely positioned or removed from the area.
- After all tools have been removed from the machine or equipment, guards have been reinstalled, and employees are in the clear, remove all lockout or tagout devices. Verify that the controls are in neutral and reenergize the machine or equipment. *Note:* The removal of some forms of blocking may require reenergization of the machine before safe removal. Notify affected employees that the servicing or maintenance is completed and the machine or equipment is ready for use.

EXAMPLE OF A LOCKOUT/TAGOUT CHECKLIST

Notification

I notified all affected employees that a lockout is required and the reason for the lockout.

Date _____ Time _____ Signature _____

Shutdown

I understand the reason the equipment is to be shutdown following the required procedure.

Date _____ Time _____ Signature _____

Disconnection of Energy Sources

I disconnected or isolated each energy source from the machinery or equipment as specified by the procedure. I have dissipated or restrained all stored energy such as springs, elevated machine members, capacitors, rotating flywheels, pneumatic and hydraulic systems, and so on.

Date _____ Time _____ Signature _____

Lockout

I locked out the energy-isolating devices using my assigned locks.

Date _____ Time _____ Signature _____

Safety Check

After ensuring that no personnel were exposed to hazards, I operated the start button and other normal operation controls to ensure that all energy sources were disconnected and that the equipment would not operate.

Date _____ Time _____ Signature _____

OUTSIDE PERSONNEL WORKING WITHIN THE FACILITY

When outside maintenance or servicing personnel are to be engaged in activities covered by the lockout/tagout standard, the on-site employer and the outside employer must inform each other of their respective lockout or tagout procedures. This would include contracted employees. The on-site employer must ensure that his or her employees understand and comply with the restrictions and prohibitions of the outside employer's energy control program.

QUESTIONS

1. Who developed the lockout/tagout standard?
2. What is the purpose of the standard?
3. List at least two examples of stored energy.
4. List at least three sources of energy that are typically found in an industrial environment.
5. Define the term *lockout*.
6. Define the term *tagout*.
7. What is an authorized employee?
8. What is an affected employee?
9. What is an energy-isolating device?
10. Are normal production operations covered by the standard?
11. What is a hasp used for?
12. Describe the typical steps in a lockout/tagout procedure.
13. Write a lockout/tagout procedure for a cell that contains electric and pneumatic energy.

APPENDIX A

Starting a New Project in ControlLogix

There are three basic steps in starting a new project. First you name the project and configure the project for the correct CPU and slot, software revision, and chassis type. Next you should set a path to the CPU. Third you add the required I/O modules. RSLinx should be configured with the correct Communications Driver to be able to communicate with the CLX controller you would like to program.

Open RSLOGIX 5000 and Select File and then New Project. The screen shown in Figure A-1 should appear. First you must choose the correct type of processor. In this

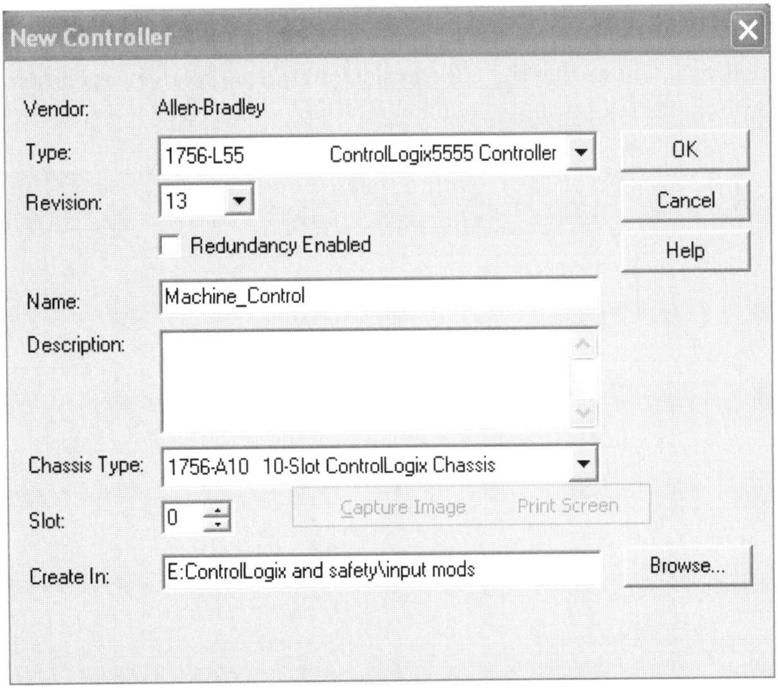

Figure A-1 New project configuration screen.

example the processor is a 1756-L55. Next choose the correct RSLogix 5000 Revision (software level). It is 13 for this example. Next you must name the project. Machine_Control was the name chosen in this example. You must also be sure to choose the correct chassis type. You may change the location where it will be created. In this example the path chosen was E:\ControlLogix and safety\input mods.

A path to the controller's CPU should be set next (see Figure A-2). In this figure the path is blank yet. You can choose the down arrow of the RSWho icon to the right to choose a path. If you select the RSWho icon, the screen in Figure A-3 will appear.

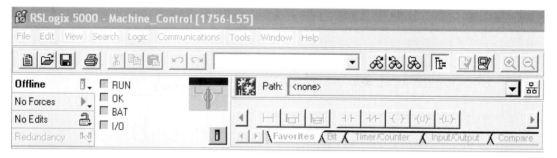

Figure A-2 Project screen. Note that the path has not been set yet.

Click on the CPU you would like to program and then choose Set Project Path. Figure A-3 shows that the CPU in slot 0 in the backplane was chosen for the CPU. In this example the path went through an RSLINX Ethernet devices driver.

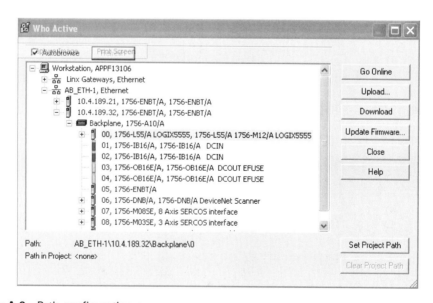

Figure A-3 Path configuration screen.

Then you can select Close to close the RSWho screen and you will see that the path now appears as shown in Figure A-4.

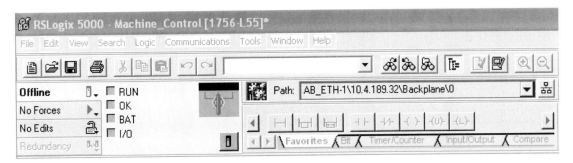

Figure A-4 Note that the path has been set.

Next you should add any modules that the application will require. Figure A-5 shows the Controller Organizer. Note that I/O Configuration is shown on the bottom but no modules have been added yet.

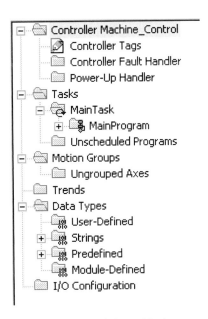

Figure A-5 Controller Organizer with no modules added yet.

To add a module, you must right-click on the I/O Configuration icon and choose New Module. The screen shown in Figure A-6 will then appear. You can choose the module from the list of available modules. In this example a 1756-IB16 was chosen for the input module. Select OK, and the screen in Figure A-7 will appear.

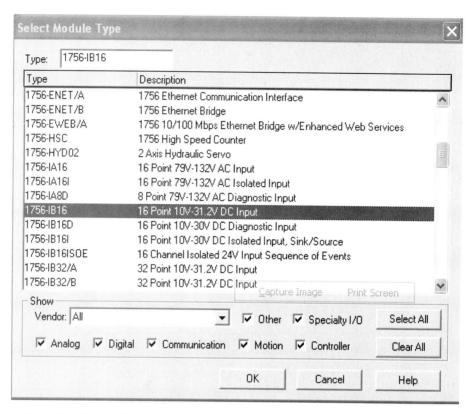

Figure A-6 List of modules that can be added. Note that a 1756-IB16 input module is highlighted in the list.

Next you must choose the Major Revision for the module you are adding. In this example the Major Revision was 2. The revision level is shown on the side of the module. You can also use RSLinx to find the revision level of the module. The use of RSLinx will assure that you have the right revision in case the module firmware was upgraded.

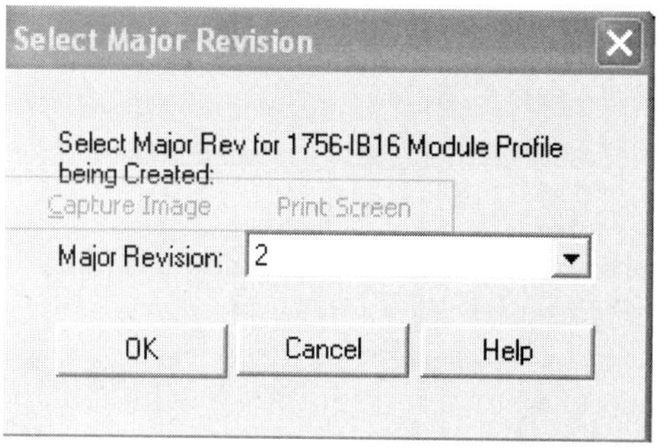

Figure A-7 Select Major Revision screen.

Figure A-8 shows the Module Properties screen for the input module that is being added. The name entered was Input_Mod_1. It is located in slot 1. The keying method chosen was Compatible Module.

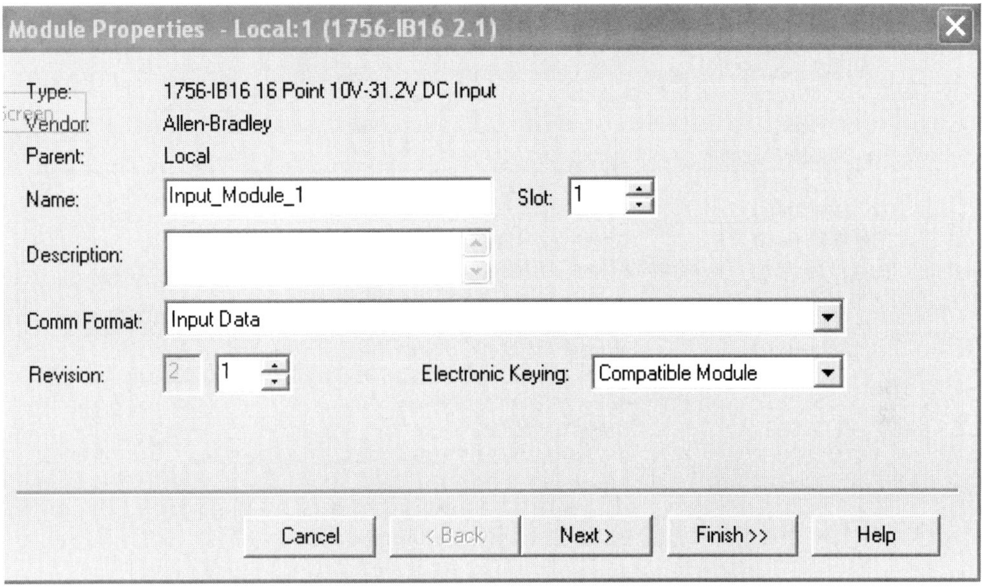

Figure A-8 Module Properties screen for the input module.

Figure A-9 shows the Controller Organizer after the input module was added.

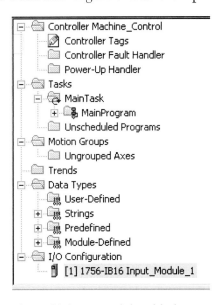

Figure A-9 Controller Organizer with input module added.

Next the output module will be added. Right click the I/O Configuration icon in the Controller Organizer and choose New Module. The screen shown in Figure A-10 shows the module selection screen that will appear. The output module for this example is a 1756-OB16E output module. This is a 16-output electronically fused module. Then select OK.

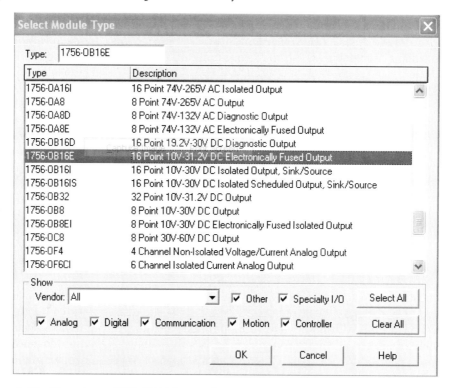

Figure A-10 Note that a 1756-OB16E was chosen in this example.

Next the Select Major Revision screen will appear (see Figure A-11). The Major Revision level for this module was 2. Select OK.

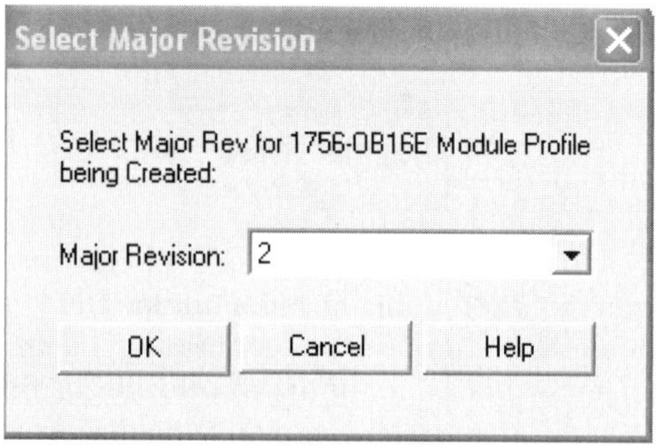

Figure A-11 Select Major Revision screen.

CAPPENDIX A—STARTING A NEW PROJECT IN CONTROLLOGIX   429

The Module Properties screen then appears (see Figure A-12). Enter a name for the module, choose the slot where it is located, enter the minor Revision level, and choose the keying method for this module. Then select Finish.

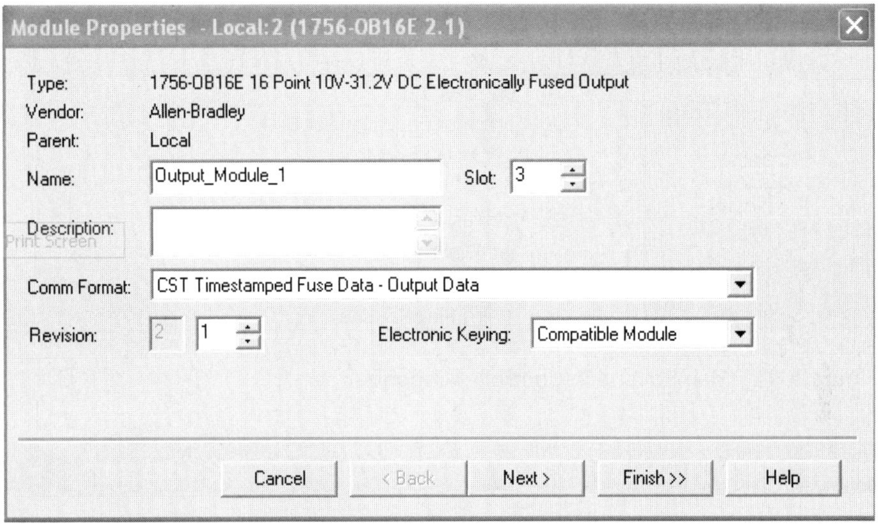

Figure A-12 Module Properties screen.

Note that this module was named Output_Module_1 in this example. The module is in slot 2. Compatible Module was chosen for the method of keying.

Figure A-13 shows the Controller Organizer after the modules have been added.

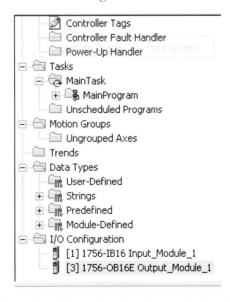

Figure A-13 Controller Organizer after an input and an output module were added.

You are now ready to program. Click on the + next to the MainProgram folder in the Controller Organizer shown in Figure A-13. Program Tags and MainRoutine will appear in the Controller Organizer as shown in Figure A-14. Double-click on MainRoutine, and the programming screen will open up as shown in Figure A-15. You are now ready to program.

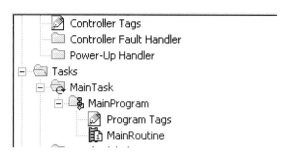

Figure A-14 MainRoutine in Controller Organizer.

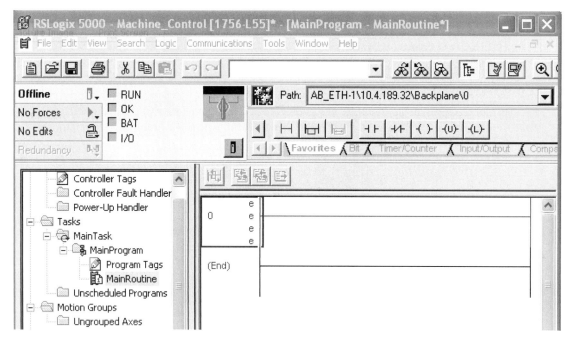

Figure A-15 Programming screen.

APPENDIX B
Configuring I/O Modules in a Remote Chassis

A controller in one chassis can own modules in another chassis. Remember that output modules can only have one owner but multiple controllers can own the same input module as long as they are configured exactly the same.

Figure B-1 shows an example of a simple system. There are two CLX systems in the figure. This appendix will show how to configure a project so that the controller in one chassis will own and control an output module in a different chassis. The project will be configured so that the controller on the left of Figure B-1 will own the output module in slot 3 of the CLX system on the right. Remember that no other controller can own the module. Note also that there is an Ethernet Bridge module in each chassis. ControlNet would also work, but this example uses Ethernet.

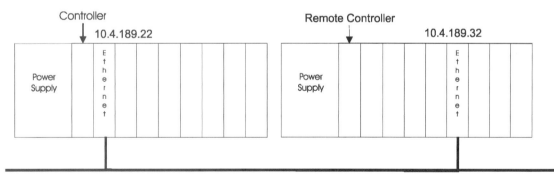

Figure B-1 Two CLX systems. Note that each chassis has an Ethernet module.

It is a relatively straightforward process to configure a project to use remote I/O. It really just involves adding cards that will provide a path under the I/O configuration for the project. Consider the system in Figure B-1. The path from the controller would be through the backplane of the Ethernet module that resides in the same chassis, then out the Ethernet module to the Ethernet module in the remote chassis, and then through the backplane to the module in slot 3. The path involves three modules and begins with the Ethernet module in the controller's chassis.

Figure B-2 shows the Project Organizer screen. You need to add modules under the I/O Configuration folder. To do this, you right-click on I/O Configuration and choose New Module. The screen shown in Figure B-3 will then appear.

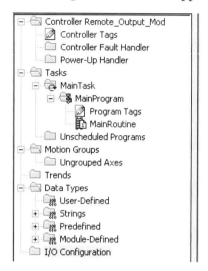

Figure B-2 Controller Organizer before any modules were added.

Select the communications module. In this example it is a 1756-EBET/B module. Then click on OK.

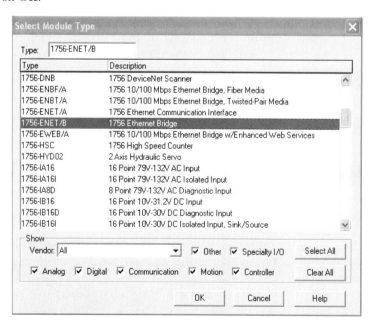

Figure B-3 Select Module Type screen.

Next the module must be configured (see Figure B-4). The IP address for this module is 10.4.189.22. The module is in slot 1 of the chassis so 1 was entered for Slot. The correct Revision level should be chosen for the module. A name is also entered for the module. In this example the module was named Ethernet_Local. Compatible Module was chosen for the Electronic Keying type. Then Finish can be selected, and the screen in Figure B-5 will appear.

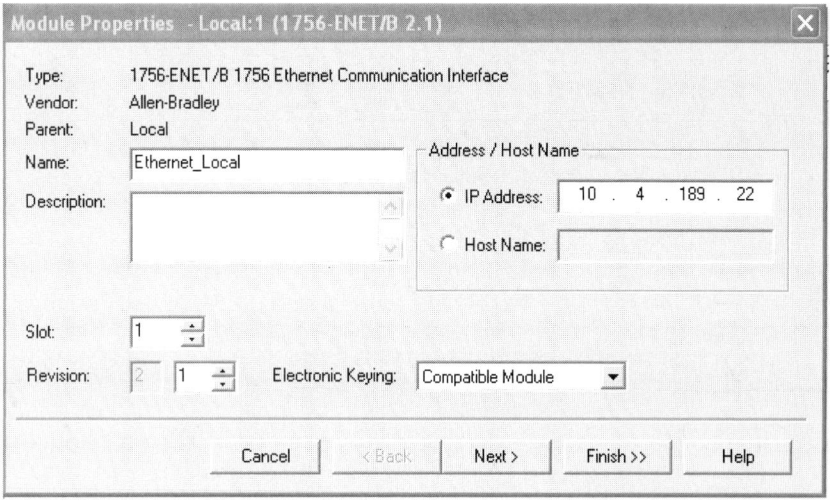

Figure B-4 Module Properties screen.

Figure B-5 shows the Controller Organizer after the Ethernet module was added. Notice the name of the module, Ethernet_Local.

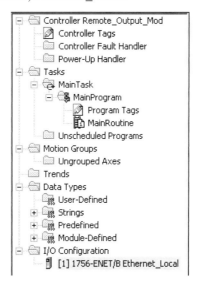

Figure B-5 Controller Organizer after the Ethernet module was added.

Next you must add the remote Ethernet module. This must be added under the Local Ethernet module that was just added. Right-click on the Ethernet module (Ethernet_Local) that was just added and choose New Module. The screen in Figure B-6 will appear. Choose the correct module. In this example the 1756-ENET/B module was chosen. Choose OK and the screen in Figure B-7 will appear.

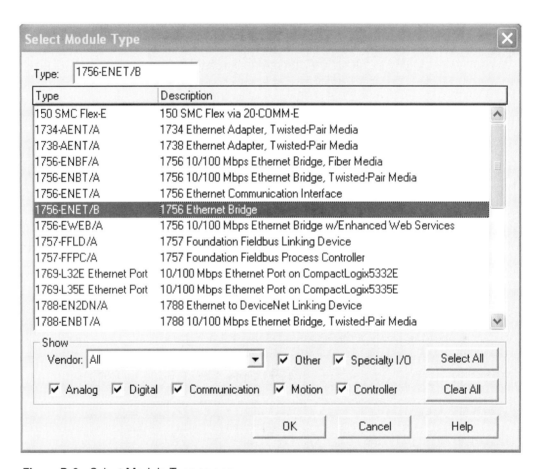

Figure B-6 Select Module Type screen.

Next the module's properties must be configured (see Figure B-7). The name Ethernet_Remote was entered for the name. The IP address was entered (10.4.189.32). The module is in slot 5, so 5 was entered for Slot. Make sure the correct Revision level is set. Compatible Module was chosen for the Electronic Keying type. Select Finish, and the screen in Figure B-8 will appear.

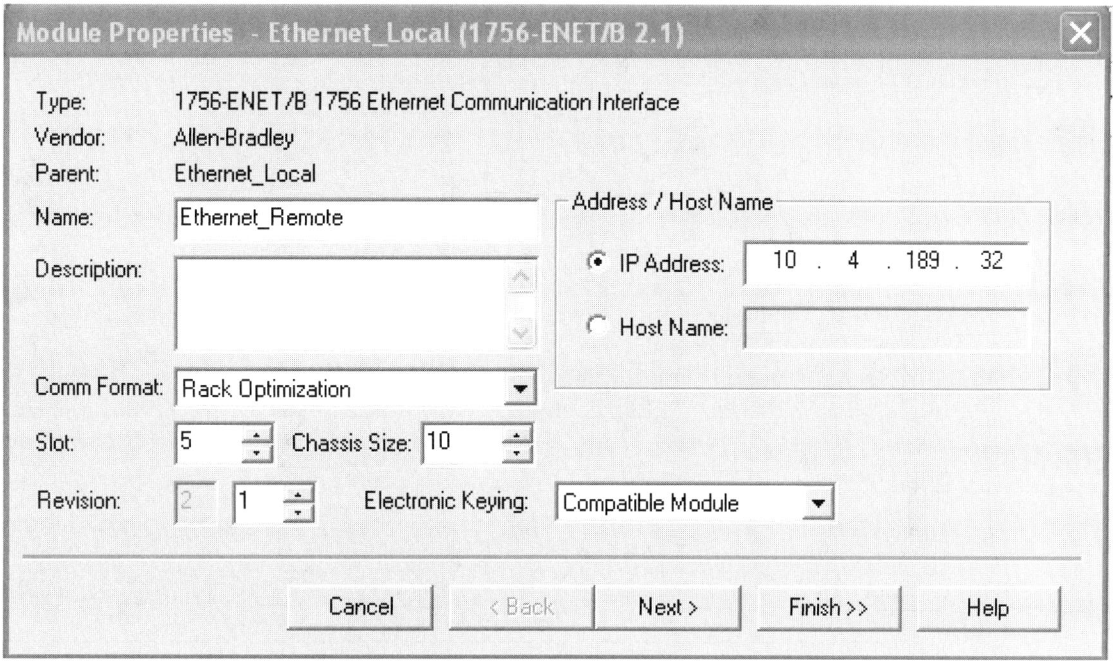

APPENDIX B—CONFIGURING I/O MODULES IN A REMOTE CHASSIS 435

Figure B-7 Module Properties screen.

Figure B-8 shows the Controller Organizer after the remote Ethernet module was added. Note the name on the module is Ethernet_Remote. Also note that it was added under the first Ethernet module.

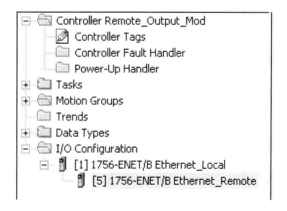

Figure B-8 Controller Organizer after the remote Ethernet module was added.

When this module is added, controller tags are automatically generated as shown in Figure B-9. They can be seen in the tag editor.

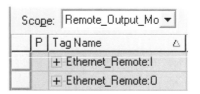

Figure B-9 Controller tags for the Ethernet module.

Next you need to add the actual I/O module, in this example an output module. Right-click on the second Ethernet module that was added. Choose New Module and the screen shown in Figure B-10 will appear. In this example a 1756-OB16E output module was chosen. Then select OK.

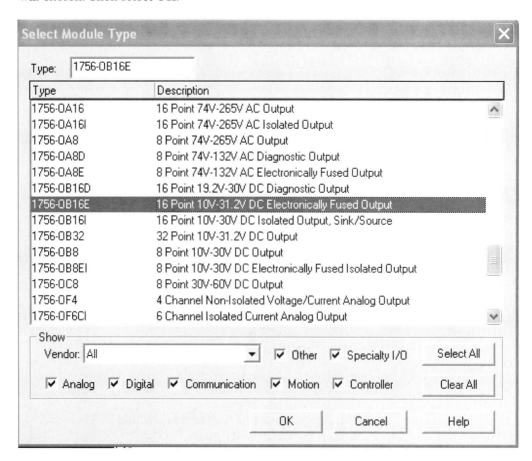

Figure B-10 Select Module Type screen.

APPENDIX B—CONFIGURING I/O MODULES IN A REMOTE CHASSIS   **437**

The correct Major Revision level is set next (see Figure B-11). Choose OK, and the screen shown in Figure B-12 will appear.

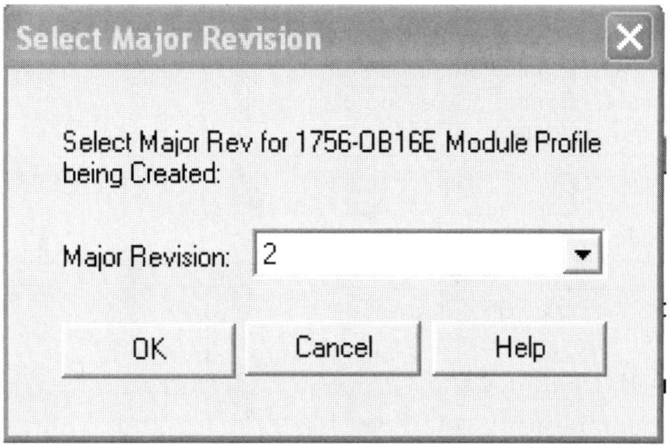

Figure B-11 Major Revision level entry screen.

In Figure B-12 the module was named Remote_Output_Module. The module resides in slot 3. Compatible Module was chosen for the Electronic Keying type. A Comm (communications) Format must be chosen. In this example Rack Optimization was chosen. Rack Optimization and the other choices are covered in Figure B-13. Choose Finish, and the screen in Figure B-14 will appear.

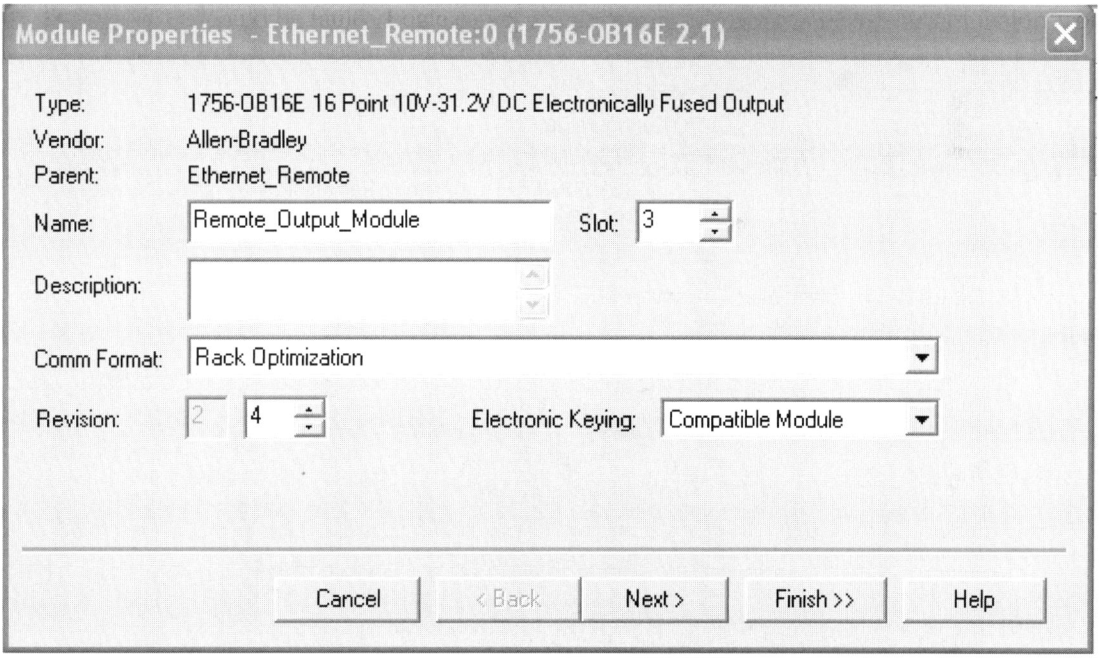

Figure B-12 Module Properties screen.

APPENDIX B— CONFIGURING I/O MODULES IN A REMOTE CHASSIS

If	Select
The remote chassis contains only analog modules, diagnostic digital modules, fused output modules, or communication modules.	None
The remote chassis only contains standard, digital input and output modules (no diagnostic or fused output modules).	Rack Optimization
You want to receive I/O module and chassis slot information from a rack-optimized remote chassis owned by another controller	Listen-Only Rack Optimization

Figure B-13 Comm Format choices.

Figure B-14 shows the Controller Organizer after the output module was added. Notice the name on the module (Remote_Output_Module). The configuration is complete at this point. If you look in the tag editor screen you will see that tags have been automatically added for the module. The actual outputs are in the tag named Ethernet_Remote:3:O (see Figure B-15).

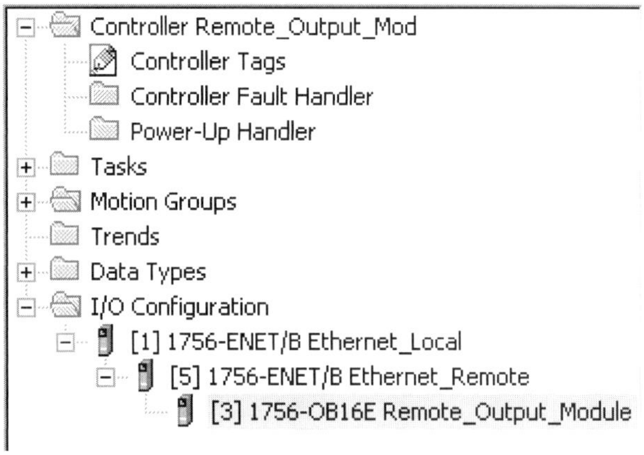

Figure B-14 Controller Organizer after the output module was added.

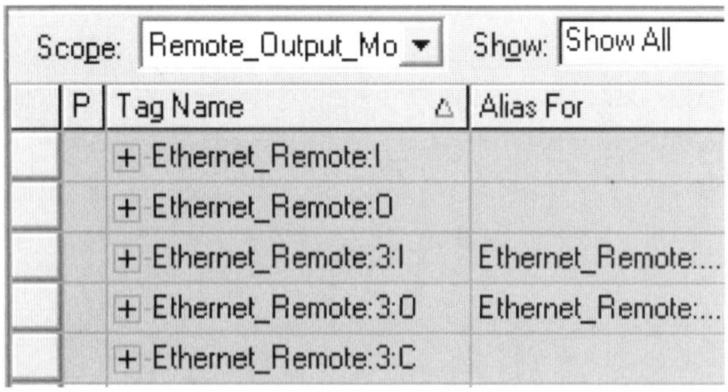

Figure B-15 Tag editor screen showing the tags that were added when the module was added.

APPENDIX C

The Use of Producer/Consumer Tags

Producer/consumer tags provide an easy means for a processor to provide data to other controllers. By using the producer/consumer model, data can be transferred between processors without any logic. The user can choose the Requested Packet Interval (RPI) for the rate at which data should be updated. Data can be transferred between DINT-type tags, an array, or a User-Defined data type.

THE PRODUCED TAG

The produced tag is the easiest part to configure. In the CLX controller that will be the producer, the user simply creates a controller Scope tag and configures it as produced. The name of the produced tag in this example is Produced_Tag (see Figure C-1). It is a DINT. Note the check mark in the P column. P stands for Produced.

Marking a tag produced enables the tag to be available to consumed tags in other controllers.

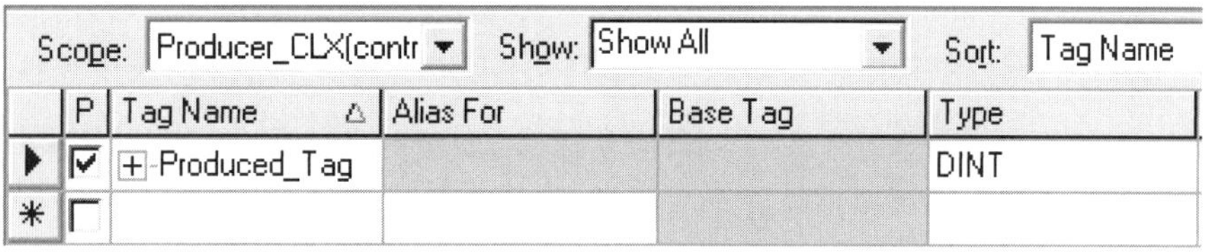

Figure C-1 A controller Scope tag named Produced_Tag. Note the check mark in the P column. This makes the tag a produced tag.

Figure C-2 shows how it appears in the Tag Properties screen.

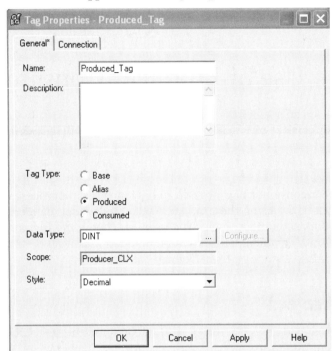

Figure C-2 Tag Properties screen for the tag. Note that the Tag Type is Produced. Note also the Connection tab on top of the screen.

If you click on the Connection tab, you can set the maximum number of consumers that are allowed to connect to the produced tag (see Figure C-3). The allowed values are 1 to 256. One other controller will be allowed access to this tag in this example. You can also send event triggers to consumers using the IOT instruction in logic.

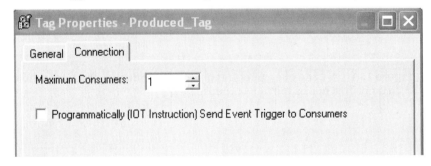

Figure C-3 Tag Properties, Connection parameters. In this example only one other controller (consumer) will be allowed to get the data for this tag because 1 was entered in Maximum Consumers.

That is all you have to do in the Producer controller.

THE CONSUMED TAG

Next a project must be developed for the controller who will be the consumer of tag data from the produced tag. Create a new project in a different controller. Set a path and add any I/O modules you need for your application. In this example only communication modules will be required.

First, you must create a communications path to the processor where the tag is being produced. Then the consumed tag can be created.

Figure C-4 shows an illustration of the two CLX systems. Note that each chassis has a controller and an Ethernet module. They are both attached to the Ethernet network. Information for this example will use the Ethernet communications network. Other networks such as ControlNet could have been used. Note that the Ethernet module is in slot 7 in the Producer Controller and in slot 6 in the Consumer Controller.

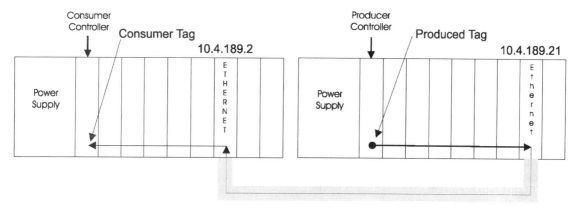

Figure C-4   The CLX systems used in this example.

In the CLX project organizer (see Figure C-5), a path will need to be configured to the processor that has the produced tag. Referring back to Figure C-4, you see the module path from the consumer CLX must be established to the producer CLX. The path will be through the Ethernet module in the consumer CLX chassis, through the Ethernet module in the producer chassis, and finally to the controller (CPU) in the producer chassis.

To add the Ethernet module in the consumer chassis, right-click on the I/O Configuration folder in the Controller Organizer (see Figure C-5). After you right-click, you will choose New Module and the screen shown in Figure C-6 will appear.

APPENDIX C—THE USE OF PRODUCER/CONSUMER TAGS

Figure C-5 Controller Organizer before the modules have been added to create a path to the Produced tag.

Next you must choose the correct module. The module in the consumer chassis is a 1756-ENET/B module. This is a 1756 Ethernet Bridge module. Then select OK. The screen shown in Figure C-7 will appear so that you can configure the module.

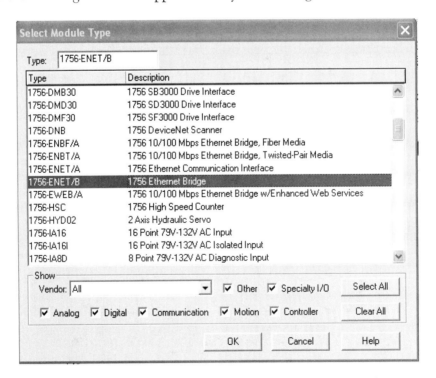

Figure C-6 The screen to select the module type.

In the Module Properties screen in Figure C-7 you enter a name for the module. You must enter the IP address of the Ethernet module. You also enter the Slot that the module resides in and the Revision. You can choose the keying method for the module.

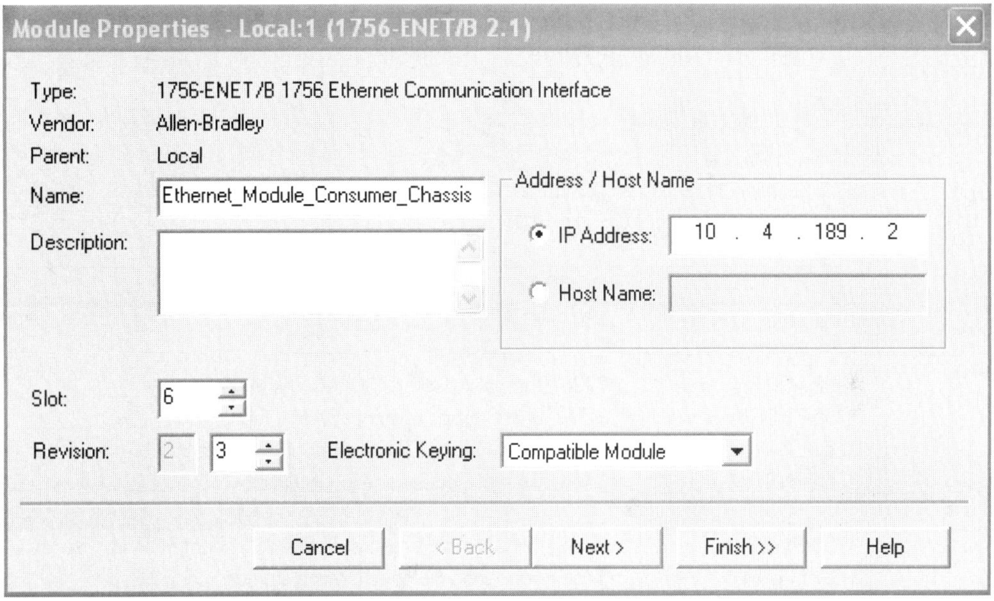

Figure C-7 Module Properties screen.

When you choose Finish, the Controller Organizer displays the module as shown in Figure C-8.

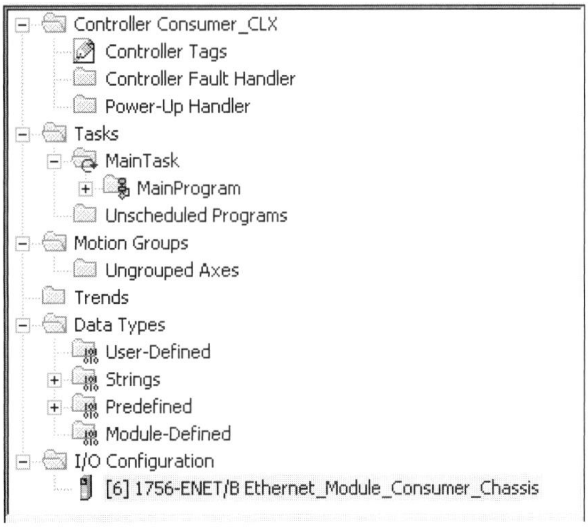

Figure C-8 Controller Organizer after the module was added.

Next you must add the Ethernet module in the other Chassis (producer CLX chassis). This one will be added under the first Ethernet module. Right-click on the Ethernet module that you just added. Choose New Module.

Next you will choose the correct module as shown in Figure C-9. The module in this example is a 1756-ENET/B module. This is a 1756 Ethernet Bridge module. Then select OK. The screen shown in Figure C-10 will appear.

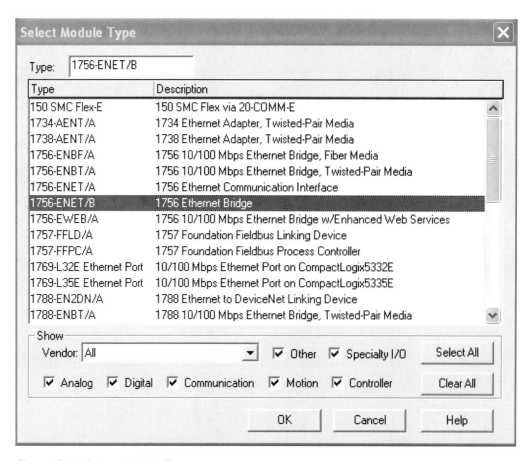

Figure C-9 Select Module Type screen.

In the Module Properties screen in Figure C-10, you enter a name for the module. You must enter the IP address of the Ethernet module. You also enter the Slot that the module resides in and the Revision level. You can choose the Electronic Keying method for the module. The last thing to choose is the Comm Format. Rack Optimization was chosen in this example.

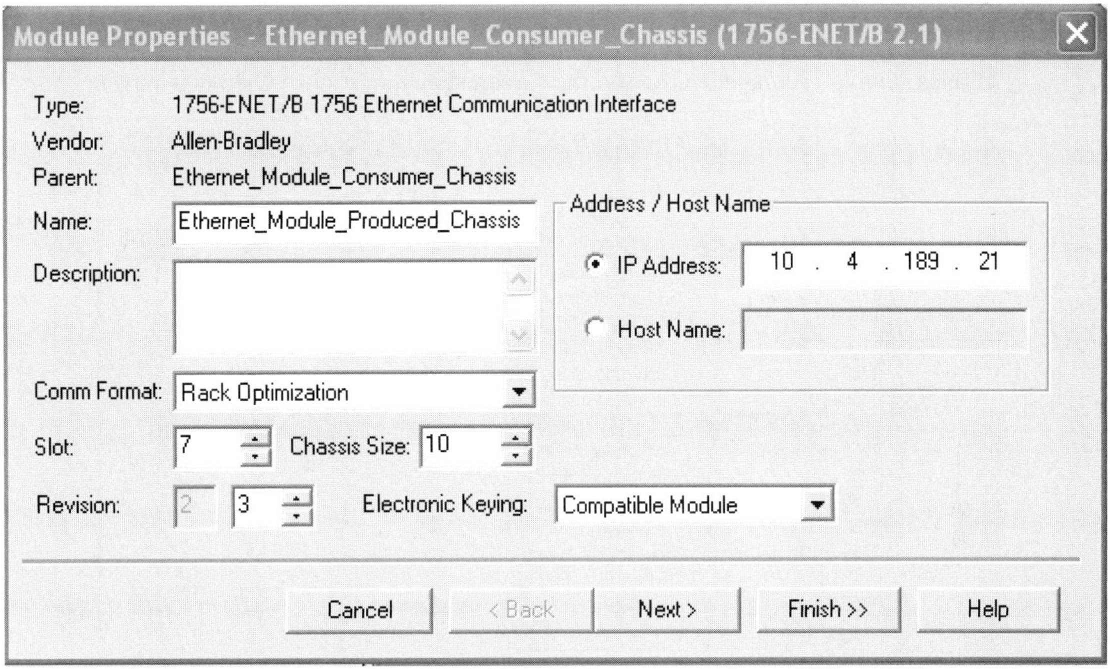

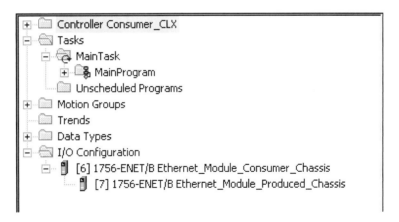

Figure C-10 Module Properties screen.

Figure C-11 shows the Project Organizer screen after the Ethernet module in the producer chassis was added.

Figure C-11 Project Organizer screen after the Ethernet module in the producer chassis was added.

So far you have a path from the Ethernet module in the consumer CLX chassis to the Ethernet module in the producer chassis. Lastly you must add the controller (CPU) that produces the tag. Right-click on the Ethernet module you just added and choose New Module. The screen in Figure C-12 will appear. Choose the type of controller, 1756-L55 in this example. Then click OK, and the screen shown in Figure C-13 will appear.

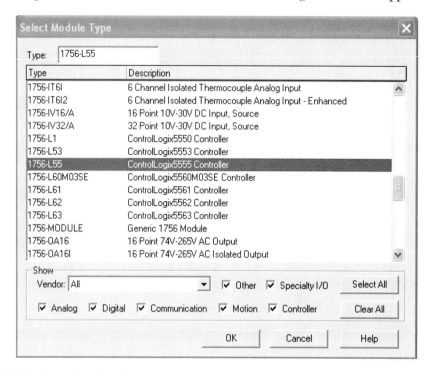

Figure C-12   Select Module Type screen.

Choose the correct Major Revision level for the producer controller.

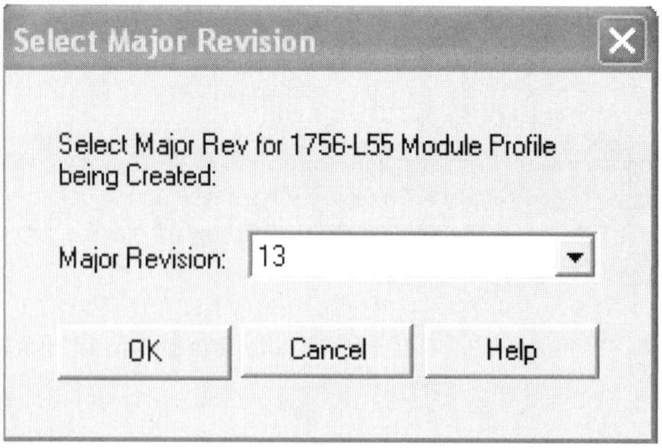

Figure C-13   Select Major Revision level screen.

After you enter the Major Revision level, click OK, and the screen shown in Figure C-14 will appear. Enter a name for the producer controller as well as the Slot that it resides in. Then click the Finish button.

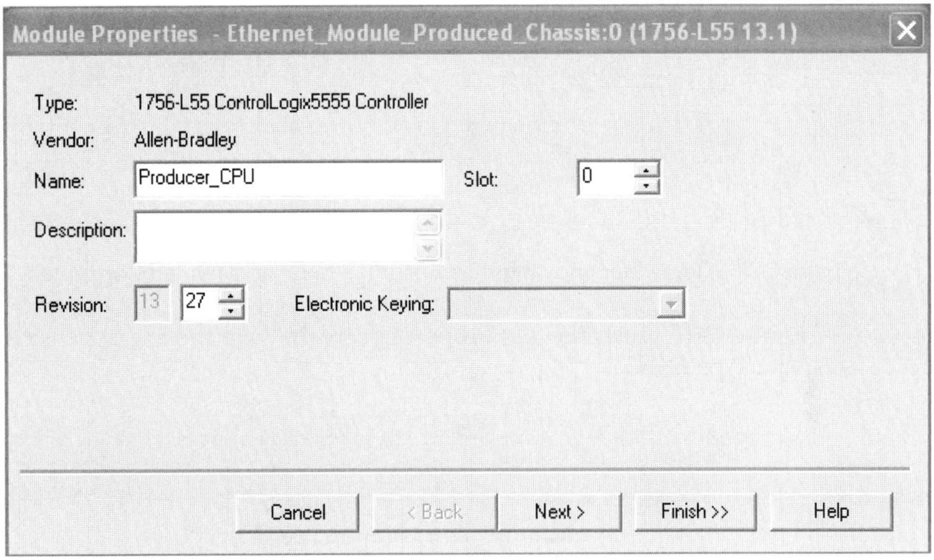

Figure C-14 Module Properties screen.

Now the Project Organizer screen looks like Figure C-15. Note that this now provides the path to get the data.

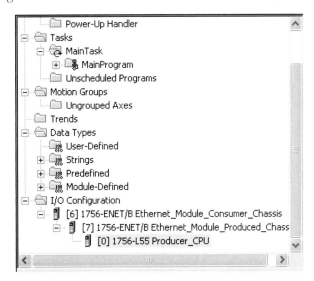

Figure C-15 Project Organizer screen after the producer controller module was added.

The last thing to do is create the consumed tag. In this example the consumed tag was named Consumed_Tag (see Figure C-16). It is a DINT type for this example.

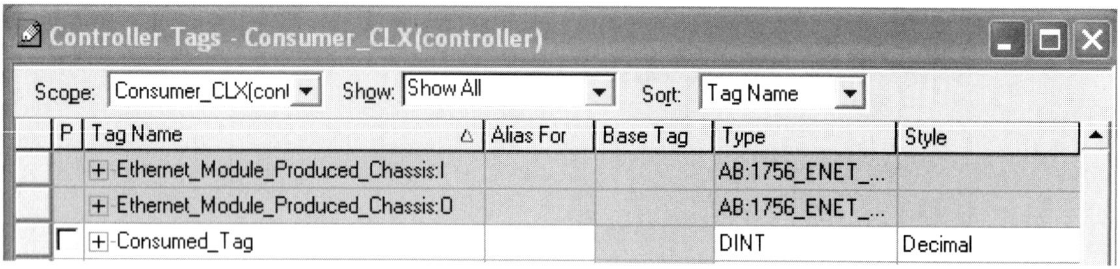

Figure C-16 Tag editor showing the consumed tag that is named Consumed_Tag.

Figure C-17 shows the Tag Properties screen. Note the Tag Type is Consumed and the Data Type DINT.

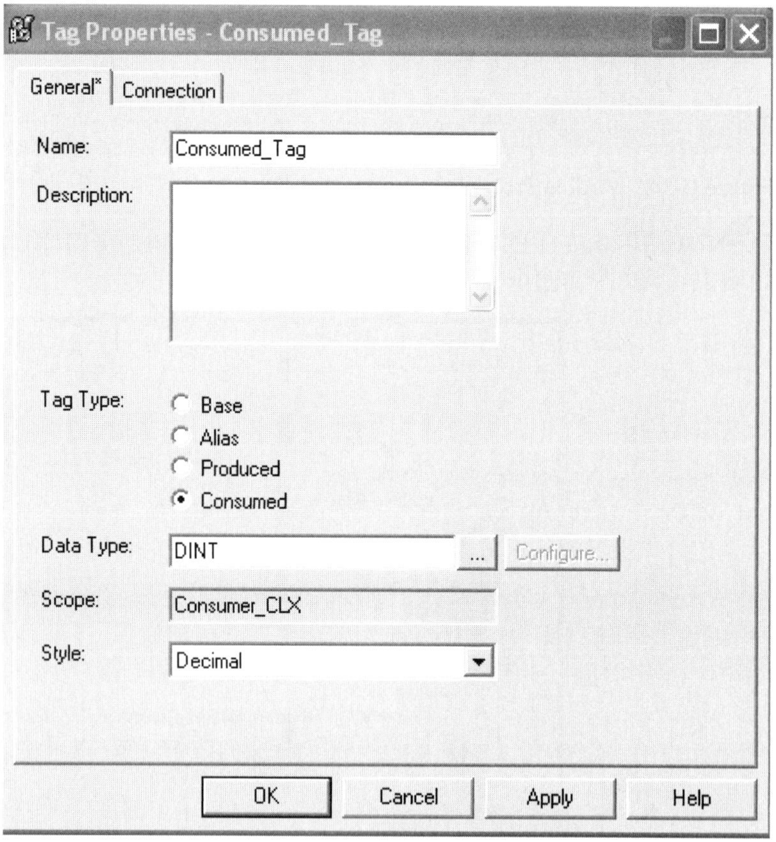

Figure C-17 Tag Properties screen.

Next click on the Connection tab on the Tag Properties screen. The screen shown in Figure C-18 appears. In this screen the Producer controller is chosen from the drop-down list (Producer_CPU in this example). Then the name of the produced tag in the remote controller is entered (Produced_Tag in this example). A value is also entered for the RPI (5 in this example). This determines how often the tag is updated.

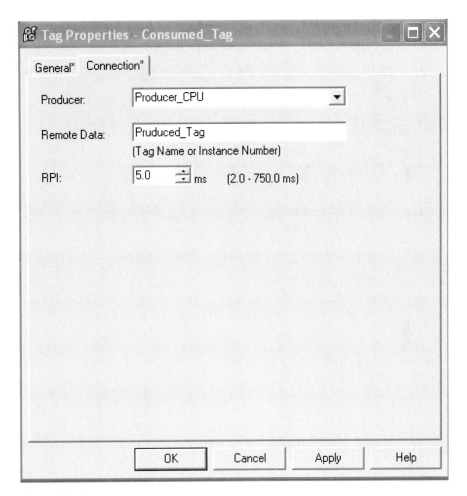

Figure C-18 Tag Properties screen.

At this point both controllers should be put into run mode. Type a number into the produced tag. Figure C-19 shows the number 55 has been entered into the Produced_Tag in the tag editor. The number should then appear in the consumed tag in the other controller. Figure C-20 shows the tag editor screen in monitor mode.

450 APPENDIX C—THE USE OF PRODUCER/CONSUMER TAGS

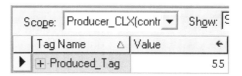

Figure C-19 Monitor mode in the tag editor screen in the producer controller.

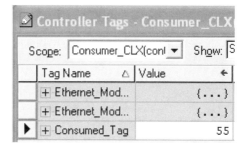

Figure C-20 Monitor mode in the tag editor screen in the consumer controller.

APPENDIX D

ControlLogix Messaging

This appendix will show how messaging can be configured and accomplished between two CLX controllers.

The producer/consumer model is very efficient for transferring data between processors, but if the data transfer does not need to occur at periodic intervals, you may be able to conserve network bandwidth using the message (MSG) instruction. Using the MSG instruction, data can even be received (or sent) from another processor, even if that processor is not present in the I/O Configuration tree.

For this example, consider an MSG instruction that will read data from another ControlLogix processor and store that data in a memory location in your controller.

Below, you can see the path to take to connect to the target processor. Once the connection is made, the Temp_In array tag in the controller on the left will receive data from the Temp_Remote array tag in the controller on the right each time the MSG instruction is executed.

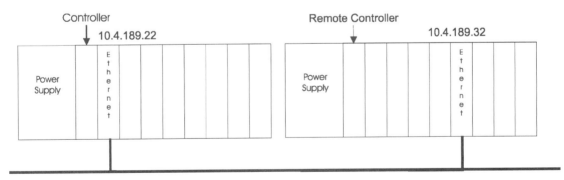

Figure D-1 The two CLX systems.

CREATE REMOTE TAGS

The first step is to create the tag in the remote CLX controller. In this example the tag is in the controller on the left of Figure D-1. Figures D-2 and D-3 show a DINT-type array tag named Temp_Remote was created and has ten members. It was created as a controller Scope tag. The Temp_Remote[0] array will be the memory location that another controller can read from. Data will be put in this tag's members, so the controller that reads from the tag can test the connection.

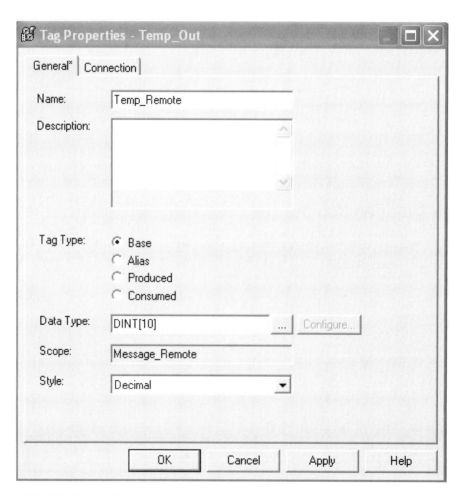

Figure D-2 Tag Properties screen.

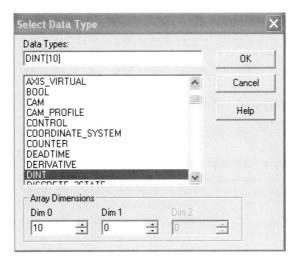

Figure D-3 Select Data Type screen.

SETTING UP THE CLX CONTROLLER FOR MESSAGING

Next you need to configure the CLX controller that will be utilizing the MSG command to get information. You need to establish a communications path for it to access the information from the remote CLX controller. You will utilize the Ethernet modules in the two controller chassis for the communications path. Figure D-4 shows the Controller Organizer before any modules have been added.

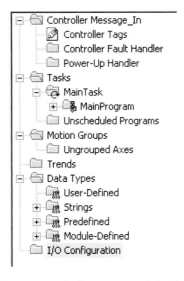

Figure D-4 Controller Organizer screen before any modules have been added.

The first module that was added in this example was the Ethernet module. To do this, right-click on the I/O Configuration shown in Figure D-4. Choose New Module. The screen shown in Figure D-5 appears. A 1756-ENET/B module is chosen. Click on OK, and the screen in Figure D-6 appears.

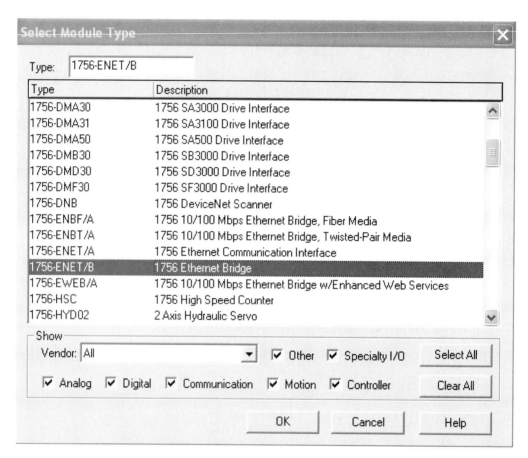

Figure D-5 Select Module Type screen.

In the screen shown in Figure D-6 the IP address was entered (10.4.189.22). The Ethernet module is located in slot 1 so 1 was entered in the Slot. Compatible Module was chosen for the Electronic Keying method. Click on the Finish button, and the screen in Figure D-7 appears.

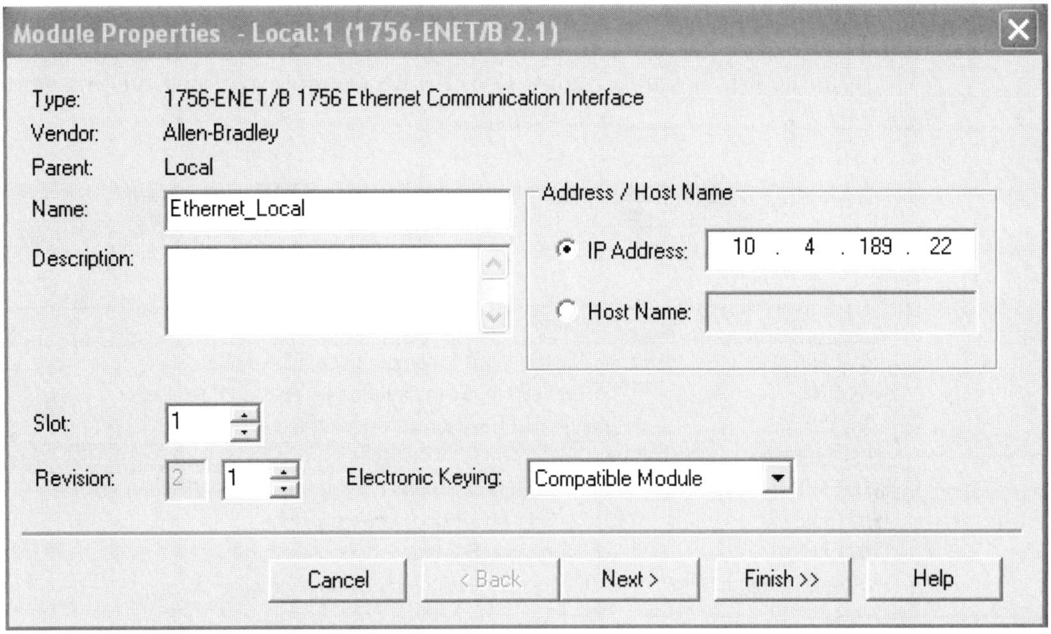

APPENDIX D— CONTROLLOGIX MESSAGING

Figure D-6 Module Properties screen.

Figure D-7 shows the Controller Organizer after the local Ethernet module was added.

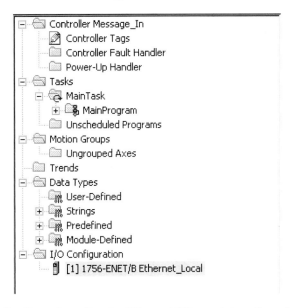

Figure D-7 Controller Organizer after the Ethernet (Ethernet_Local) module was added.

Next you must add the Ethernet module in the remote chassis. Right-click on the Ethernet module that was just added and choose New Module. Choose the correct module from the list shown in Figure D-8. In this example a 1756-ENET/B module was chosen.

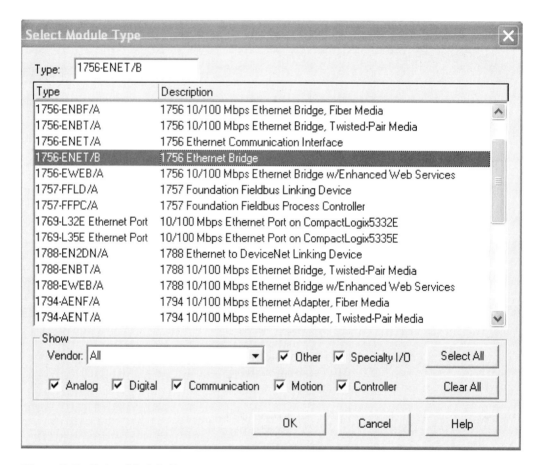

Figure D-8 Select Module Type screen.

The IP address for the remote Ethernet module was entered (10.4.189.32) in this example (see Figure D-9). A 5 was entered for the Slot, and Compatible Module was chosen for the Electronic Keying method.

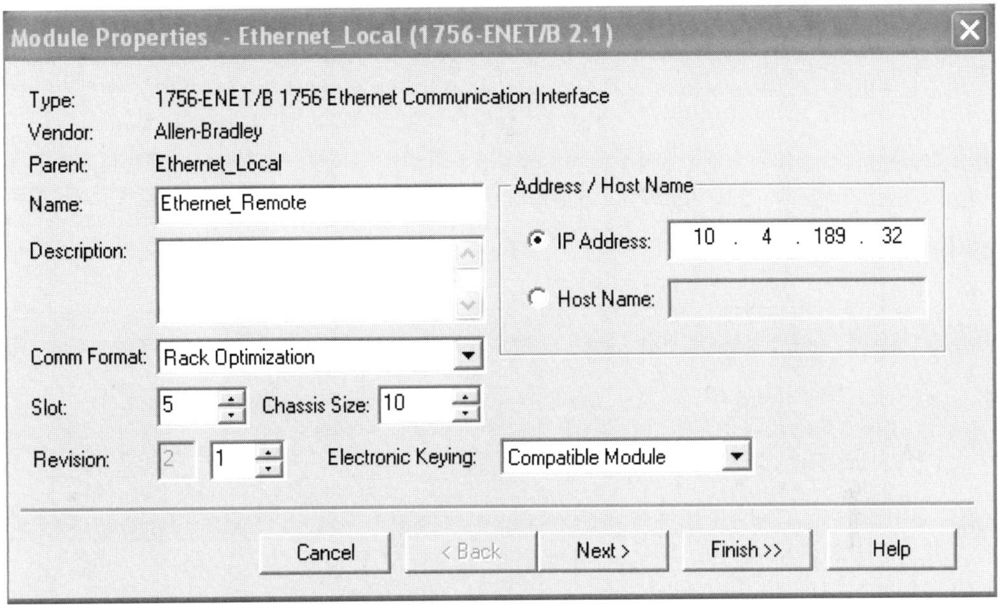

APPENDIX D— CONTROLLOGIX MESSAGING 457

Figure D-9 Module Properties screen.

Figure D-10 shows the Controller Organizer after the second Ethernet module was added. Notice it was added under the first Ethernet module.

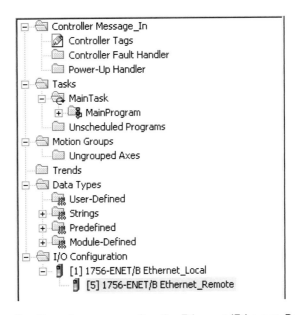

Figure D-10 Controller Organizer screen after the Ethernet (Ethernet_Remote) module was added.

The last thing that needs to be done is to add the CLX controller module that is located in the remote chassis. Right-click on the Ethernet module that was just added (see Figure D-10). Choose New Module. Choose the correct controller from the list shown in Figure D-11. In this example 1756-L55 was chosen. Then select OK.

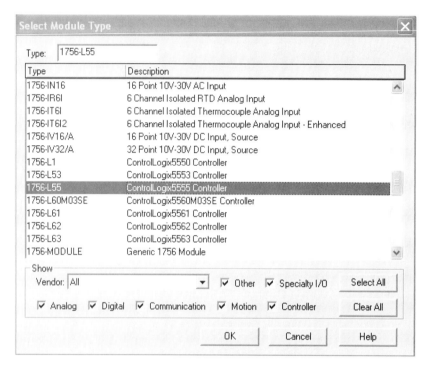

Figure D-11 Select Module Type screen.

Enter the Major Revision as shown in Figure D-12, and then choose OK.

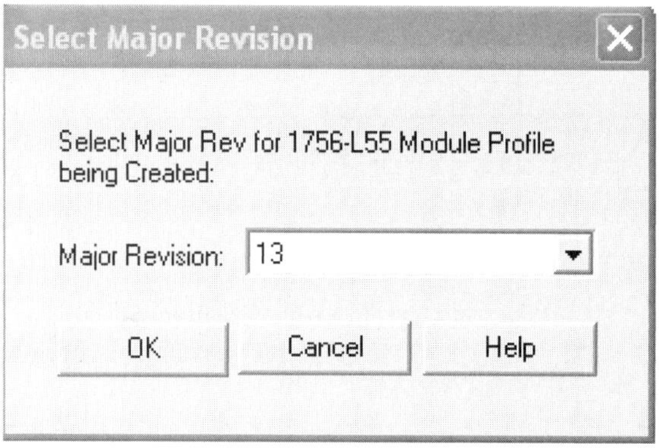

Figure D-12 Select Major Revision level screen.

Next enter the name for the module as shown in Figure D-13. Choose the correct Slot, 0 in this example. Then choose Finish.

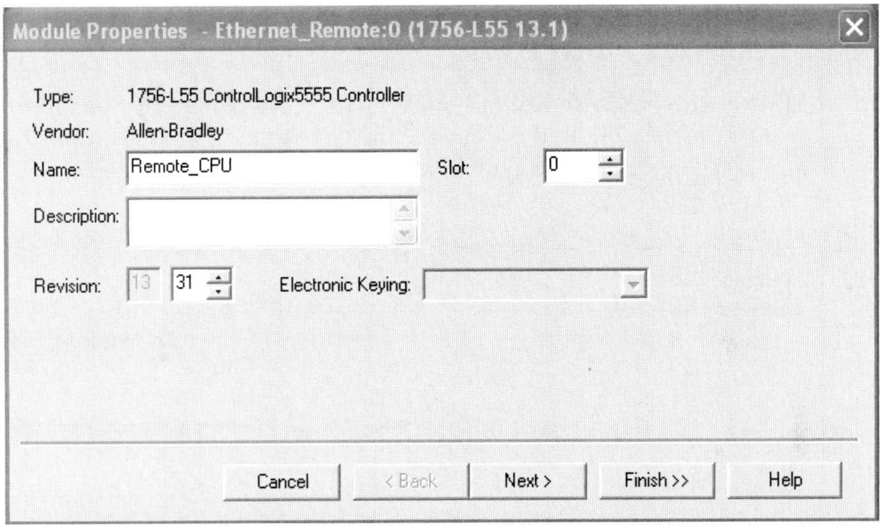

Figure D-13 Module Properties screen.

The Controller Organizer now shows the controller has been added (see Figure D-14).

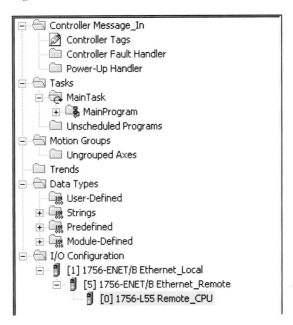

Figure D-14 Controller Organizer screen after the controller (Remote_CPU) module was added.

The path is now complete. Next the logic can be developed and a tag to receive the data from the remote tag can be entered.

MESSAGE COMMAND AND LOGIC

A MSG instruction is shown in Figure D-15. A tag name must be entered for a control tag for the instruction. In this example the tag is named MSG_Control_Tag and its type is Control. When this instruction is true, it will read the specified tag in the remote controller and put it in the local tag that will be specified.

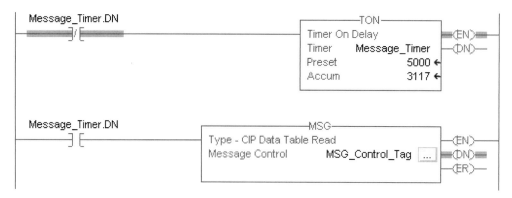

Figure D-15 Logic for the MSG instruction.

Click on the ellipsis on the MSG instruction to configure it. The screen shown in Figure D-16 will appear. CIP Data Table Read was chosen for the type. The Source Element is the name of the tag in the remote controller. The Number Of Elements to be read is 10 as the tag is an array tag with 10 members. The Destination Element is the name of the tag to which the tag data from the remote controller should be written (Temp_In[0] in this example). The table in Figure D-17 shows the possible choices for the MSG instruction.

APPENDIX D— CONTROLLOGIX MESSAGING

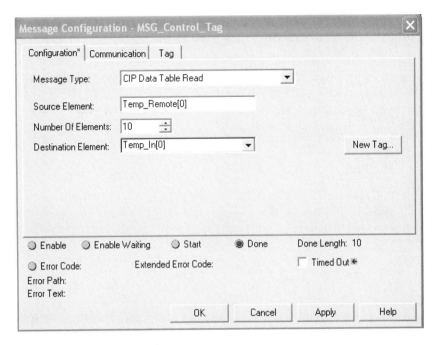

Figure D-16 Message Configuration screen.

Target Device to Communicate With	Select
Logix5000 Controller	CIP Data Table Read
	CIP Data Table Write
I/O Module that you configure using RSLogix 5000 software	Module Reconfigure
	CIP Generic
PLC-5 Controller	PLC5 Typed Read
	PLC5 Typed Write
	PLC5 Word Range Read
	PLC5 Word Range Write
SLC Controller	SLC Typed Read
MicroLogix Controller	SLC Typed Write
Block Transfer Module	Block-Transfer Read
	Block-Transfer Write
PLC-3 Processor	PLC3 Typed Read
	PLC3 Typed Write
	PLC3 Word Range Read
	PLC3 Word Range Write
PLC-2 Processor	PLC2 Unprotected Read
	PLC2 Unprotected Write

Figure D-17 Message Type choices.

462 APPENDIX D— CONTROLLOGIX MESSAGING

Remote_CPU was entered for the path in Figure D-18. This is the name of the controller in the remote chassis (see Figure D-14).

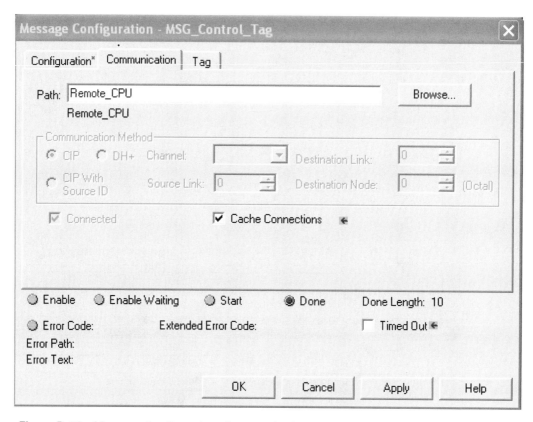

Figure D-18 Message Configuration, Communication screen.

The Tag tab was then chosen in the Message Configuration screen. MSG_Control_Tag was entered for the name of the control tag (see Figure D-19).

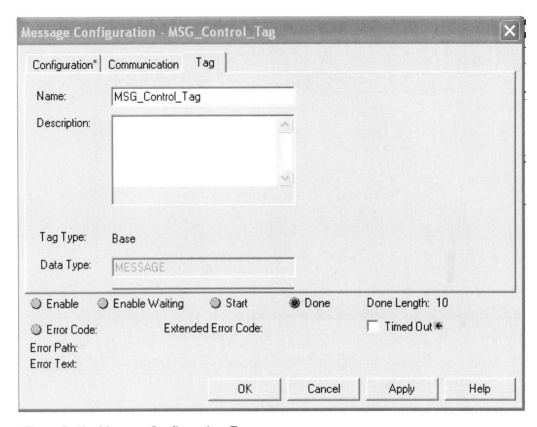

Figure D-19 Message Configuration, Tag screen.

CREATING THE TAG IN THE CONTROLLER

A controller Scope tag was created in the tag editor (see Figure D-20). The tag name is Temp_In. This is the tag to receive the data read with the MSG instruction. Note that the tag is an array of 10 members of Data Type DINT.

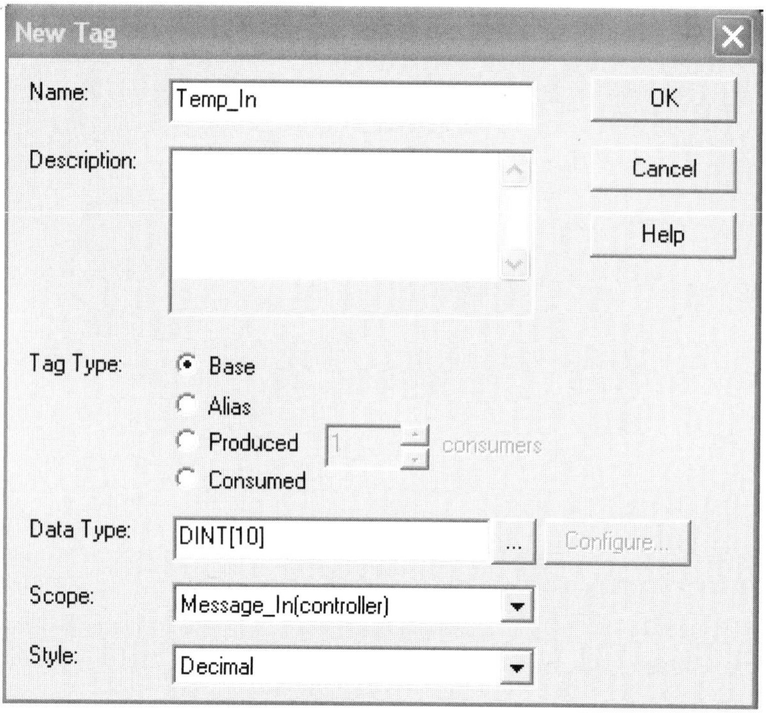

464 APPENDIX D— CONTROLLOGIX MESSAGING

Figure D-20 New Tag configuration screen.

MSG LOGIC

Next the logic is created. Figure D-21 shows the logic that was used for this example. In this example the timer DN bit is used to execute the instruction once every 5 seconds.

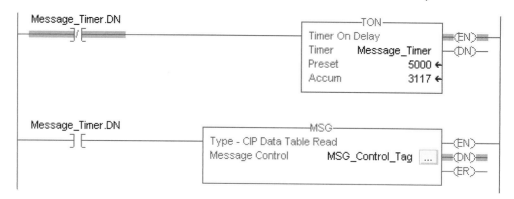

Figure D-21 Logic for the MSG routine.

Figure D-22 shows the tag array in the remote controller. Note the numbers in the 10 tag members.

Tag Name	Value
− Temp_Remote	{...}
+ Temp_Remote[0]	220
+ Temp_Remote[1]	210
+ Temp_Remote[2]	200
+ Temp_Remote[3]	190
+ Temp_Remote[4]	180
+ Temp_Remote[5]	170
+ Temp_Remote[6]	160
+ Temp_Remote[7]	150
+ Temp_Remote[8]	140
+ Temp_Remote[9]	130

Figure D-22 Remote tag in the tag editor.

Figure D-23 shows the tag array in the controller that is running the MSG instruction. Note that the numbers that the instruction read from the remote tag members are written in the tag members' Value in this controller.

Tag Name	Value
+ Ethernet_Remote:I	{...}
+ Ethernet_Remote:O	{...}
+ MSG_Control_Tag	{...}
− Temp_In	{...}
+ Temp_In[0]	220
+ Temp_In[1]	210
+ Temp_In[2]	200
+ Temp_In[3]	190
+ Temp_In[4]	180
+ Temp_In[5]	170
+ Temp_In[6]	160
+ Temp_In[7]	150
+ Temp_In[8]	140
+ Temp_In[9]	130

Figure D-23 Tag in the tag editor in the controller executing the MSG Read instruction.

APPENDIX E

Configuring ControlLogix for Motion

This information is written to help configure a motion project for RSLogix 5000. The objective is to correctly configure all the parameters for the drives and motors when a new motion project is developed. The configuration of the drives and motors should all be correct before beginning to develop the logic.

Start a new ControlLogix project. Click on Controller Properties, click on Properties, click on the Date/Time tab (see Figure E-1), click on Make this controller the Coordinated

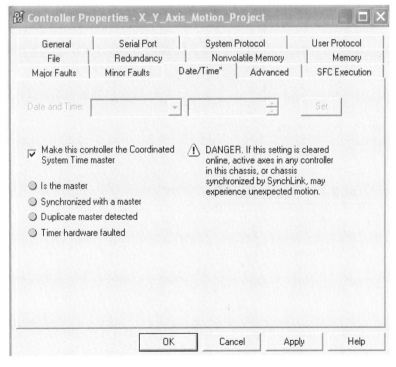

Figure E-1 Configuring the controller to be the Coordinated System Time (CST) master.

System Time master, and click OK. This makes sure that this CPU will coordinate the motion of all the axes.

Next we need to add the SERCOS card and the drives. In this example there are two drives that control two axes of motion. Figure E-2 shows the controller organizer before the modules have been added.

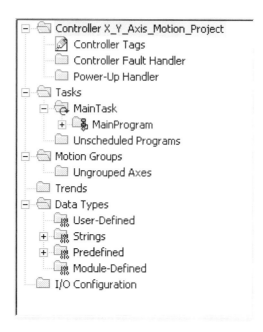

Figure E-2 Controller Organizer before any modules have been added.

Here is a quick overview of what needs to be done.

I/O modules need to be added to the project as well as the SERCOS module. Then the axes have to be configured to operate correctly in the application.

First add the SERCOS module. Right click on I/O Configuration in the Controller Organizer. Choose New Module, and the screen shown in Figure E-3 will appear. In this example a 1756-M08SE module was chosen. Select OK.

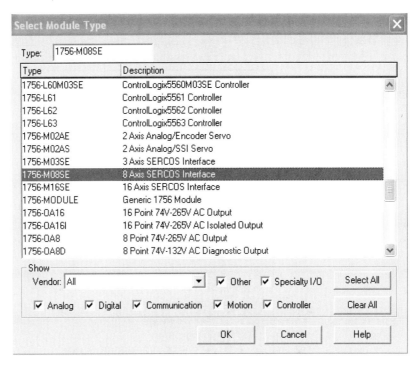

Figure E-3 Select Module Type screen. Note that a 1756-M08SE module was chosen.

Next the Module Properties screen will appear. SERCOS_Module was entered for the name. A 4 was entered for the Slot of the module. The Revision level was also entered. Compatible was chosen for the Electronic Keying method. Then Finish was selected.

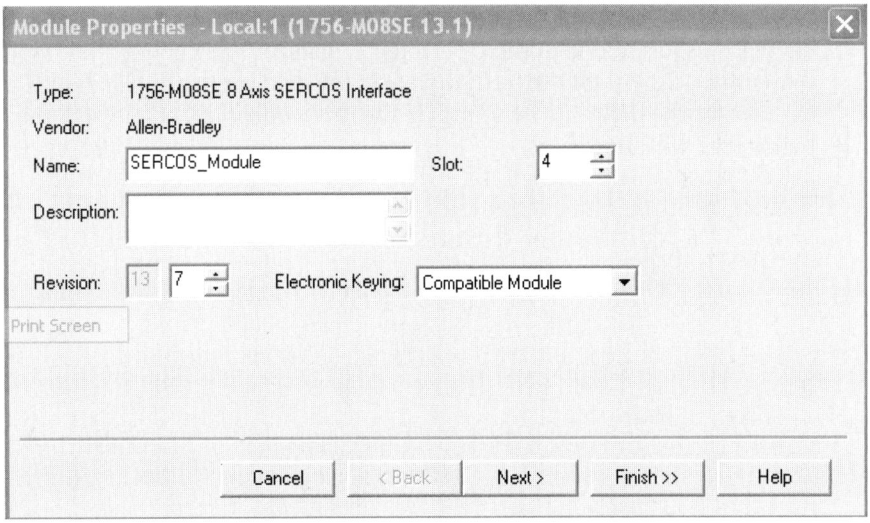

Figure E-4 Module Properties screen.

The SERCOS module is now shown under the I/O Configuration folder in the Controller Organizer in Figure E-5.

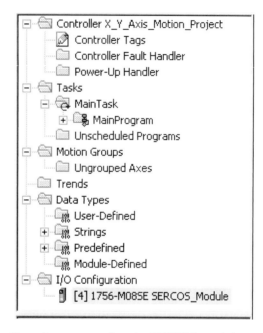

Figure E-5 Controller Organizer screen after the SERCOS module was added.

Drives must be added under the SERCOS module. The drives must be named and configured. A tag must also be created for each axis. Right-click on the SERCOS module that was just added in the Controller Organizer and choose New Module. Figure E-6 shows that a 2098-DSD-010-SE drive was selected for this application. The correct model number for the drive can be found on the drive. After OK is selected, the screen shown in Figure E-7 will appear.

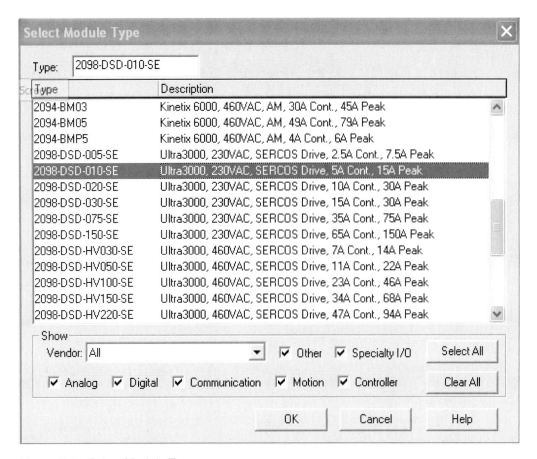

Figure E-6 Select Module Type screen.

Figure E-7 shows how the drive is configured. This drive was named X_Axis and 11 was entered for its Node address. The Node address is set on the front of the drive for this model on two dials. The Node address must be set to the same Node address on the actual drive. The drives are addressed to 11 (X_Axis) and 22 (Y_Axis). The correct Revision level was chosen and Compatible Module was chosen for the Electronic Keying method. The Next button was chosen, and the screen shown in Figure E-8 appeared.

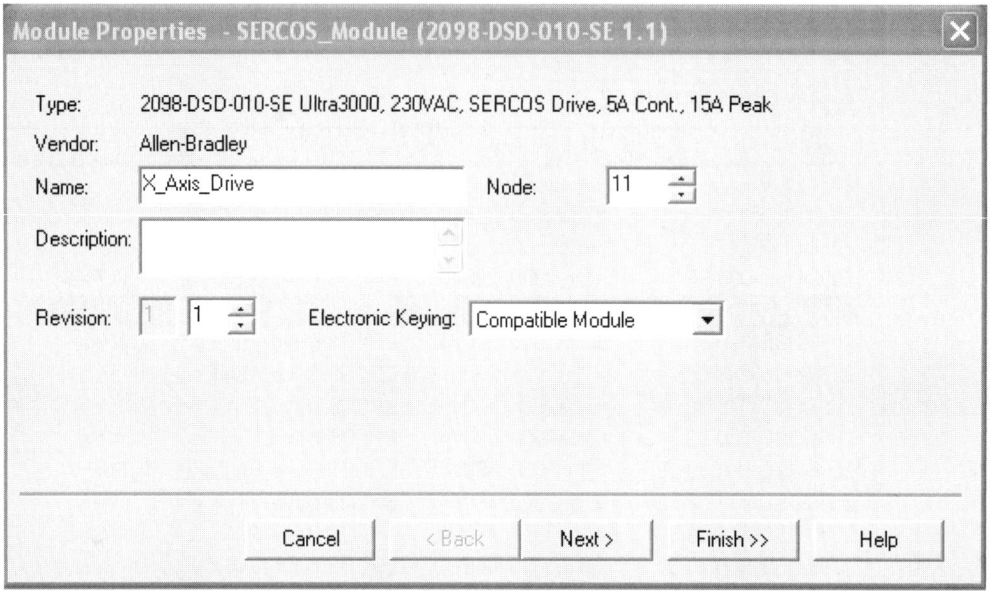

Figure E-7 Drive Name and Node number configuration.

In the screen shown in Figure E-8 you must create a New Axis tag as there is none in the selection list that have been previously created. Click on New Axis and the screen in Figure E-9 appears.

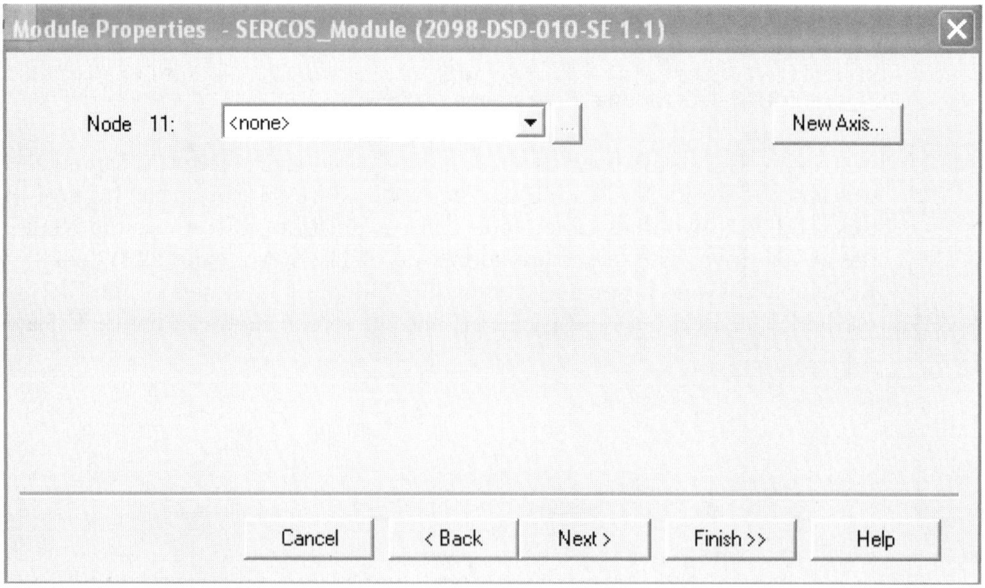

Figure E-8 Module Properties tag selection screen.

The tag name for the X_Axis is entered as X_Axis in the screen shown in Figure E-9. Its data type is Axis_Servo_Drive. OK is then selected, and the screen in Figure E-10 appears.

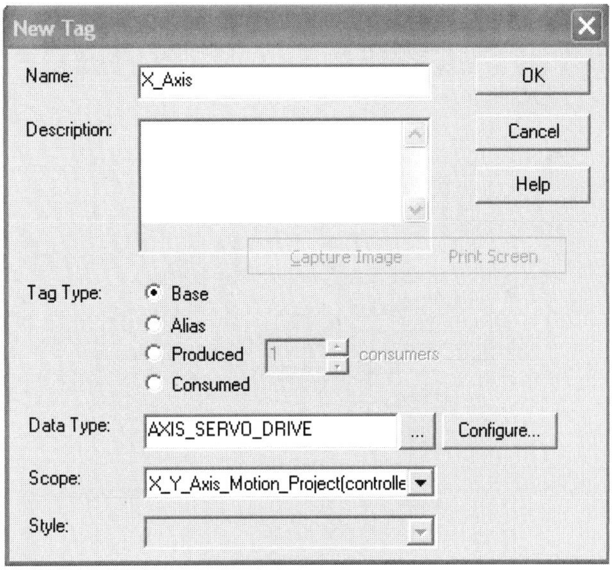

Figure E-9 New Tag screen.

Figure E-10 shows that the New Tag X_Axis was chosen for the tag for this axis. Then Finish was selected.

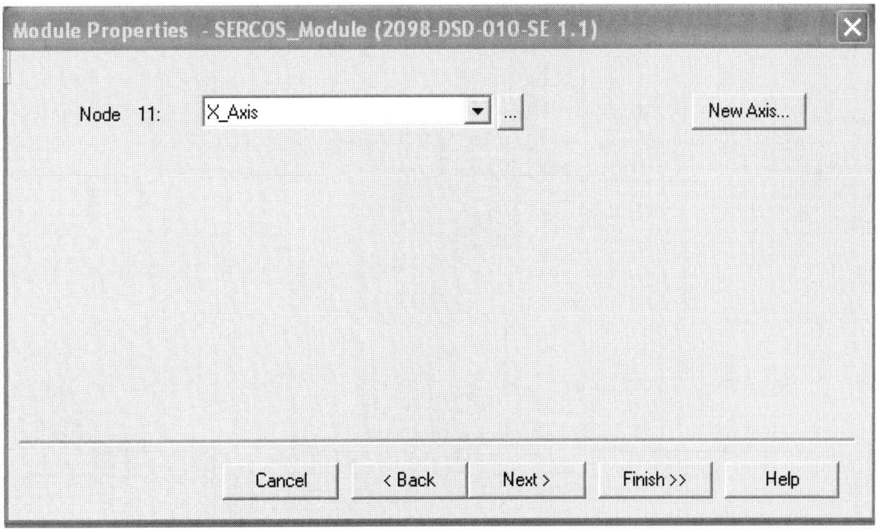

Figure E-10 Module Properties screen.

Figure E-11 shows the Controller Organizer after the first drive and its tag was added. Note the X_Axis tag under the Ungrouped Axes folder and the X_Axis_Drive under the SERCOS_Module.

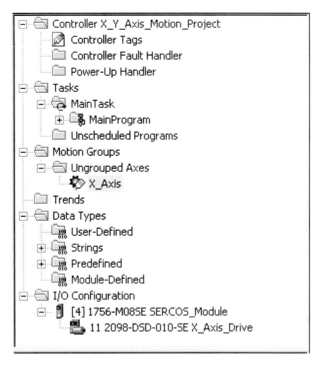

Figure E-11 Controller Organizer screen after the drive and tag were added for the X_Axis.

Next you must add the second drive. Right-click on the SERCOS module in the controller organizer and choose New Module. Choose the correct drive from the list in Figure E-12. A 2098-DSD-010-SE drive was selected for this application.

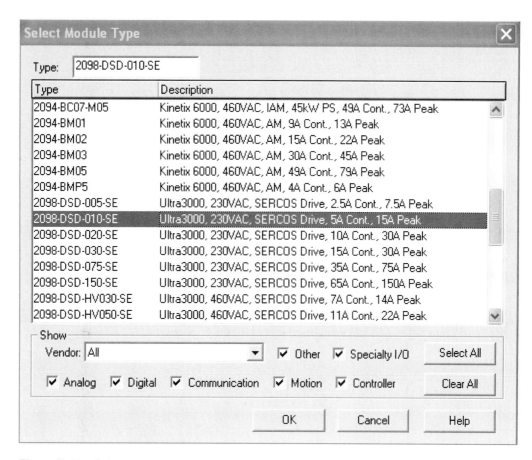

Figure E-12 Select Module Type screen.

Figure E-13 shows how this drive is configured. This drive was named Y_Axis and 22 was entered for its Node address. The Node address must be set to the same Node address on the actual drive. The correct Revision level was chosen and Compatible Module was chosen for the Electronic Keying method. Finish was chosen, and the screen shown in Figure E-14 appeared.

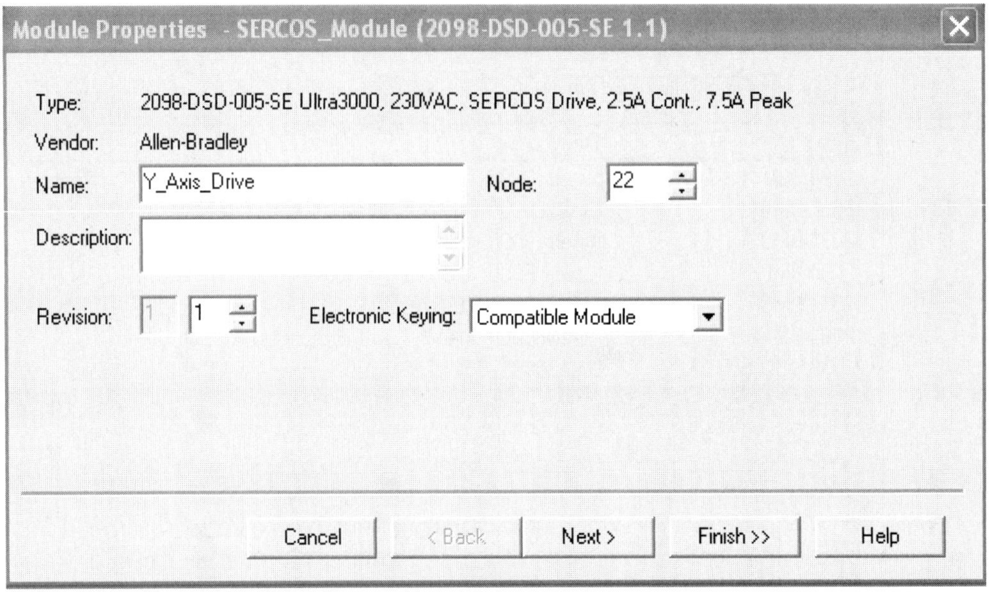

Figure E-13 Module Properties screen for the Y_Axis drive.

Next a tag must be created for the Y_Axis. The New Axis button was chosen and the screen in Figure E-15 appeared.

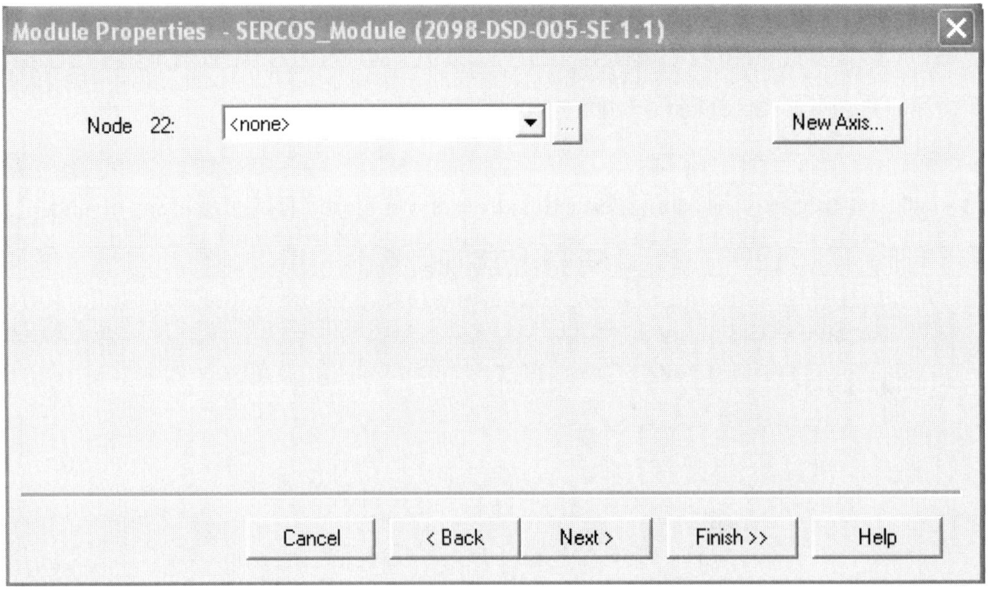

Figure E-14 Module Properties screen for the drive.

APPENDIX E—CONFIGURING CONTROLLOGIX FOR MOTION

Y_Axis was entered for the Y_Axis tag name. The Data Type for the tag is AXIS_SERVO_DRIVE. Then OK was selected.

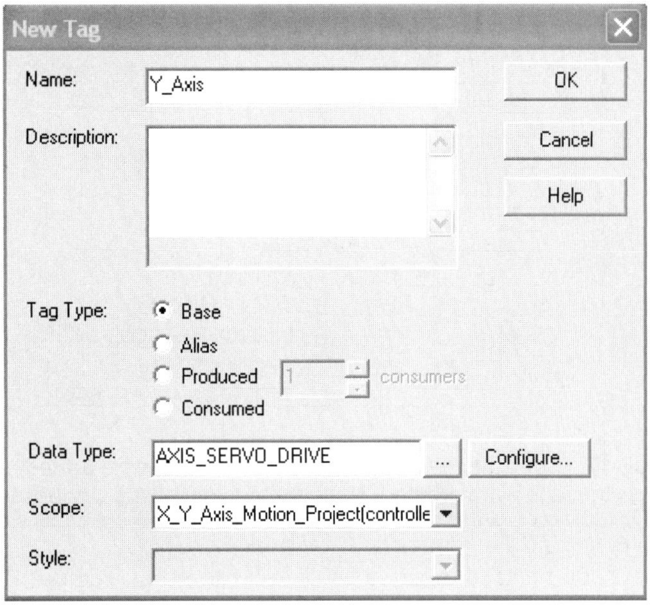

Figure E-15 New Tag screen for the axis drive configuration.

The Y_Axis tag was selected for the axis in Figure E-16. Then Finish was chosen.

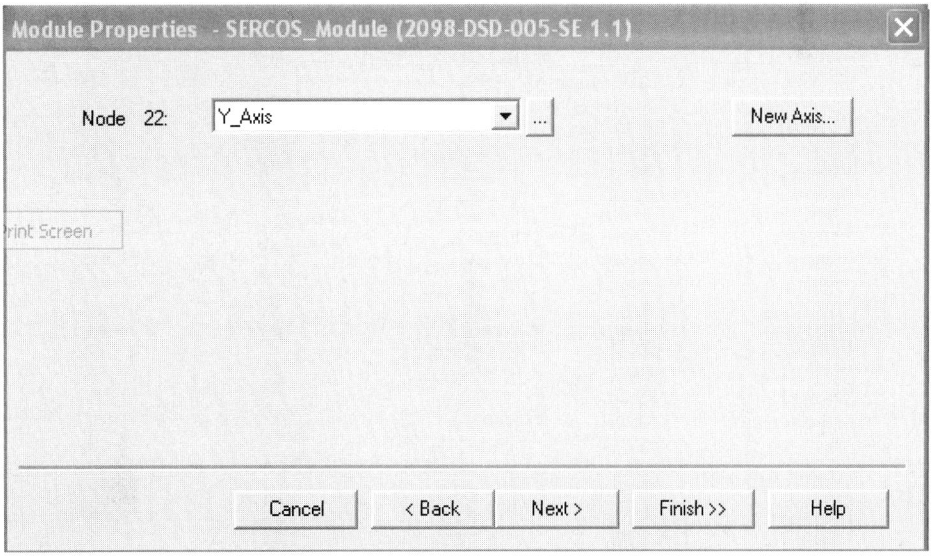

Figure E-16 Module Properties screen for the Y_Axis drive tag.

Figure E-17 shows what the Controller Organizer looks like at this point.

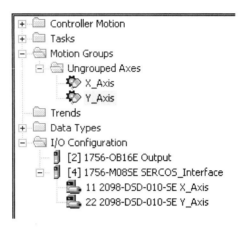

Figure E-17 Controller Organizer after the drives were added under the SERCOS module and after the tags were created. Note the tags under the Ungrouped Axis folder.

Next you need to create a Motion Groups tag. Right-click on the Motion Groups icon in the Controller Organizer and add a new group. The screen shown in Figure E-18 should appear. The motion group tag was named X_Y_Motion_Group.

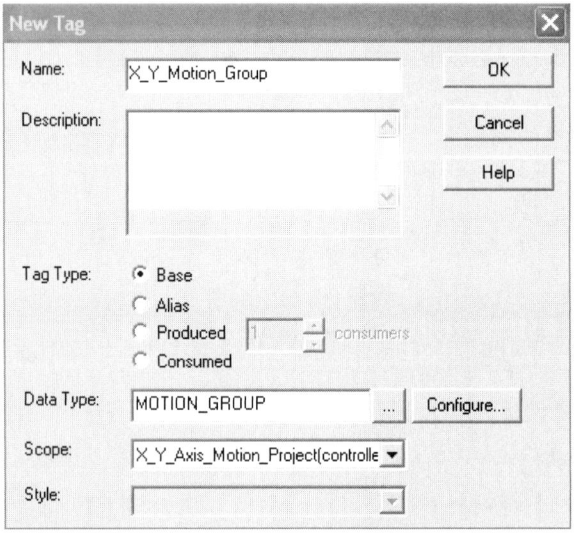

Figure E-18 New Tag screen for creating a new motion group.

Right-click on the new motion group you just created and choose Properties. The screen shown in Figure E-19 should appear.

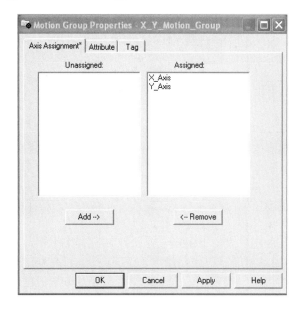

Figure E-19 Motion Group Properties screen.

Next click on the Attribute tab at the top of the Motion Group Properties screen and the screen shown in Figure E-20 will appear. Set the Coarse Update Period to 4. This number represents the number of 0.5 ms used to update the motion. You should have a minimum of 2 per axis used. In this application you have 2 axes so the Coarse Update Period should be set to 4 minimum.

Figure E-20 Motion Group Properties screen. This is where the Coarse Update Period is set. Then select Apply and OK.

Figure E-21 shows what the Controller Organizer looks like after the motion group X_Y_Motion_Group was added.

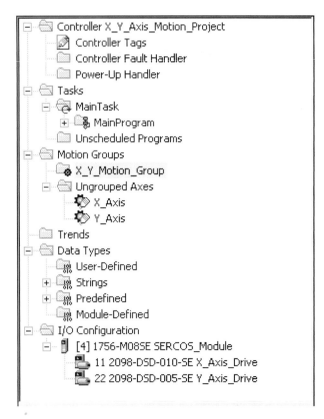

Figure E-21 The new motion group named X_Y_Motion_Group under Motion Groups in the Controller Organizer.

Next you need to move the axis tags you created into the new motion group. Left-click and drag the axes tags (X_Axis and Y_Axis) you created (under the Ungrouped Axes tag) and drop them into the motion group folder you created for this application (see Figure E-23). In this example the name of the motion group folder is X_Y_Motion_Group. The Controller Organizer is shown in Figure E-22.

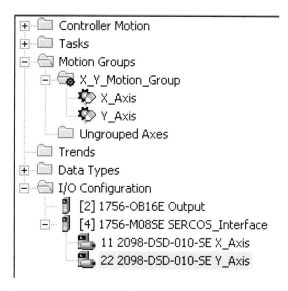

Figure E-22 Controller Organizer screen after drives, motion tags, and a motion group folder X_Y_Motion_Group were added. Note that the axes tags were dragged to the new motion group folder X_Y_Motion_Group.

CONFIGURING THE AXES OF MOTION

Next the axes need to be configured. This is done in the tag for each axis. Right-click on the X_Axis tag under Motion Group and select Axis Properties, and the screen shown in Figure E-23 will appear.

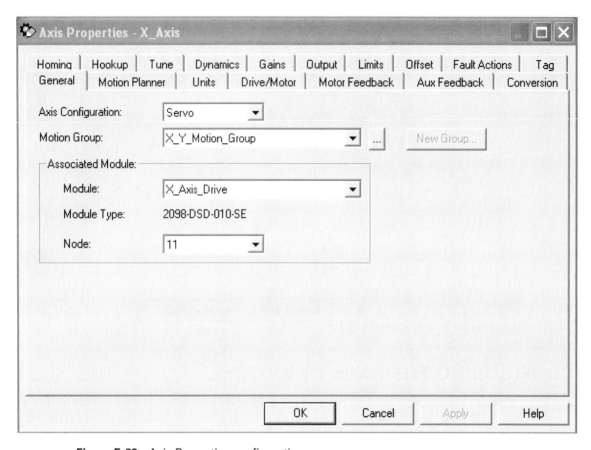

Figure E-23 Axis Properties configuration screen.

Click on the Drive/Motor tab and the screen in Figure E-24 will appear. The drive in this example is a 2098-DSD-010-SE model. The drive model number was entered into the Amplifier Catalog Number. Next the motor model is entered. Click on the Change Catalog button and select the correct motor model. The Motor Catalog Number is found on the motor. The motor model in this example was an N-3412-2-H.

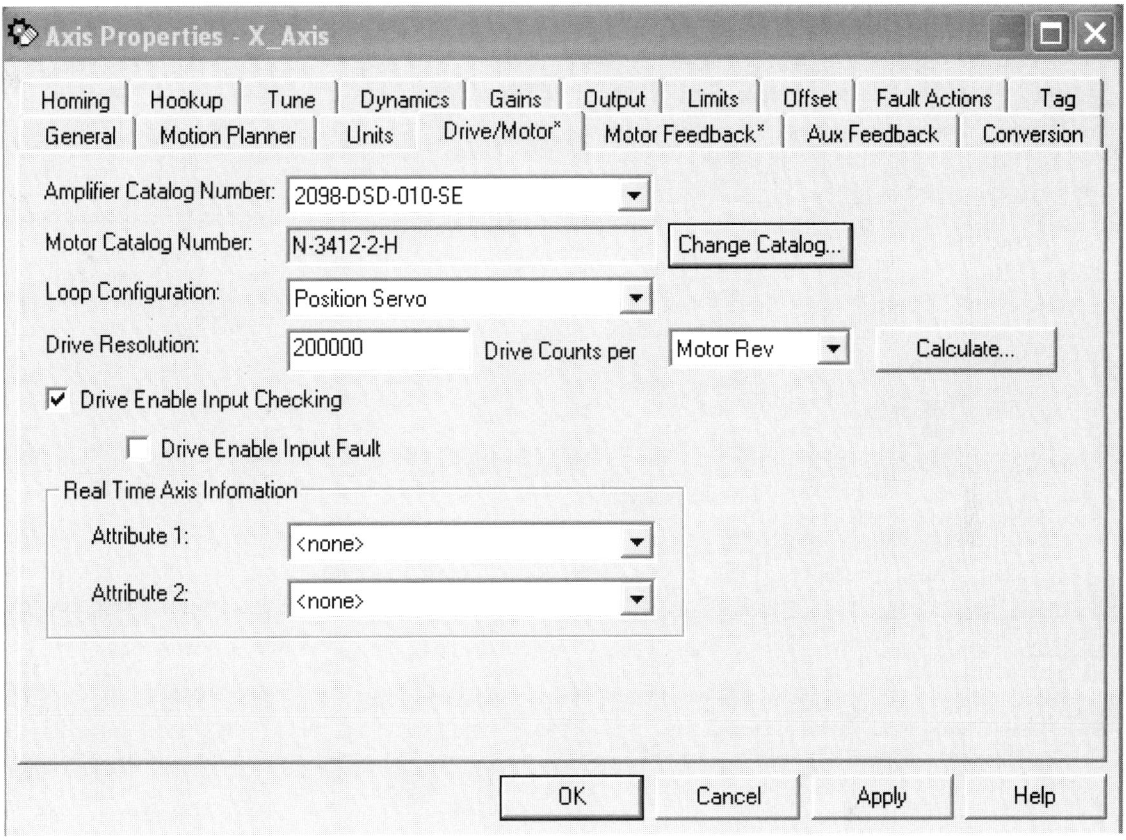

Figure E-24 Drive/Motor selection.

There are several more parameters that need to be configured to prevent damage to the drive or motor. You will be configuring several important parameters for how the drives will function. The most important to prevent damage to the axis in this application is the parameter that sets hard limits for the application. In this application if hard limits are not set, a crash could occur and do great damage. To set hard limits, click the Limits tab and check the box next to Hard Travel Limits (see Figure E-25).

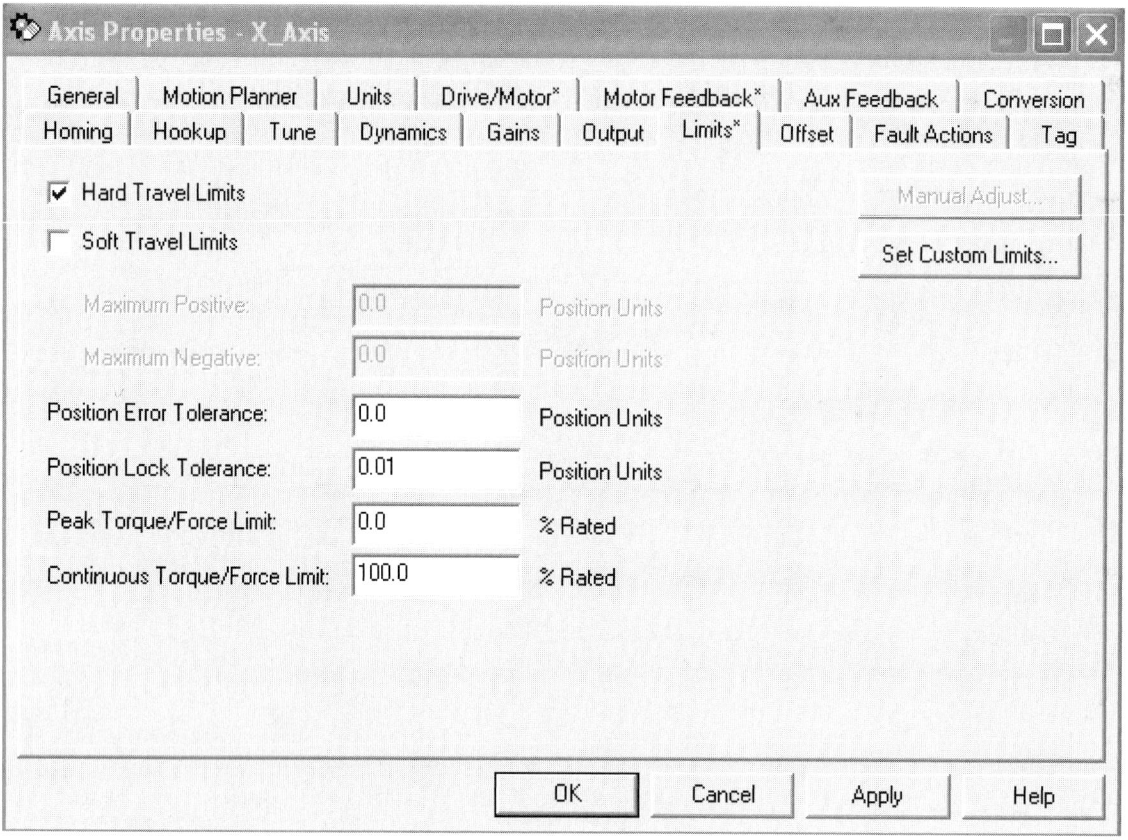

Figure E-25 How Hard Travel Limits are set for a drive.

Next choose the Fault Actions tab and select Stop Motion in the Hard Overtravel checkbox (see Figure E-26). This tells the system to stop motion if an overtravel switch goes false. Note that the end limit switches on this system are normally closed switches. This must be done individually for each axis.

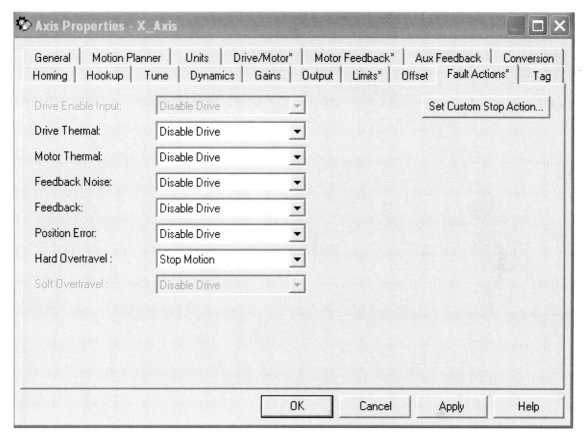

Figure E-26 Setting the Hard Overtravel configuration. In this example you want the drive to stop the motion if the axis moves too far and hits an end limit switch.

Next homing will be configured. This is configured for each axis in the axis tag. Choose the Homing tab in the Axis Properties screen (see Figure E-27). You will utilize the homing method that uses the home sensor (switch) and the index pulse on the encoder (marker pulse). Choose Switch/Marker for the type of homing, set the Limit Switch type to Normally Open, and set the homing Direction to Reverse Bi-directional. This will tell the drive that when a home command is executed it should move in a negative direction to find the home switch. The home switches in this example system are on the negative end of each axis. Speed and Return Speed homing velocities must be entered. Homing is normally done at a low speed. A 5 was entered in each for this example; 5 represents 5 inches per minute in this application.

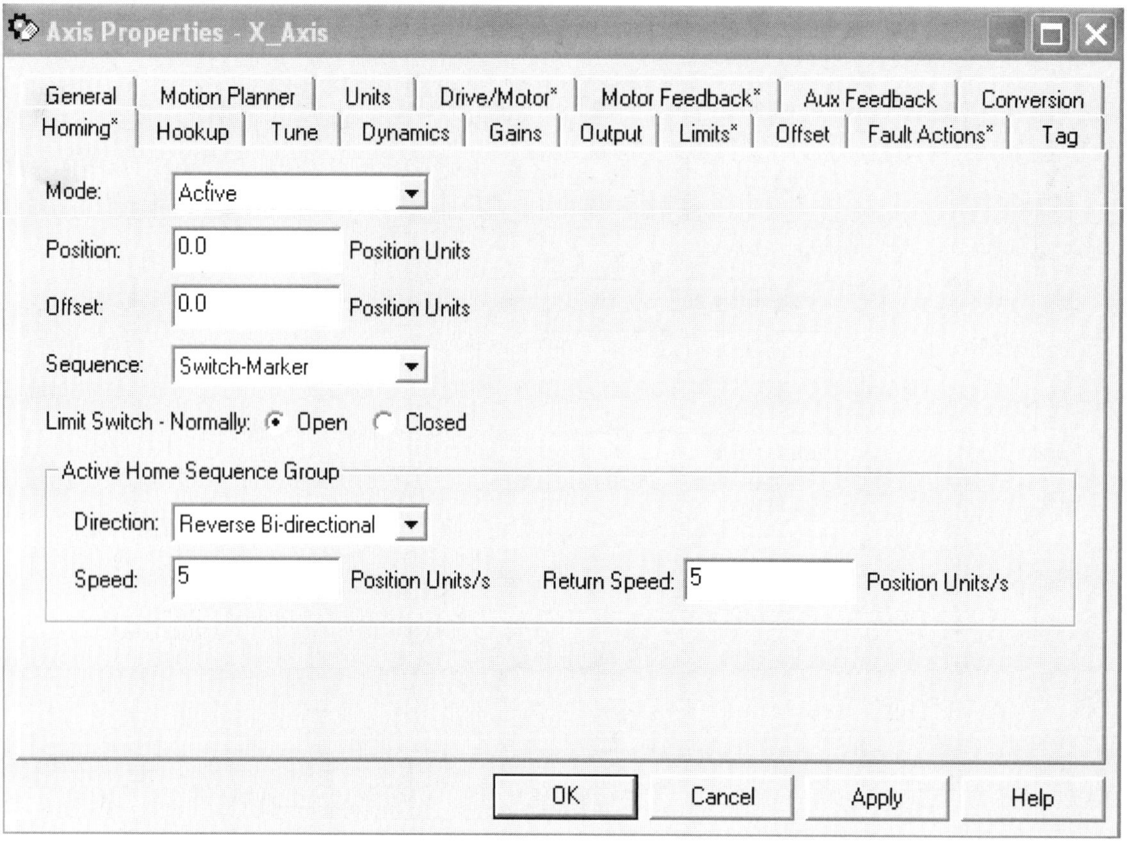

Figure E-27 Homing configuration screen.

RESOLUTION OF THE AXES

Figure E-28 shows the Axis Properties screen for the X_Axis. Note that Conversion Constant 200000 was entered in this example. This number represents Drive Counts/1.0 Position Units. In this example the desired unit of measurement for the system is inches, so 200000 Drive Counts is equal to 1 inch of travel.

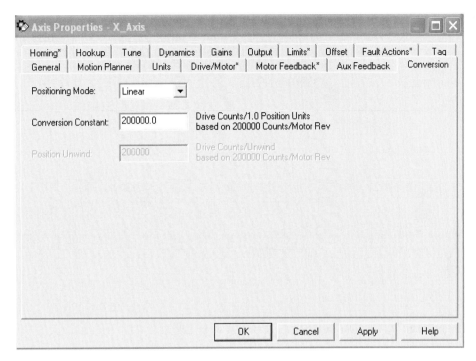

Figure E-28 Axis Properties screen for X_Axis.

A discrete output module will be required for this application. Outputs from the output module will be used to enable each drive. Right-click on I/O Configuration and choose New Module. Add an output module. Figure E-29 shows the Controller Organizer after the output module was added. Note that you could have added the output module before you added the SERCOS card and drives.

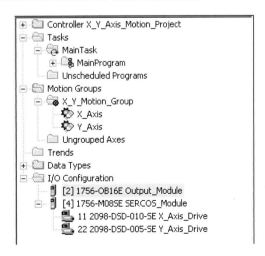

Figure E-29 Controller Organizer after the output module was added.

At this point the configuration is complete. Next you need a simple program to enable the drives. In this example the drive enable input for the X_Drive was connected to output 0 on the output module in slot 2. The enable input for the Y_Drive was connected to output 1 on the output module in slot 2. Figure E-31 shows a simple ladder diagram that could be used to enable the drives.

Figure E-30 Logic to enable the two drives. Note that the outputs from the output module are connected to the drive enable input on each drive.

At this point logic can be written to command motion or the drives can be tested with motion direct commands.

MOTION DIRECT COMMANDS

Motion direct commands can be used to test your axes before you write the logic. You must be online in order for any of these commands to work. Right-click the axis icon for the axis you want to test, then click on Motion Direct Commands. The Motion Direct Commands window opens; first highlight MSO, the Motion Servo On instruction, and then hit Execute. The MSO instruction closes the servo loop and puts the drive in control. From here there are several commands that can be used.

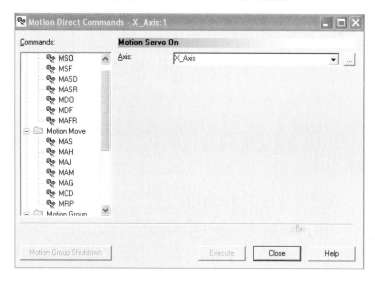

Figure E-31 Motion Direct Commands screen.

First the drives must be enabled. Figure E-30 shows a simple ladder diagram that turns on the drive enable input for each axis. A BOOL tag was used to control the output to each drive's enable. This is done to electrically enable and disable each drive.

If the CLX is put in Run mode and the BOOL tags are energized to enable the drives, motion direct commands can be used to test the axes.

Right-click on one of the drives and choose Motion Direct Commands.

The first command that must be executed is an MSO. This closes the servo loop for the drive. This must be done, or the drive will not execute any commands.

When an axis is first powered up, it does not know its current position. It needs to be *homed* to establish its current position. The motion direct command MAH will use the parameters that were set up above in the homing properties. Figure E-32 shows the Motion Direct Commands screen and the MAH command in the list. Note that the X_Axis was chosen in this example. If the Execute button is chosen, the drive should initiate the homing routine.

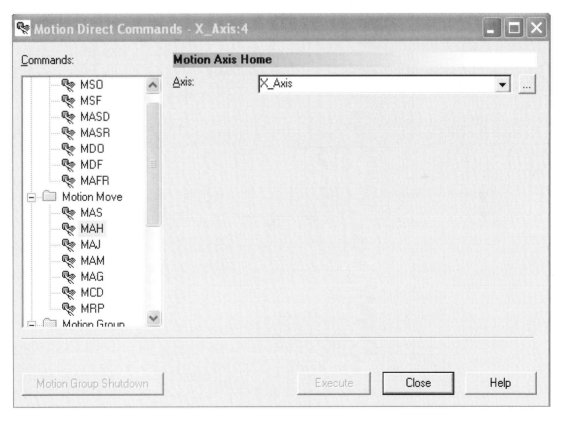

Figure E-32 Motion Direct Commands screen.

Once the axis is homed, other commands may be tried. A motion axis jog (MAJ) instruction is used to jog an axis. With a MAJ command you have to use a motion axis stop (MAS) command to stop the axis. When the MAJ command is executed in incremental mode, the drive will continue to move until a MAS command is executed for that axis. To use an MAJ instruction, you must choose a Direction and enter a Speed. Forward was chosen for direction and 2 was entered for the MAJ motion direct command in Figure E-33. Note the X_Axis was chosen for the Axis. When the Execute button is chosen, the axis will move in a positive direction until a limit is encountered or a MAS instruction is executed.

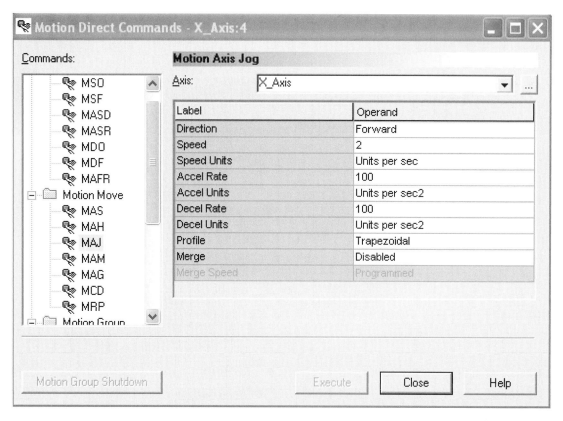

Figure E-33 MAJ command.

A MAS instruction is shown in Figure E-34. Note that X_Axis was chosen. If the Execute button is chosen, the MAS command will stop motion of the X_Axis.

APPENDIX E—CONFIGURING CONTROLLOGIX FOR MOTION **491**

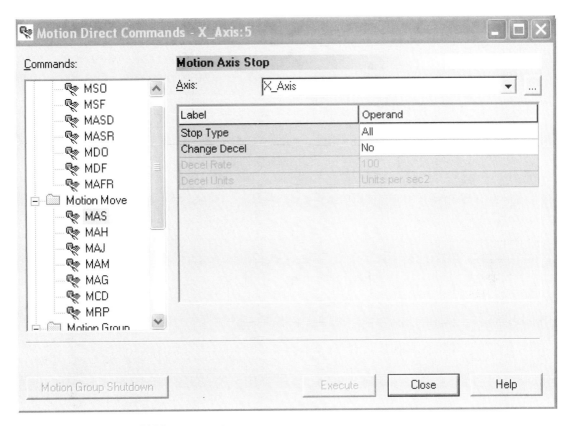

Figure E-34 MAS command.

There are many other motion direct commands that can be executed. The use of motion direct commands can help understand motion programming instructions available in CLX.

GLOSSARY

A

AC input module A module that converts an AC input signal to a low-level DC signal logic level required by the PLC CPU.

AC output module A module that converts the CPU's low-voltage DC level to an AC output signal to control a device.

Accumulated value The present time or count. Applies to the use of timers and counters.

Accuracy The difference between the actual position and the programmed position.

Actuator An output device normally connected to an output module. An example would be an air cylinder and valve.

Analog A signal with a smooth range of possible values. For example, a pressure that varies from 3 to 15 psi is sensed by a pressure sensor that outputs a signal between 4 and 20 mA.

ANSI American National Standards Institute.

Array A systematic arrangement of numbers or symbols in rows or rows and columns. In a CLX array all members must have the same data type.

American Standard Code for Information Interchange (ASCII) A system used to represent letters and characters. Seven-bit ASCII can represent 128 different combinations. Eight-bit ASCII (extended ASCII) can represent 256 different combinations.

Asynchronous communications Communications that use a stream of bits to send data between devices. There is a start bit, data bits (7 or 8), a parity bit (odd, even none, mark, or space), and stop bits (1, 1.5, or 2). Only one character is transmitted at a time.

Awareness barrier A physical barrier like a railing, chain, or cable suspended at waist height. Requires an intentional effort to get beyond it, making it better than just a yellow line on the floor.

B

Backplane Bus in the back of a PLC chassis. It is a printed circuit board with sockets that accept various modules. It powers the modules and connects the modules for communication.

Ball screw A mechanical device that is used to change rotational motion to linear motion. A threaded shaft provides a spiral raceway for ball bearings which act as a precision screw. The ball assembly acts as the nut while the threaded shaft is the screw.

Barrier A device or object that provides a physical boundary to a hazard.

Baud rate The speed of serial communications. The number of bits per second transmitted. For example, RS-232 is normally used with a baud rate of 9600. This would be 9600 bits per second. It takes 10 bits in serial to send an ASCII character so a baud rate of 9600 would transmit about 960 characters per second.

Binary Base 2 number system in which 1s and 0s are used to represent numbers.

Binary-coded decimal (BCD) number system A number system in which each decimal number is represented by four binary bits. For example, the decimal number 341 would be represented by 0011 0100 0001 in BCD.

Bit One binary digit, the smallest element of binary data. A bit can have a value of 0 or 1.

Blanking Bypassing a portion of the sensing field of a presence-sensing safeguarding device such as a light curtain.

BOOL A binary digit. A BOOL-type tag can have a value of 1 or 0. Abbreviation of Boolean.

*Bootstrap Protocol (BootP) A protocol that assigns the same IP address to a device every time it connects to the network. Its server has a list of hardware addresses and IP addresses that belong to each device. When a device (such as a PLC) connects to the network, it will give the BootP server its hardware address. The BootP server will then look up the hardware address in a list and see which IP address belongs to the PLC. It will then return the IP address (and other information such as the subnet mask) to be used by the device. Similar to Dynamic Host Configuration Protocol.

*Bounce The erratic make and break of electric contacts as they close or open. Branch: Parallel logic in a ladder diagram. Used to create OR logic.

Byte 8 bits or 2 nibbles.

C

Cascade Programming technique that is used to extend the range of counters and timers.

European Committee for Electrotechnical Standardization (CENELEC) An organization that develops standards for dimensional and operating characteristics of control components.

Change of state Property of a device such that it reports only when the data changes.

*Chassis A frame of an electronic device. Also called a rack. Available in various sizes. The user chooses the size chassis that is needed to hold the number of modules required for the application. Chassis have slots that locate and power the modules. The slots connect the module to the backplane. The backplane passes power to operate modules and also enables the modules to communicate with the controller and other modules.

CLX ControlLogix.

Color mark sensor Sensor that was designed to differentiate between different colors.

Complement The inverse of a binary number.

Contact A symbol used to represent inputs. There are two types: normally open and normally closed.

Contactor A special-purpose relay that is used to control large electric current.

ControlNet An open industrial network protocol for industrial automation. It is normally used for communication between controllers.

Control reliability The capability of the machine control system, the safeguarding, other control components, and related interfacing to achieve a safe state in the event of a failure within their safety-related functions.

Central Processing Unit (CPU) The microprocessor portion of the programmable logic controller (PLC) that handles the logic.

Carrier Sense Multiple Access/Collision Detection (CSMA/CD) A network on which each device monitors the wires for a carrier frequency (carrier sense) and can talk if the line is not being used. If the line is being used, the device must wait for the line to become clear before it can gain access to send a message. The weakness in Ethernet is that two devices may try to transmit at exactly the same time. This would cause a data collision, and neither device's message would get through. To take care of this problem, Ethernet uses Collision Detection (CD).

Current sinking A NPN output device that allows current flow from the load through the output to ground.

Current sourcing A PNP output device that allows current flow from the output through the load and then to ground.

Cyclic Redundancy Check (CRC) A calculated value, based on the content of a frame of communication. It is inserted in the frame to enable a check of data accuracy after receiving the frame across a network.

D

Dark-on sensor A photosensor in which the output is on when the receiver is not seeing any light.

Data Highway Plus (DH+) A communications network that allows programmable logic controllers (PLCs) to communicate.

Debugging The process of finding and fixing problems (bugs) in a system.

Deterministic A property of a process that a result can be determined. For example, in communications systems, devices must gain access to the network to be able to communicate. Token passing is one access method. In the token-passing access method, only the device that has control of the token can talk. The token is passed on from device to device until one of them wants to talk. The device then takes control of the token and is free to talk. With token passing, access times for a device are predictable. This can be very important in a manufacturing environment. The access times in a token-passing access method are called deterministic because actual access times can be calculated on the basis of the actual bus and nodes.

DeviceNet A communications protocol used in automation to connect field devices such as sensors, motors, valves, and so on. It uses Controller Area Network (CAN chip) as its backbone and defines an application layer to cover a range of device profiles. Typical applications include safety devices and I/O devices.

DINT Double integer.

Dynamic Host Configuration Protocol (DHCP) When a device such as a computer is configured to use DHCP, a DHCP server should be available on the network. As soon as the device connects to the network, it will ask the DHCP server to automatically assign an IP address, subnet mask, DNS servers, gateway addresses, and so on. This IP address is dynamic, so the device could get a different IP address each time it's connected to the network.

Diagnostics Software routines devices often have that aid in identifying and finding problems in the device. They identify fault conditions in a system typically with a readout or LEDs.

Digital output An output that can have two states: true (on) or false (off). They are also called discrete outputs.

Discrete The property of having two states: on or off.

Domain Name Server (DNS, Primary and Secondary) The server that resolves host names into an IP address. When you enter an address such as Amazon.com into your Web browser, your PC does not understand where to go. It must ask the DNS server to look up the IP address of the name you entered.

Downtime The time a production system is not available for production or operation.

E

EEPROM Electrically erasable programmable read only memory.

Emergency stop (E-stop) A manually actuated control device that can be used to initiate an emergency stop function. E-stops should be a red mushroom button with yellow background.

Encoder A transducer that converts rotary motion or position to a code of electronic pulses.

Energy source Any electric, mechanical, hydraulic, pneumatic, chemical, thermal, potential, kinetic, or other source of power or movement.

Examine if closed (XIC) A normally open contact used in ladder logic. This instruction is true (logic 1) when the hardware input (or internal relay equivalent) is energized.

Examine if open (XIO) A normally closed contact used in ladder logic. This instruction is true (logic 1) when the hardware input (or internal relay equivalent) is *not* energized.

Examine off A normally closed contact used in ladder logic. The contact is true (or closed) if the real-world input associated with it is off.

Examine on A normally open contact used in ladder logic. This type of contact is true (or closed) if the real-world input associated with it is on.

Expansion chassis A chassis added to a PLC system when the application requires more modules than the main rack can contain. Sometimes used to permit I/O to be located away from the main chassis.

F

False 0 or off.

Firmware A combination of software and hardware to download new software to upgrade the system. A CLX controller and modules can be upgraded by downloading new firmware.

Force Change the state of I/O by changing the bit status in the controller. You can force an output on by changing the bit associated with the real-world output to a 1. Forcing is normally used to troubleshoot a system.

Full duplex Communications scheme where data flows in both directions simultaneously.

Function block One of the languages specified in IEC 61131-3. Can be used to program CLX.

G

Gateway address The IP address of a server or hardware router that connects a device to other networks such as the Internet.

Ground A direct connection between equipment (chassis) and earth ground.

Guard A barrier that prevents a person from reaching over, under, around, or through it, avoiding both intentional and unintentional access to hazardous areas.

H

Half-duplex The property of a communication such that data flows in both directions but in only one direction at a time.

Hasp A safety device that enables the use of several locks to lock out energy.

Hazard A potential source of harm to individuals.

High-Level Data Link Control (HDLC) Standard protocol of communication orientation in message transmission (frames). The Serial Data Link Control (SDLC) is a subset of the HDLC that defines the whole protocol in more detail and is byte oriented.

Hexadecimal Base 16 number system.

Hysteresis A dead band that is used to prevent false reads in the case of a sensor.

I

International Electrotechnical Commission (IEC) An organization that develops and distributes recommended safety and performance standards.

IEC 61131-3 An international standard for PLCs. Actually a collection of standards for PLCs and their associated peripherals. The standard consists of eight parts: Part 1: General information, Part 2: Equipment requirements and tests, Part 3: Programming languages, Part 4: User guidelines, Part 5: Communications, Part 6: Reserved for future use, Part 7: Fuzzy control programming, and Part 8: Guidelines for the application and implementation of programming languages. Part 3 (IEC 61131-3) is the most important to the PLC programmer. It specifies the following languages: ladder diagram, instruction list, function block diagram, structured text, and sequential function chart.

IEEE Institute of Electrical and Electronic Engineers.

Incremental The property of an encoder that uses pulses to establish position and direction.

Integer A whole number.

Interfacing Connection of a PLC to other industrial devices.

Interlocked barrier guard A barrier or section of a barrier interfaced with the machine control system to prevent inadvertent access to the hazard.

I/O Input/output.

IP rating A rating system established by the IEC that defines the protection offered by electrical enclosures.

Isolation A technique used to separate real-world inputs and outputs from the CPU. Isolation assures that even if there is a problem with real-world I/O, the CPU will be protected.

K

K An abbreviation for the number 1000. In computer language it is equal to 2 to the 10th power (1024).

Keying A method to ensure that modules are not put in the wrong slots of a PLC. Done by mechanical means in most PLCs, but electronically in CLX.

L

Ladder diagram A programmable controller language that uses contacts and coils to develop the logic for an application.

Latch An instruction used in ladder diagram programming to retain a coil's state even if the rung controlling it becomes false.

Leakage current A small amount of current that flows through load-powered sensors. The small current is used to operate the sensor. The small amount of current flow is normally not enough to be sensed by the PLC input.

Light-emitting diode (LED) A solid-state semiconductor that emits visible light or invisible infrared light.

Light-on sensor A photosensor in which the output is on when the receiver sees light.

Linear output Analog output.

Line-powered sensor Normally a three-wire sensor. The line-powered sensor is powered from the power supply. The third wire is used for the output.

Load-powered sensor A two-wire sensor. A small leakage current flows through the sensor even when the output is off. The leakage current is required to operate the sensor. The leakage current is too small to activate a PLC input if the sensor output is off.

Lockout The placement of a lockout device on an energy-isolating device, in accordance with an established procedure, to ensure that the energy-isolating device and the equipment being controlled cannot be operated until the lockout device is removed.

Lockout device A device such as a lock, either key or combination type, to hold an energy-isolating device in the safe position and prevent the energizing of a machine or equipment.

LSB Least significant bit.

M

Master The device that controls the communication traffic in a network. The master polls every slave to check if it has something to transmit. In a master-slave system, only the active master can place a message on the bus. The slave can reply only if it receives a logical token that explicitly enables it to respond.

Master control relay (MCR) A hardwired relay that can be deenergized by any series-connected switch. Used to deenergize all devices. If one emergency switch is hit, it causes the MCR to drop power to all devices.

Memory map A drawing showing the areas, sizes, and uses of memory in a computer or PLC.

Mesh network A network in which each device passes the message to its neighboring device until it reaches the destination device. If a neighboring device is damaged, another neighbor is used.

Microsecond (μs) One-millionth (0.000001) of a second.

Millisecond (ms) One-thousandth (0.001) of a second.

MSB Most significant bit.

Muting The automatic temporary bypassing of safety-related function(s) of the control system or safeguarding device.

N

National Electrical Manufacturers Association (NEMA) An organization that develops standards that define a product, process, or procedure. The standards consider construction, dimensions, tolerances, safety, operating characteristics, electrical rating, and so on.

Nibble 4 binary bits. One half of a byte.

Noise Unwanted electrical interference. It is caused by motors, high voltages, coils, welding, and so on. It can interfere with communications and control.

Nonretentive coil A coil that will turn off on removal of power to the CPU.

Nonretentive timer A timer that loses the accumulated time if its rung goes false.

Nonvolatile memory Memory in a controller that does not require power to be retained.

NOR A logic gate that results in zero unless both inputs are zero.

NOT A logic gate that results in the complement of the input.

NUT Network update time.

O

Octal Base 8 number system. Uses digits 0 through 7.

Off-delay timer (TOF) A timer whose output is on immediately when it is enabled. The output turns off after it reaches its preset time.

On-delay timer A timer whose output does not turn on until its accumulated time has reached the preset time value.

One-shot contact A contact that is only on for one scan when energized.

Optical isolation A technique used in I/O module design that protects the CPU from signals from the outside world.

OR A logic gate that results in one unless both inputs are zero.

OTE Output energize coil.

OTL Output latch coil.

OTU Output unlatch coil.

P

PAC Programmable automation controller.

Parity Bit used to help check for data integrity during communications.

Peer to peer The type of communication that occurs between similar devices. For example, two PLCs communicating would be peer to peer.

Proportional, integral, derivative (PID) control A control algorithm that is used to closely control temperature, position, velocity, and so on. The proportional portion corrects for the magnitude of the error. The integral corrects small errors over time. The derivative compensates for the rate of change in the error.

Pitch The distance between two adjoining threads.

PLC Programmable logic controller.

Polled The property of communications of individual devices that the scanner polls for their status.

Producer/consumer A communication system in which devices can send and receive data independently, communication is optimized, and data is collected directly from devices without the need for complicated programming. In a producer/consumer system a controller can produce certain data and other controllers can consume the data. Used in such networks as DeviceNet, ControlNet, and EtherNet/IP.

Project The overall application you develop in CLX. Contains all of an application's elements and is broken into tasks, programs, and routines.

Protective device A device, other than a guard, that reduces a risk, either alone or associated with a guard but not including personal protective equipment.

Pulse modulated The property of a device in which it is turned on and off at a very high frequency. Used with LEDs. In sensors the light source is modulated; the receiver only responds to that frequency. Used to make photosensors immune to ambient lighting.

Q

Quadrature The situation where two output channels are out of phase with each other by 90 degrees. Used in encoders to determine direction of rotation.

R

Rack A PLC chassis. Normally holds the CPU and PLC modules and is usually attached to the power supply.

Random access memory (RAM) Volatile memory. Normally considered user memory.

REAL A real or decimal number.

Residual risk The risk that remains after protective measures have been taken. Every industrial machine, regardless of how well safeguarded it appears to be, has some degree of residual risk.

Resolution A measure of how closely a device can measure or divide a quantity. For an analog to digital card, resolution would be the number of bits of resolution. For example, for a 16-bit card the resolution would be 65536.

Retentive coil A coil that will remain in its last state, even if power is removed.

Retentive timer A timer that retains the present accumulated time even if the input enable signal is lost. When the input enable is active again, the timer begins to time again from where it left off.

Retroreflective A photosensor that emits light that is reflected from a reflector back to the receiver. When an object passes through the light, it breaks the light beam.

Risk A combination of the probability and the degree of the possible injury or damage to health in a hazardous situation. Used to select appropriate safeguards.

Risk assessment The process by which the intended use of the machine, the tasks and hazards, and the level of risk are determined.

Read-only memory (ROM) The nonvolatile operating system memory. Memory is not lost when the power is turned off.

Routine The entities in which an application's logic is created. Each CLX program can have one or more routines. In most PLCs they are called programs or subprograms.

RS-232 A serial communications standard that specifies the purpose of each of 25 pins. It does not specify connectors or which pins must be used.

RS-422 and RS-423 Standards for two types of serial communication. RS-422 is the standard for a balanced serial mode in which the transmit and receive lines have their own common instead of sharing one. This allows for higher data transmission rates and longer transmission distances. RS-423 is for the unbalanced mode, whose speeds and transmission distances are much greater than with the RS-232 but less than with the RS-422.

RS-449 Electrical standard for RS-422/RS-423.

RS-485 A derivation of the RS-422 standard. The standard is now officially known as EIA/TIA-485, although it is still commonly referred to as RS-485. RS-485 is for a multidrop protocol, in which many devices can be connected on the same network. The standard limits the number of stations to 32. This allows for up to 32 stations with transmission and reception capability, or 1 transmitter and up to 31 receiving stations. The maximum distance for RS-485 is 1200 meters. The total number of devices and maximum distance can be extended if repeaters are used.

Rung A horizontal line in a ladder diagram that has the contacts and coils that form the logic.

S

Safeguarding Protecting workers and equipment with guards, safeguarding devices, awareness devices, safeguarding methods, and safe work procedures.

Safety distance The calculated distance between a hazard and its associated safeguard.

Safety function The function of a machine that provides safety; the malfunction of the machine would increase the risk of harm.

Scan time The amount of time it takes a programmable controller to evaluate logic once. It is typically in the low-millisecond range. The PLC continuously scans the logic. The time it takes to evaluate it once is the scan time.

Scope Who has access to a tag. There are two scopes for tags in CLX: controller scope and program scope. A controller scope tag is available to every program in the project. It is also available to the outside world, such as SCADA systems. A program scope tag is only available within the program it is created in. Programs cannot access or use a different program's tag if it is program scope.

Serial Data Link Control (SDLC) A subset of the HDLC protocol that is used in a large number of communications systems like Ethernet, ISDN, BITBUS, and others. The SDLC protocol defines the structure of the frames and the values of a number of specific fields in these frames.

Sensitivity The property of a device to discriminate between levels. Sensitivity for a sensor relates to the finest difference the sensor can detect. On some sensors the sensitivity adjustment is used to set the level at which the output should energize.

Sensor A device used to detect change. The outputs of sensors change state when they detect the correct change. Sensors can be analog or digital.

Sequencer An instruction type that is used to program a sequential operation. Similar to a drum controller.

Sequential function chart (SFC) A PLC language that is similar to a decision tree or flowchart. It organizes an application into steps for programming.

Serial Real-Time Communications System (SERCOS) A digital control bus that is used to connect motion controllers, drives, and I/O for motion control applications such as numerically controlled machines. It is very widely used in motion control applications.

Serial communication The sending of data one bit at a time. The data is represented by a coding system such as ASCII.

Shield A barrier that prevents unintentional contact with hazardous machine areas.

SINT Single integer.

Slave The devices on a master-slave network that can transmit information to the master only when they are polled (called) by the master. There is usually one master and several slaves.

Spread spectrum technology (SST) The technology in which the message is modulated across a wide bandwidth, over many different frequencies. This ensures that interference on a single frequency cannot prevent the data from reaching its destination. A special code determines the actual transmitted bandwidth. Authorized receivers use the code to extract the message from the signal. The transmission looks like noise to unauthorized receivers. This makes SST very noise immune. Used for wireless communications. Developed by the U.S. military during World War II to prevent jamming of radio signals. It also helped make signals harder to intercept.

Step A programming element used in SFC programming to divide an application into logical blocks (steps).

Stop functions Functions that bring a machine to a stop. There are three types. Category 0 is an uncontrolled stop by immediately removing power to the machine actuators. Category 1 is a controlled stop with power to the machine actuators available to achieve the stop then removed when the stop is achieved. Category 3 is a controlled stop with power left available to the machine actuators.

Structures Entities that enable a programmer to create a structure-type tag, available in CLX, that can hold multiple data types. An array can only hold one data type.

Structured text (ST) A PLC language that is very similar to regular computer programming language such as C or Pascal.

T

Tagout The placement of a tag on an energy-isolating device, in accordance with an established procedure, to indicate that the energy-isolating device and the equipment being controlled may not be operated until the tag is removed.

Tagout device A very visible warning device, such as a tag and a means of attachment, that can be securely fastened to an energy-isolating device in accordance with an established procedure to indicate that the energy-isolating device and the equipment being controlled may not be operated until the tagout device is removed.

Task A CLX project can have one or more tasks. Tasks can be used to divide an application (project) into logical parts. Tasks have a couple of important functions. The task is used to schedule the execution of programs in the task. A CLX project can have up to 32 tasks. A task's execution can be configured to be executed continuously, periodically, or on the basis of an event.

Thermocouple A temperature-sensing device. It changes a temperature to a current. The current can then be measured and converted to a number by a PLC input module.

Thumbwheel A device used by an operator to enter a number between 0 and 9. Thumbwheels can be combined to enter numbers of more than one digit. Thumbwheels typically output BCD numbers.

Token passing A communications access method in which only the device that has the token can talk. It is considered a deterministic access method.

Tolerable risk What a company considers acceptable for a given task-hazard combination.

Transition An element used in SFC programming that determines when the next step can be executed.

Transitional contact A contact that changes state for one scan when energized.

True A 1 or high state.

Two-hand control device An actuating control that requires the concurrent use of the operator's hands to initiate machine motion during the hazardous portion of the machine cycle.

U

Underwriters Laboratory (UL) An organization that operates laboratories to investigate systems with respect to safety.

User Defined tag A tag in CLX that is defined by a user. A User Defined tag type can contain multiple tag members that can be of different types.

V

Validation The confirmation by examination and testing that the particular requirements for a specific intended use are met.

Verification The process or act of confirming that a device or function conforms or performs to its design.

Volatile memory Memory that is lost when power is lost.

W

Watchdog timer Tasks have a watchdog timer. A major fault occurs if the scan time exceeds the watchdog value. It can be set under task properties.

Word The length of data in bits that a microprocessor can handle. For example, a word for a 32-bit computer would be 32 bits long, or two bytes. A 64-bit computer would have a 64-bit word.

INDEX

1-dimensional arrays, 29, 154
2-dimensional arrays, 30
3-dimensional arrays, 30
7-bit ASCII, 292–293
8-bit extended ASCII, 292–293
556x processors, 9
1756-OB16E output module, 93
5555 ControlLogix processors, 9
60204-1 standard, 367

A

ABS (absolute) instruction, 269
ABS (Absolute value) function, 206
absolute encoders, 125–126
absolute move, 333–334
absolute position, 319
AC devices, 93
AC output module and wiring, 95
AC outputs and surge suppression, 399
ACC (preset count), 273
ACC value, 74, 77
acceleration move profile, 331–332
acceleration parameter, 331–333
acceptable risk, 345
accumulating timers, 70, 76
ACOS (Arc cosine) function, 206
ACS instructions, 165
Action Properties screen, 236
action tags structure, 235
actions, 221
    assignment statement (;) and, 236
    Boolean, 236–238
    calling subroutine, 239
    comments, 254
    non-Boolean, 236
    order of execution, 238
    qualifiers, 240
    setting bit member, 237
Add (+) operator, 205
ADD (add) instruction, 152
ADD function block, 260–261, 266
Add-On instructions, 279, 341
    benefits, 280
    creation, 285–287
    data types, 282
    defining primary functionality, 282
    description, 285
    developing, 280
    developing logic, 286–287
    documentation, 279, 284–285
    help, 284
    Help file instructions, 285
    local tags, 281
    logic routine, 282
    parameters, 280
    scan mode routines, 283
    storing, 287
Add-On Instructions command, 285
adjustable guards, 362

Agere, 299
Aironet, 299
alarm area, 372
AlarmHigh member, 232
alarms
    analog modules, 104–105
    PID (Proportional, Integral, Derivative) instructions, 179
alias-type tags, 27
American Society of Safety Engineers, 343
analog data and integer mode, 110
analog I/O module, 4
analog input modules, 97–112
    inputs, 14, 176
    module properties screen, 99–100
    troubleshooting, 402–403
analog modules
    alarms, 104–105
    current-sensing, 99
    data format, 108
    differential inputs, 106–107
    field calibration, 102–103
    freezing state of channels, 103
    high-speed-mode differential wiring method, 107
    resolutions, 101
    RPI (Requested Packet Interval) parameter, 98–99
    RTS (Real Time Sample) parameter, 98
    scaling, 100–101
    sensor offset, 104
    single-ended inputs, 105
    timing parameters, 97–99
    updating controllers, 103
analog output, 176
    ramping signal, 110–111
analog output devices, controlling, 15
analog output modules, 15–16
    RPI value, 108–109
    troubleshooting, 404
    wiring, 111
analog resolution, 101
analog sensors output, 137
analog voltage input modules, 99
AND condition, 51
AND instruction, 163
AND logic block, 383–384
AND operation, 163
ANSI (American National Standards Institute)
    ANSI B11.TR3-2000, 344
    control reliability in Standard B11.192003 (3.14), 381
ANSI B11-2008 safety standard, 344, 359–360, 362
ANSI B11.TR3-2000 standard
    acceptable risk, 345
    frequency of exposure to hazard, 346
    probability of injury occurring, 346

responsibility on machine manufacturers and users, 344–345
risk estimation example, 347–348
safeguarding, 348
severity, 345
tolerable risk, 345
ANSI PMMI B1555.1-2006, 344
ANSI/RIA-1999. "Industrial Robots and Robot Systems" standard, 345
ANSI/RIA 15.06-1999 standard, 344
    risk classification, 351
ANSI/RIA R16.06-1999 (4.5.4) standard, 382
applications and modules, 6
arcs, 322–323
arithmetic conversion operations, 156–157
arithmetic functions, 205–206
arithmetic operators
    order of precedence, 209
    ST (structured text) programs, 204–205
arrays
    creation, 30–31
    dimensions, 29–30
    elements, 29
    filling range of memory locations in, 174
    holding values of multiple tags, 29
    indexing, 29
    taking one order out of, 192–193
    User-Defined structure tags, 35
ASCII, 292–293
ASCII/RTU mode, 296–297
ASIN (Arc sine) function, 206
ASN instructions, 165
assignment statements (;), 201, 203
Assume Data Available Marker, 263–264
asynchronous serial communications, 293
ATAN (Arc Tangent) function, 206
ATN instructions, 165
automated systems
    clearance distances around devices, 389
    disconnects, 392
    documentation, 388
    enclosures, 388–391
    fusing, 388
    installation, 387–393
    math instructions, 151
    MCR (master control relay), 392–393
    metal chips or components, 391
    reliability, 387
    temperatures, 389
    transformers, 393
    wiring, 388
automating processes, 2
automation, 113
    hardwired relays, 2
    safety, 113
AVE (average) instruction, 154–155
averaging values, 154–155

INDEX

axes
    calibrating position to known reference, 318
    closing servoloop, 328
    direction, 330
    homed, 326
    interpolation, 320
    jogging, 326, 329–333
    known position (home position) for, 326
    moving and controlling, 328–329
    moving in absolute or incremental mode, 333–334
    multidimensional linear coordinated move, 335–339
    resolution, 319
    speed, 330–331
    stopping, 326
    testing, 326
axis of motion
    single-axis of motion, 316–319

B

backbone cable, 298
background tasks, 20
backplane, 5–6
backplane connector, 12–14
backplane CPU, 9
ball screws, 317–318
BAND (Boolean AND) function block, 268
base-type tags, 25–27
BAT, 7
battery backup, 9–10, 400
BCD value, converting integer to and from, 271–272
bifurcated cables, 118
bipolar modules, 99
bit file, comparing values, 174–175
bits, 163–164
black wiring, 394
blanking, 376
bleeder resistor, 92
blue wire, 394
Boolean actions, 236–238
Boolean (BOOL)-type tag, 26
Boolean transitions, 241
BOR (Boolean OR) function block, 268
branches, 52–53, 56
branching outputs, 53
BRK (break) instruction, 195
BSL (bit shift left) instruction, 191–192
BSR (bit shift right) instruction, 191–192

C

cables, 118
calibration of analog modules, 102
calorimetric principle flow sensors, 133–134
CAN chip, 301
capacitive field sensors, 126
Cartesian coordinate system, 319
    multidimensional linear coordinated move for axes, 335–339
cascade control, 182
CASE OF statement, 212–213
CCW limit switch, 318
certified safety controller, 381
Change History entry screen, 283
channels and encoders, 124
chassis, 5–6
    CST (Coordinated System Time), 89
    grounding, 395
circular interpolation, 321–323, 339–341
circular path, 321–323
Cisco, 299
clockwise motion, 321
CLX DC input modules, wiring, 89
CMOS RAM, power for, 9
CMP (compare) instruction, 161–162
coils, 16, 45, 48–49
color mark sensors, 115, 122
colors, differentiating between, 122
3Com, 299
combination modules, 99
comments, 203–204, 254
communications
    control-level, 291, 306–310
    enterprise-level, 310–312
    serial, 292–296
    wireless, 299–300
CompactFlash card, 10
comparing
    bits, 164
    discrete inputs, 268
    large blocks of data, 174–175
    values, 161–162
Compatible keying, 88
computers, 4, 7
conditions
    locking in, 60
    true and actions, 211–212
configuration tags, 38–39
connectors
    FBD (function block diagram) programs, 265–266
    names, 266
constant as masks, 186
constraints, moving to another location, 172–173
constructs, 209
    CASE OF statement, 212–213
    FOR DO statements, 211
    ELSEIF (Else If) statements, 210–211
    REPEAT UNTIL statements, 212
    WHILE DO statements, 211–212
consumed-type tags, 32
contacts, 16, 45
    conditions on rungs, 49
    multiple, 50–53
contiguous memory, 174–175
continuous tasks, 20–21, 23, 41
continuously running timers, 89
control devices, battery backup, 400
control instructions
    BRK (break) instruction, 195
    EVENT instructions, 197
    GSV (get system values) instruction, 196
    FOR (loop) instruction, 194–195
    MCR (master control reset) instruction, 194
    RET (return) instruction, 195
    SSV (set system values) instruction, 196
control interlocking, 367
control-level communications
    ControlNet, 307–310
    SERCOS (Serial Real-Time Communications System), 306–307
control-level networks and ControlNet, 291
control reliability, 381–382
control transformers, 393
Controller Organizer, 274
Controller Organizer screen, 286
Controller Properties ST dialog box, 232–234
controller scope tags, 28
controllers
    See also CPUs (central processing units)
    Ethernet capability, 311
    getting and setting status data, 196
    Listen-Only modes, 84
    overhead operations (output processing) for I/O modules, 41
    owning I/O modules, 83
    owning output module, 84
    sharing tag information with multiple, 32
ControlLogix (CL or CLX), 1, 3
    architecture, 13
    capability as communications gateway, 308
    circular interpolation, 321
    communications modules, 308
    controlling robot, 341
    CPUs (central processing units), 4
    motion control, 323–339
    multiple types of control and communications, 308
    programming languages, 17
    projects, 20–23
ControlLogix controller, 315–316
ControlLogix CPU, 7, 9
ControlLogix Part_Timer tag members, 72
ControlLogix PLC, 2, 4
ControlLogix projects and ST programs, 201–204
ControlLogix technology, 1
ControlNet, 291
    bus, star, or tree topologies, 308
    compatibility, 308
    deterministic, 307, 310
    dual-media option, 308
    performing multiple functions, 307
    producer/consumer method, 310
    RG6-U cable, 308
    token-ring system, 309–310
ControlNet network
    devices, 309
    distance, 307
    guardband period, 309
    logically arranging devices, 310
    maintenance message, 309
    maximum speed, 308
    moderator frame, 309
    NUT (network update time), 309
    peer to peer or master-slave communication, 309–310
    scheduled time, 309
    unscheduled time, 309

INDEX

convergent photosensors, 120
Coordinate System Properties General entry screen, 336
Coordinate System Properties Units entry screen, 336
Coordinate System tag name, 336
coordinated motion, 333
coordinated motion instructions, 333
COP (copy) instruction, 171–172
copy conversion operations, 156–157
COS (change of state) instruction, 89, 165
COS (Cosine) function, 206
count-up enable input, 273
counterclockwise motion, 321
counters, 77–79
   function blocks, 273–274
CPT (compute) instruction, 156–157
CPUs (central processing units)
   adding memory, 5
   battery backup, 9–10
   ControlLogix PLC, 4
   Ethernet, 7
   explicit messaging, 305–306
   grounding, 395
   I/O messaging, 305
   memory cards, 10
   microprocessor, 7
   modular PLCs, 6–8
   power requirements, 6
   protecting from real world, 11, 84–85
   RS232 serial programming port, 7
   status LEDs, 7
   troubleshooting, 404
crash switch, 384
CRC (cyclical redundancy check), 297
cross talk, 295
CSMA/CD (Carrier Sense Multiple Access/Collision Detection), 310–311
CST (Coordinated System Time), 89
CTD (count-down) counters, 79
CTU (count-up) counters, 77–79
CTUD (count up/down) counters, 79
CTUD instruction, 274
current output, 136
current-sensing analog modules, 99
CV (control variable), 176
CW limit switch, 318
cyclic transmission, 303

D

D (derivative) gain, 177
daisy chaining, 304
dark-on outputs, 118
dark-operate output, 118
data
   comparing large blocks, 174–175
   copying bit by bit, 173
   floating point, 108
   integer, 108
   masks, 173
data format for analog modules, 108
data types, 35
   Add-On instructions, 282
   tags, 328
DC devices and discrete output modules, 93
DC output module wiring, 94
DC outputs
   noise suppression, 399
   transistor output, 93
DDT (diagnostic detect) instruction, 174–175
dead-band parameter, 181
deceleration move profile, 331–332
deceleration parameter, 331–333
decimal (floating-point) values, 27
DEG instruction, 167, 271
DEG (radians to degrees) function, 206
degrees, converting radians to and from, 166, 271
device-level networks, 300–301
   DeviceNet, 291
DeviceNet, 291
   broadcast-based communications protocol, 301
   CAN chip, 301
   change of state, 302–303
   communications flow, 305–306
   communications module, 305
   components, 303
   cyclic transmission, 303
   devices, 301–302
   drop lines, 303
   flat cable, 303
   Flex I/O modules, 305
   flexibility in wiring topology, 304
   node, 303
   open network standard, 301
   polled communications, 302
   scanlist, 305
   scanner, 305–306
   strobed communications, 302
   thick wire, 303
   troubleshooting, 306
   trunk lines, 303–304
   wiring, 303–305
devices
   clearance distances around, 389
   collision, 311
   ControlNet network, 309
   DeviceNet, 301–302
   gaining access to network, 299
   polling for status, 302
   reporting only when data changes, 302–303
   strobing for status, 302
   transmitting, 310–311
   turning off at end of step, 232–234
   two states, 46
DF1 protocol, 296
DH+ (Data Highway Plus), 298–299
diagnostic digital I/O modules LED indicators, 85–86
diagnostic modules and tags, 37–38
diagnostic output modules, 96
differential encoders immunity to noise, 124–125
differential inputs and analog modules, 106–107
differential wiring, 106–107
   incremental encoders, 124–125
diffuse sensors, 115
digital I/O module, 4
digital input modules, 84–97
   accessing CST, 89
   input, 84–85
   LED status information, 85–87
digital output modules, fusing, 97
digital sensors, 114–115
digital signals, 14
digital values, converting analog output signals to, 108–109
Disable keying, 88
disconnects and automated systems, 392
discrete devices states, 46
discrete input modules, 13
   troubleshooting, 402
discrete inputs, comparing, 268
discrete modules. *See* digital modules
discrete output modules, 93–94
   troubleshooting, 403
discrete sensors, 114
disturbances, dealing with known, 181–182
DIV (divide) instruction, 153
DIV function block, 267
Divide (/) operator, 205
division, remainder of, 155, 269
DN bit, 74, 247, 273
documentation
   Add-On instructions, 279, 284
   automated systems, 388
   SFC (sequential function chart) programs, 254
double-integer (DINT) -type tag, 26
down counters, 77, 79
drive application, sequence for starting, 325–326
drives
   controlling position and velocity, 326
   enabling, 326, 328
   enabling input, 319
   homed, 326
   motion direct commands, 326–328
drop lines, 303–304
drum controllers, 183–184
dual-channel gate switches, 374
dual-channel safety switch wiring diagram, 380

E

E-stop button, 365
E-stop condition, 383
E-stop input circuit safety relays, 371–372
E-stop pushbuttons, 367–368
E-stop switch, 372, 393
Edit tag command, 250
Edit Tags command, 30
EDM (external device monitoring) contacts, 374
efector dualis sensors, 121
EIA/TIA-485, 295
electrical noise, 397–399
electromechanical relays, 371
electronic fuses, resetting, 97
electronic keying, 88
electronic sensors, 114

INDEX

electronically fused output module and LED troubleshooting panel, 93
elements, 29
ELSEIF (Else If) statements, 210–211
emergency stop switches, 392–393
EMI (electromagnetic interference), 395
employee notification of lockout and tagout devices, 414
EN 953. "Safety of Machinery Guards, General Requirements for the Design and Construction of Fixed and Movable Guards" standard, 365
EN 60 204-1 standard, 363
EN 60947-5-1 standard, 363
EN 954-1 standard
   categories to describe level of safety requirements, 352
   safety system classifications, 352–353
EN 954-1 standards, 351–352, 355, 368
EN 1088 standard, 363
EN (enable) bit, 71
EN ISO 13849-1 standard, 351, 354–355
EN ISO 12100 standard, 348–349, 355
enclosure fans, 400
enclosures
   automated systems, 388–391
   IP (Ingress Protection) rating, 388
   NEMA's standard, 389
   power supply to convert AC to DC, 393
encoders
   absolute, 125–126
   channels, 124
   incremental, 122–125
   index pulses, 318
   LEDs, 122
   position feedback, 122
   quadrature, 124
   resolution, 122
   velocity feedback, 122
endpoint, 322
energy-isolating devices, 413–414
energy lockout/tagout standard, 411–419
enterprise-level communications, 310–312
EOT (End of Transition), 242, 243
EQU (equal) math instruction, 78–79
EQU (equal to) instruction, 158
Equal (=) operator, 206
equipment
   returning to service, 420
   risk assessments, 344
equipment protection earth ground, 397
ESEL (enhanced select) function block, 277–278
ESPE (electrosensitive protective equipment) standards, 361
Ethernet, 7, 310–312
Ethernet/IP
   connecting remote analog output modules, 109
   connecting remote input modules, 99–100
Ethernet module, 99
Ethernet network
   devices, 310
   underloading, 311

Ethernet TCP/IP, 312
European safety standards
   EN ISO 12100, 348–349
   ISO 14121-1:2007 standard, 349
event-based tasks, 20–21, 23
EVENT instructions, 197
event-type tasks, 197
Exact Match keying, 88
execution types characteristics, 40
explicit messaging, 305–306
Exponent (°°) operator, 205
extruders, 419–420

F

fail-safe stop switch, 54
Faraday's law of electromagnetic induction, 133
fault status LED, 86
FBC (file bit comparison) instruction, 174–175
FBD (function block diagram) programming
   ADD function block, 260–261
   extensive information/data flow, 259
   ladder diagram, 285
FBD (function block diagram) programs
   connectors, 265–266
   elements of function blocks, 261–262
   feedback, 263–265
   order of execution, 262–263
   starting, 274
   wire type, 262
FCC (Federal Communications Commission), 300
feedback and FBD (function block diagram) programs, 263–265
feedforward, 181–182
FFL (FIFO load) instruction, 192
FFL (file fill) instruction, 174
FFU (FIFO unload) instruction, 192–193
fiber-optic cables types, 118
fiber-optic sensors, 117
fiber-optic thru-beam sensors, 147
field buses, 300
field calibration, 102–103
field devices, 301
field sensors
   capacitive, 126
   hysteresis, 126–127
   inductive, 126
   mounting, 129–130
   preventing object from teasing, 127
   sensing range, 128–129
   shielding, 129
file instructions
   COP (copy) instruction, 171–172
   DDT (diagnostic detect) instruction, 174–175
   FBC (file bit comparison) instruction, 174–175
   FFL (file fill) instruction, 174
   MOV (move) instruction, 172–173
   MVM (masked move) instruction, 173
file shifts, 192
fixed blanking, 376

fixed guards, 362
floating blanking, 376
floating-point data, 108
floating-point mode, 110
floating-point value, 272
flow control instructions, 62
flow sensors
   calorimetric principle flow sensors, 133–134
   magnetic inductive flow meters, 133
   ultrasonic flow sensors, 134–135
FLT LED, 7
flush sensors, 129
FOR DO statements, 211
FOR (loop) instruction, 194–195
FORCE LED, 7
forward kinematics, 341
FRD instruction, 167, 271–272
full duplex, 295–296
Function Block command, 274
function block diagram, 17
function block instruction, 260
function blocks
   Assume Data Available Marker, 263–264
   counters, 273–274
   data flow between, 262
   elements, 261–262
   feedback, 263–265
   ICONs (input connectors), 262
   IREF (input reference), 261, 262
   loops, 263
   mathematical, 266–269
   mathematical conversion, 271–272
   multiple connections between, 264
   OCONs (output connectors), 262
   one-scan delay between, 264
   order of execution, 262–263
   OREF (output reference), 261, 262
   outputs, 259
   program/operator control, 277–279
   programming routines, 274–276
   providing input data, 265
   SEL (select) function block, 276
   setting parameters, 261
   statistical, 269–271
   switching between program and operator control, 278
   tag members, 260
   tags, 260
   timers, 272–273
   transferring data between output and input pins, 265–266
   trigonometric, 269
   wire type, 262
   wiring, 262–263
functional ground, 397
fuse status LED, 86
fusing automated systems, 388

G

gate switches, 369–371
   dual- or single-channel, 374
gates, 369–371
gateways, 308
generic CL I/O module, 12–13

INDEX

generic output module, 14
GEQ (greater than or equal to) instruction, 159
GM (General Motors Corporation), 2
Greater than (>) operator, 206
Greater than or equal (>=) operator, 206
green wire, 394
ground, 397
ground loops, 398
grounding, 395–397
group lockout/tagout, 418
GRT (greater than) instruction, 159
GSV (get system values) instruction, 196
guard-limit switch, 393
GuardPLC, 386
guards
   adjustable, 362
   ergonomic considerations, 364
   fixed, 362
   interlocked, 363
   self-adjusting, 362–363
   standards, 365

H

half duplex, 295–296
handshaking, 294, 408
hardware relays wiring, 2
hardwired relays, 2
Harris Semiconductor, 299
hazards and risk estimation analysis, 350–354
Help tab screen, 284
High alarm, 104
high-density I/O modules, 97
high-density modules, 97
High-High alarm, 104
high-speed-mode differential wiring method, 107
high-voltage spikes, 398
higher-priority tasks, 41
HMI (human-machine interface) monitors, 21
home command, 326
home switch, 318–319
homing, 318–319
hot tap operations, 412
hysteresis, 126–127

I

I (integral) gain, 177
I/O addressing, 36
I/O LED, 7
I/O messaging, 305
I/O module tags, 37–39
I/O modules, 83–84
   high-density, 97
   inserting and removing from chassis, 85
   mechanically keying, 87
I/O status LEDs, 86
I/O table, updating, 50
ICONs (input connectors), 262, 265–266, 275
IEC 61508, "Functional Safety of Electrical/Electronic/Programmable Electronic Safety-Related Systems" standard, 381

IEC 61131-3 international standard, 17
IEC 61131-3 languages, 259
IEC 61496 international standard, 361
IEC 62061 standard, 381
IEC/EN 62061 standard, 354–355
IEC (International Electrotechnical Commission), 388
IEEE 802.3 standard, 310
IEEE 802.11 standard, 299
IF statements, 209–210
immediate outputs, 59
In signal, 270
incremental encoders, 122–126
incremental moves, 333–334
incremental position, 319–320
index pulses, 318
inductive devices, 398
inductive field sensors, 126
inductive RFID technology, 131
inductive sensors, 129, 148
industrial buses, 300
industrial controllers, 291
industrial devices, 291
Industrial Ethernet, 307, 311
industrial networks categories, 291
information networks, 291
injuries, 343, 351
input modules
   analog, 14
   calibration, 102
   converting electric signal to binary equivalent, 12
   diagnostics, 11
   discrete, 13
   LEDs for monitoring, 86
   remote, 99–100
   replacing, 13
   rolling time stamp, 89
   sensor connected to, 11
   sinking or sourcing, 90–92
   states for LED status indicator, 87
   tags, 38
   time-stamping inputs, 89
   wiring, 89–90
input monitoring, 378
input signal, 269–271
input tags, 38
input value, converting to floating-point value, 272
inputs
   analog input modules, 14, 176
   continuously monitoring, 50
   counters, 273–274
   digital modules, 84–85
   highest, lowest, median, or average, 277
   LEDs for monitoring, 86
   PLCs, 11–14, 47
   program or operator control, 277
   selecting, 276
   time-stamping, 89
   troubleshooting, 402
inspection photosensors, 121
instruction list, 17
instructions
   Add-On, 279

control, 194–197
displaying information about, 196
file, 171–175
jump, 193
languages used in, 213
load, 192–193
logical, 162–164
math, 151–157
math conversion, 167
MSG (Message), 175
PID (Proportional, Integral, Derivative), 175–182
relational, 158–162
retentive, 61
sequencer, 184–190
shift, 190–192
trigonometric, 165–166
integer data, 108
integer (INT)-type tag, 26
integer mode, 110
integers, 271–272
interlocked guards, 363–364
interpolation, 320
   circular, 339–341
Intersil, 299
inverse kinematics, 341
IOT (immediate output) instruction, 59
IP 54 enclosure, 388
IP (Ingress Protection) rating, 388
IREF (input reference), 261, 262, 265
ISO 14121-1:2007 standard, 349
ISO EN138490-1 and -2 standards, 368
ISO EN 13849 standard, 381
isolation transformers, 393

J

JMP (jump) instruction, 193
jogging
   acceleration or deceleration parameter, 331–333
   axes, 329–333
   direction, 330
   merging, 333
   speed, 330–331
   speed units, 331
JSR (jump-to-subroutine) instructions, 23, 62–63, 193
   ST programs, 202
jump instructions, 193

K

key switch, 7
keying modules, 87

L

ladder diagrams, 17, 49–50, 285
ladder logic, 2, 7, 199
   coils, 16, 45, 48–49
   contacts, 16, 45–46
   EN (enable) bit, 71
   instructions, 45
   ladder diagrams, 49–50
   normally closed contacts, 47–48
   normally open contacts, 46–47
   outputs, 55–64

power rails, 16
PRE value, 71, 73
problems, 408
real-world switches, 46–48
RSLogix 5000 toolbar, 49
SQO (Sequencer Output) instruction, 187
start/stop circuits, 54–55
timer status bits, 71–73
laser distance sensors, 145
laser scanners, 372, 379–380
laser sensors, 120, 147
latches, 60
latching instructions, 60
LBL (label) instructions, 62
LED status indicator, 87
LEDs
   encoders, 122
   laser sensors, 120
   optical sensors, 115
LES (less than) instruction, 160
Less than (<) operator, 206
Less than or equal (<=) operator, 206
LFL (LIFO load) shift instructions, 192
LFU (LIFO unload) shift instructions, 192
light curtain, 371–373, 384
   blanking, 376
   muting, 376–378
   safety relay, 377–378
light-on outputs, 118
light-operate output, 118
LIM function block, 274
LIM (limit) function block instruction, 274
LIM (limit) instruction, 74, 160
limit switches, 90
line filters, 397
line-powered sensors, 140
linear interpolation, 320
linear output sensors, 114
Listen-Only modes, 84
LN (Natural log) function, 206
load instructions, 192–193
load-powered sensors, 139–140
lockout, 414
lockout devices, 414
   removing, 417–418
   requirements, 416
   testing machines, equipment or components, 418
lockout/tagout checklist example, 421
lockout/tagout standard
   defining energy source, 412
   energy-isolating devices, 413
   group lockout/tagout, 418
   hot tap operations, 412
   lockout, 414
   normal production operations, 412
   outside maintenance or servicing personnel, 421
   personnel or shift changes, 419
   procedures for application, 416–417
   retraining, 415
   servicing or maintenance, 412–413
   tagout, 414
   training, 415
LOG (Log base 10) function, 206
logic, evaluating, 50

Logic menu, 30
logic routine, 282
Logical AND (&, AND) operator, 207
logical conversion operations, 156–157
logical instructions
   AND instruction, 163
   NOT instruction, 163–164
   OR instruction, 164
logical operators, 207–208
Logix5000 controller, 40
Logix family of controllers, 3
looping routines, 194–195
loops
   function blocks, 263
   number of times executed, 211
Low alarm, 104
Low-Low alarm, 104
lower-priority tasks, 41
LRC (longitudinal redundancy check) calculation, 297
Lucent Technologies, 299

M

machine limits, 350
machine safety, 344
machines
   circular interpolation, 321–323
   linear interpolation, 320
   returning to service, 420
MAFR (Motion Axis Fault Reset) command, 327
magnetic inductive flow meters, 133
MAH (Motion Axis Home) instruction, 319, 326, 328–329
MAH motion direct command, 326
main device, 296
main power disconnect, 392
main-sub mode, 296
MainProgram command, 246
MAJ (Motion Axis Jog) instruction, 328–333
MAM (Motion Axis Move) instruction, 328, 333–334
MAS (Motion Axis Stop) instruction, 318, 326, 328
   MAJ instruction and, 329–330
MASD (Motion Axis Shutdown) command, 327
masks, 173, 186
MASR (Motion Axis Shutdown Reset) command, 327
master devices addresses, 297
master-slave mode, 296
math conversion instructions, 167
math instructions, 151
   ADD (add) instruction, 152
   AVE (average) instruction, 154–155
   CPT (compute) instruction, 156–157
   degrees-to-radians conversion, 166
   DIV (divide) instruction, 153
   MOD (modulo) instruction, 155
   MUL (multiply) instruction, 153–154
   NEG (negate) instruction, 156
   SQR (square root) instruction, 156
   SUB (subtract) instruction, 152
math operations
   equal order, 156–157

overriding precedence, 157
math statements and precedence, 208
mathematical conversion function blocks, 271–272
mathematical function blocks
   ABS (absolute instruction), 269
   ADD function block, 266
   BAND (Boolean AND) function block, 268
   BOR (Boolean OR) function block, 268
   DIV function block, 267
   MOD instruction, 269
   MUL function block, 267
   NEG instruction, 269
   SUB function block, 267
MAVE (moving average) instruction, 269, 270
MAX instruction, 270
MAXC (maximum capture) instruction, 270–271
maximum ramp, 111
MCCM (Motion Coordinated Circular Move) instruction, 321, 339–341
MCLM (Motion-Coordinated Linear Motion) instruction, 336–339
MCR (master control relay) instruction, 392–393
MCR (master control reset) instruction, 194
MCR zone, 194
MDF (Motion Direct Drive Off) instruction, 327
MDO (Motion Direct Drive On) instruction, 327
mechanical devices, 113–114
mechanical disconnect, 392
mechanical sensors, 113–114
mechanical switches, 46, 113–114
members, 32
memory, 4
   moving to another location, 172–173
   running programs in, 7
   setting bit, 48
   user-friendly name for location, 24
memory cards, 10
merge disabled, 333
merge parameter, 333
merge speed, 333
mesh wireless networks, 300
metalworking machine tools and ANSI B11 safety standards, 362
microprocessor, 7
MIN instruction, 269
MINC (minimum capture) instruction, 269, 271
minimum capture, 269
MOD (modulo) instruction, 155, 269
Modbus, 296–298
Modbus Plus, 296, 298
Modbus/TCP, 296
modular PLCs
   battery backup, 9–10
   chassis, 5–6
   CPUs (central processing unit), 6–8
   memory cards, 10
   power supply, 6
module status LED, 86

INDEX

modules, 5–6, 83
   analog input, 97–112
   analog output, 108–109
   analog resolution, 101
   applications, 6
   bipolar, 99
   characteristics and operation, 38
   combination, 99
   Compatible keying, 88
   configuring, 84
   digital, 84–97
   Disable keying, 88
   discrete output, 93–94
   electronic keying, 88
   Exact Match keying, 88
   faults, 85–87
   floating-point mode, 110
   integer mode, 110
   keying, 87
   LED status information, 85–87, 97
   Locking tab, 85
   modifying input or output, 100–101
   output resolution, 109
   power requirements, 6
   replacing, 88
   rolling time stamp, 89
   RTBs (Removable Terminal Blocks), 88
   scaling, 109–110
   sinking *versus* sourcing, 90
   tags automatically created for, 37–38
   unipolar, 99
modulo instruction, 205
Modulo (MOD) operator, 205
motion, programming logic for, 328
motion commands, 328–329
motion control
   ControlLogix, 323–339
   ControlLogix controller, 315–316, 341–342
   X axis, 315
   Y axis, 315
motion control applications and SERCOS (Serial Real-Time Communications System), 306–307
Motion Direct commands, 326–328
Motion Direct Commands window, 326
Motion Direct-Motion Move commands, 328
motion function block tags, 328
motion instruction, 328
motion tag members, 329
motion-type tag, 339
MOV (move) instruction, 172–173
move profiles, 331–332
moving average, 269–270
moving standard deviation processes, 269
MSF (Motion Servo Off) command, 327
MSG (Message) instructions, 175
MSO (motion servo enable) instruction, 326, 328
MSTD instruction, 269
MUL function block, 267
MUL (multiply) instruction, 153–154
multiaxis motion synchronized motion control, 323
multidimensional arrays and AVE (average) instruction, 155

multidimensional linear coordinated move, 335–339
multidrop protocol, 295
multiple contacts, 50–53
Multiply (*) operator, 205
multipoint to peer wireless networks, 300
muting, 376–378
MUX (multiplex) instruction, 279
MVM (masked move) instruction, 173

N

National Fire Protection Associates, 394
NEG (negate) instruction, 156, 269
negative output, 95
NEMA (National Electrical Manufacturers Association), 388–390
NEQ (not equal to) instruction, 161
networks
   device-level, 300–301
   devices gaining access, 299
   reducing unnecessary traffic, 303
   wiring, 301
neutral, 397
New Add-On Instruction command, 285
New Data Type command, 33
NFPA 79: Electrical Standard for Industrial Machinery standard, 367, 394
NFPA 70 National Electrical Code, 394
no-load detection, 96
node, 303
noise, 397–399
Nokia, 299
non-Boolean actions, 236
nonflush sensors, 129
nonretentive timers, 70, 274
nonshielded sensors, 129–130
normal production operations, 412
normally closed contacts, 47–48
normally closed switches, 46–48
normally open contacts, 46–47
normally open switches, 46–47
Not equal (<>) operator, 206
NOT instruction, 163–164
NOT (Logical complement) operator, 207
NPN (Sinking Type) sensors, 141–142
numbers
   adding, 152, 266
   comparing bits, 164
   conversion operations, 156–157
   dividing, 153, 267
   multiplying, 153–154, 267
   subtracting, 152, 267
NUT (network update time), 89, 309

O

object inspection, 121
object recognition, 121
objects, sensing with light, 115
OCONs (output connectors), 262, 265–266, 275
off-delay timers, 75–76
on-delay timers, 70
ONF (one-shot-falling) instructions, 60
ONR (one-shot-rising) instructions, 60
ONS (one-shot) instructions, 60

Open DeviceNet Vendor Association Inc., 301
OperOperReq input, 277
OperProgReq input, 277
optical sensors
   See also photosensors
   fiber-optic sensors, 117–118
   LEDs, 115
   reflective sensors, 115
   retro-reflective sensors, 116
   thru-beam sensors, 116–117
opto-isolation, 84
opto-isolator phototransistor, 84
optoelectronic safety devices
   blanking, 376
   input monitoring, 378
   laser scanners, 379–380
   light curtains, 375–376
   muting, 376–378
OR conditions, 52–53
OR instruction, 164
OR (Logical OR) operator, 207
OREF (output reference), 261, 262
OSHA (Occupational and Safety Health Administration) guidelines, 344
OTE (output energize) instruction, 48
OTU (output unlatch) instruction, 62
OUT (output match) instruction, 61–62
output biasing, 181–182
output coil, 75
output devices, 93–94
output instructions, 55–56
output modules
   analog, 15–16
   backplane connector, 14
   calibration, 102
   current limit specifications, 93
   digital signals, 14
   fusing, 94, 97
   limiting to single owner, 83–84
   no-load detection, 96
   on- or off-type signals, 14
   relay outputs, 15
   resolutions, 101
   rolling time stamp, 89
   RTB (Removable Terminal Block), 14
   sinking *versus* sourcing, 95–96
   solid-state outputs, 15
   status LEDs, 15
   wiring, 94–95
output resolution, 109
output tags, 39
outputs, 48–49
   analog output, 176
   branching, 53
   counters, 273–274
   energized, 57
   function blocks, 259
   immediate, 59
   inputs, 279
   logic examples, 56
   multiple conditions to control, 53
   negative, 95
   nesting branches, 56
   off and on, 56, 58–60, 70, 76
   PLCs, 14–16, 47

INDEX

positive, 95
scheduling, 89
troubleshooting, 402
outside maintenance or servicing personnel and lockout/tagout standard, 421

P

P (proportional) gain, 176–177
PACs (Programmable Automation Controllers), 2–3
parity and asynchronous serial communications, 293
passive ID (identification) tag, 131
peer-to-peer communication, 296, 298
peer to peer wireless networks, 300
perimeter guarding, 372
periodic tasks, 20–21, 23, 180
photodetector, 115
photoemitter, 115
photoreceiver, 115
photosensors
　See also optical sensors
　convergent, 120
　dark-on outputs, 118
　immune to ambient light, 115
　inspection, 121
　laser sensors, 120
　light-on outputs, 118
　photodetector, 115
　photoemitter, 115
　photoreceiver, 115
　polarizing, 119, 146
　special, 119–122
PID (Proportional, Integral, Derivative) instructions
　alarms, 179
　cascading loops, 182
　configuration parameters, 178–179
　CV (control variable), 176
　D (derivative) gain, 177
　dead-band parameter, 181
　derivative-smoothing filter, 181
　executing, 180
　feedforward, 181–182
　I (integral) gain, 177
　input and output, 176–177
　main parameters, 178
　output biasing, 181–182
　output limiting, 181
　P (proportional) gain, 176–177
　periodic tasks, 180
　PV (process variable), 176
　scaling parameters, 180
　setpoint, 176
　summing junction, 176
　update time, 180
　zero crossing, 181
Pilz GmbH and Company, 371
PLCs
　automating processes, 2
　capabilities, 2
　communicating between, 175
　conceptual view, 50
　CPUs (central processing units), 4–5
　downtime, 400

dust and dirt, 400
enclosure fans, 400
Ethernet communication modules, 311
history of, 2–3
IEC 61131-3 international standard, 17
input, 2, 4, 11–14, 47
ladder logic, 2, 7
maintenance, 400
memory, 4
modular, 5–10
output, 2, 4, 14–16, 47
PACs (Programmable Automation Controllers), 2–3
programming languages, 7, 16–17
replacing components, 400
replacing hardwired relays, 2
run mode, 50
safety-rated, 381
similarities with personal computers, 4
wiring devices, 2
pneumatic cylinders and sensors, 146
PNP sensors, 140–141
point to multipoint wireless networks, 300
polarizing photosensors, 119, 146
polarizing reflector, 119
poor maintenance procedures, 343
positioning, incremental and absolute, 319–323
positive output, 95
positively guided contacts, 369
potential safety hazards, 344
power interlocking, 366–367
power line disturbances, 397
power rails, 16
power supply, 4, 6
PRE (preset) value, 70–71, 77
PRE timers tag member, 70
precedence, 208
predefined structures, 32
pressure sensors, 131–132, 148
process sensors and transmitters, 136
processes
　controlling, 175–182
　moving standard deviation, 269
　sequential, 183–190
produced-type tags, 32
producer/consumer technology, 13
products, monitoring size, 270
PROG position, 7
ProgOperReq input, 277
ProgProgReq input, 277
program flow instructions, 62
program mode, 7, 102
program scope tags, 28
programming
　additional languages for, 199
　circular path, 321–323
　function block routines, 274–276
　ladder logic, 199
　masks, 173
　selection branches, 252–253
　simultaneous branches, 250–251
programming languages, 7, 199
programs, 21, 23
　controlling execution sequence, 62

documenting with comments, 203–204
documenting with tag names, 24
jumping between areas in, 62
jumping to subroutine, 193
routines, 21, 23
Project Explorer window, 285
projects, 20–23
　configuring modules, 84
　programs, 21
　routines, 21
　ST programs, 201–204
　tasks, 20–21
Properties command, 42
pullback devices, 366
PV (process variable), 176

Q

Q bit, 236
quadrature encoders, 124

R

racks. See chassis
RAD (Degrees to radians) function, 206
RAD instruction, 167
radians, converting degrees to and from, 271
radio frequency noise, 398
ramping and analog output signal, 110–111
rate limiting, 110–111
REAL control variable, 181
REAL-type tag, 27
real value, truncating, 272
real-world I/O addressing, 36–39
real-world input states, 56–59
real-world switches, 46–48
redundant PLC, 385
Reference bit, 175
reflective fiber-optic sensors, 117
reflective sensors, 115, 120
relational instructions
　CMP (compare) instruction, 161–162
　EQU (equal to) instruction, 158
　GEQ (greater than or equal to) instruction, 159
　GRT (greater than) instruction, 159
　LES (less than) instruction, 160
　LIM (limit) instruction, 160
　NEQ (not equal to) instruction, 161
relational operators, 206–207
relay outputs, 15
relays, 2, 371–374
REM position, 7
remainder, 155
remote analog output modules, 108–109
remote chassis
　containing I/O module, 84
　Ethernet module, 99
remote input modules, 99–100
remote mode, 7
remote output modules, 109
renaming tasks, 20
REPEAT UNTIL statements, 212
RES (reset) instruction, 74
reset switch, 371–372, 383–384
resistors, 137
resolution, 101

INDEX

axis, 319
encoders, 122
incremental encoders, 126
restraint devices, 366
retentive instructions, 61
retentive output instructions, 62
retentive timers, 70, 274
retraining and lockout/tagout standard, 415
retro-reflective fiber-optic sensors, 117
retro-reflective sensors, 116
RFID (Radio Frequency Identification) sensors, 130–131
risk assessment
   ANSI/RIA-1999. "Industrial Robots and Robot Systems" standard, 345
   categories, 351
   EN 954-1 safety standards, 351–352
   EN ISO 13849-1 safety standards, 351, 354
   IEC/EN 62061 standards, 354–355
   machine limits, 350
   necessary safety level, 350–354
   standards, 344
   task-based, 345
Risk Assessment and Risk Reduction-A Guide to Estimate, Evaluate and Reduce Risks Associated with Machine Tools, 344–348
risk assessment fundamentals, 344
risk classification, 351
risk estimation analysis, 350–354
risk estimation example, 347–348
risk reduction, 344
   EN ISO 12100 standard, 355
   information on residual risks, 355–356
   safe design, 355–356
   strategies, 356–357
   task-based, 345
   technical protective measures, 355–356
robot gripper, 384
robot kinematics, 341
robots, controlling, 341–342
Rockwell Automation, 386
   DF1 protocol, 296
   explicit messaging, 306
   input module wiring, 90
   Logix family of controllers, 3
   subroutine instructions, 62–63
rolling time stamp, 89
routines, 21
   executing, 23
   function block instruction, 260
   interrupting execution, 195
   looping, 194–195
   multiple sheets for, 276
   program tags, 28
   programming for function blocks, 274–276
RPI (Requested Packet Interval) parameter, 98–99
RS232, 7
RS-232, 294–295
RS-422, 294–295
RS-423, 294–295
RS-485, 295–296
RSLogix 5000
   organizing tags, 24
   toolbar, 49
RTBs (Removable Terminal Blocks), 13–14, 88
RTDs (resistive temperature devices), 138
RTO (retentive-timer-on) instruction, 76
RTOR instruction, 274
RTS (Real Time Sample) parameter, 98
run mode, 50, 61, 102
RUN position, 7
RUN status LED, 7
rungs
   disabling, 194
   jumping over multiple, 193
   output instructions in series, 55

S

S-curve profile, 331–332
S-type action, 244
safe design, 355–356
safeguarding methods, 361
   control and interface requirements, 381
   guards, 362–365
   location, 365
safety
   ANSI B11 2008 safety standard, 359–360
   applications and sensors, 113
   considerations, 361
   control reliability, 381–382
   costs, 343
   design, 361
   determining necessary level, 350–354
   E-stop pushbuttons, 367–368
   ESPE (electrosensitive protective equipment) standards, 361
   gates, 369–371
   IEC 61496 international standard, 361
   importance, 343
   interlocking, 361, 366–367
   safety relays, 371–374
   standards, 359–361
   two-hand switches, 368–369
safety bumper, 366
safety controllers, 382–384
safety controls, 361
safety devices, 365–366
safety edge, 366
safety outputs, 384
safety PLCs, 361, 385–386
safety-rated PLCs, 381
safety relays, 371–374, 383
   light curtain, 377–378
   series circuit, 377
   two-hand control, 380
   types, 377
safety standards, 351
safety switches, 373
safety trip controls, 366
safety trip wires, 366
scaling
   analog modules, 100–101
   modules, 109–110
scan mode routines, 283
Scan Modes configuration screen, 283
scan time, 50
scanning, 50
SCL (scale) instruction, 272
scope and tags, 28
Seedbeck, Thomas J., 135
SEL (select) function block, 276
selection branches, 252–253
SelectorMode input, 277
self-adjusting guards, 362–363
sensing range, 128–129
sensor offset, 104
sensors
   analog, 114–115
   choosing, 149
   color mark, 115, 122
   connected to input module, 11
   diffuse, 115
   discrete, 114
   efector dualis, 121
   electric mounting, 144
   electronic, 114
   external power supply, 11
   fiber-optic, 117–118
   fiber-optic thru-beam, 147
   field, 126–129
   flow, 133–135
   flush, 129
   inductive, 148
   installation considerations, 144
   laser, 120, 147
   laser distance, 145
   line-powered, 140
   linear output, 114
   load-powered, 139–140
   mechanical, 113–114, 144
   negative output, 141–142
   nonflush, 129
   nonshielded, 129
   normally open or closed outputs, 142–143
   NPN (Sinking Type), 141–142
   optical, 115
   output current limit, 142
   pneumatic cylinders and, 146
   positive output, 140–141
   pressure, 131–132, 148
   reflective, 115
   response time, 143
   retro-reflective, 116
   RFID (Radio Frequency Identification), 130–131
   safety applications, 113
   shielded, 129
   sinking, 92, 140–143
   smart level, 144
   sourcing, 92, 140–143
   switching frequency, 143
   temperature, 135–139
   three-wire, 140
   thru-beam, 116–117
   timing functions, 142
   two-wire, 92, 139
   ultrasonic, 135
   wiring, 139–140, 144
sequence of events and time-stamping, 89
sequencer array, 190

sequencer instructions
  SQI (Sequencer Input) instructions, 188–190
  SQO (Sequencer Output) instruction, 184–187
sequencers, 183–190
sequential function chart, 17
sequential processes, 183–190
SERCOS I, 307, 324
SERCOS II, 307, 324
SERCOS III, 307
SERCOS (Serial Real-Time Communications System), 17, 306–307
  device standards, 307
  drives, 324–325
  fiber-optic connections, 325
  motion control, 4
  Motion Direct commands, 326–328
  node addresses, 324
  ring topology, 323
  sequence for starting drive application, 325–326
serial communications
  ASCII, 292–293
  asynchronous, 293
  RS-232, 294
  RS-422, 294–295
  RS-423, 294–295
  synchronous, 293–294
serial master-slave system (RS-232/485), 296
serial protocols, 295–296
series logic, 50–52
series output instructions, 55
servos, 17
SFC (sequential function chart) programming, 219–220
SFC (sequential function chart) programs
  action tags, 250
  actions, 221, 235–240, 250
  adding elements, 247–250
  comments, 254
  concurrent step processing, 225–226
  documenting, 254
  ending, 245
  heating application, 222–223
  keeping outputs on, 243–244
  looping back, 224
  restarting after stop, 246
  sample, 222–223
  selection branch, 226–227, 252–253
  SFC_Routine, 246
  simultaneous branch, 226, 244, 250–252
  step and transition icon, 247
  step tags, 250
  steps, 221, 223–234, 246
  Stop element, 245
  stop icon, 248
  text boxes, 254
  transitions, 221, 241–244, 250
  wiring steps, 224
SFR (SFC reset) instruction, 246
shielded sensors, 129–130
shift instructions, 190–192

shift registers, 190–191
Sick safety devices and controllers, 383
signal wiring, 394
signals, improving noise immunity, 106–107
SILs (safety integrity levels), 354
simultaneous branches, 250–252
SIN instructions, 165–166
SIN (Sine) function, 206
single-axis motion instructions, 333
single-axis motions, 333
single-axis of motion, 316
  ball screws, 317–318
  homing, 318–319
  resolution of system, 319
single-channel gate switches, 374
single-ended inputs, 105
single-ended wiring, 105, 124–125
single-integer (SINT)-type tag, 26
sinking input module, 90–92
sinking output modules, 95–96
sinking sensors, 140–143
slave devices, 297
slot fillers, 4
smart level sensors, 144
snubber, 399
solid-state outputs, 15
solid-state relays, 371
source addresses, 163
sourcing input module, 90–92
sourcing output modules, 95–96
sourcing sensors, 140–143
special photosensors, 119–122
SQI (Sequencer Input) instructions, 188–190
SQL (Sequencer Load) instruction, 190
SQO (Sequencer Output) instruction, 184–187
SQR (square root) instruction, 156
SQRT (Square root) function, 206
SST (Spread Spectrum Technology), 300
SSV (set system values) instruction, 196
ST (structured text), 17
  benefits, 200
  executing instructions or calling subroutine, 236
  ladder diagram, 285
  overview, 199–200
  programming fundamentals, 200–201
  timers, 213–215
ST (structured text) programs
  adding routine, 201
  arithmetic operators, 204–205
  assignment statement (;), 201, 203
  case insensitivity, 201
  constructs, 209–213
  ControlLogix projects, 201–204
  entering code, 201
  JSR instruction, 202
  logic, 200–201
  modulo instruction, 205
  preplanning, 203
  sentence-like structure, 200
  step tags, 234
  terminating with semicolon (;), 203

start bit, 293
Start contact, 54
start/stop circuits, 54–55
start switch, 54
statistical function blocks, 269–271
status bits
  counters, 77
  timers, 71–73
status LEDs, 7, 15
Step Properties dialog box, 248–249
Step Properties setup screen, 238
step tags, 234
steps, 221, 246
  AlarmHigh member, 232
  concurrent (simultaneous) processing, 225–226
  DN bit, 247
  identifying, 227–228
  keeping outputs on during multiple, 243–244
  length of time executing, 230
  linear sequence, 223–224
  organizing execution, 223–227
  physical change, 227–228
  PRE (preset time) value, 230
  properties, 248–249
  renaming, 228
  selecting branching, 226–227
  tag members, 228–230
  turning devices off at end, 232–234
  wiring (connecting), 224
Stop contact, 54
Stop element, 245
stop switch, 54
strain gauges, 131–132
strings, comparing, 206–207
structure-type tag, 32
structures, 32, 35
sub devices, 296
SUB function block, 267
SUB (subtract) instruction, 152
subroutine instructions, 62–63
subroutines
  actions calling, 239
  EOT (End of Transition), 242, 243
  exchanging data with, 62
  jumping to, 193
  nesting, 62
  storing sections of logic, 62
Subtract (-) operator, 205
surge suppression, 398–399
surge suppressors, 397
switch/marker method, 318–319
switches
  Ethernet, 312
  gate, 369–371
  guard interlocking, 363–364
  limit, 90
  mechanical, 113–114
  normally closed or open, 46–48
  two-hand, 368–369
Symbol Technologies, 299
synchronous communications, 294
synchronous serial communications, 293
system clock, 89

INDEX

T

tag arrays, copying values between, 171–172
tag data types
  alias-type tags, 27
  arrays, 29–31
  assigning dimensions, 30
  base-type tags, 25–27
  Boolean (BOOL)-type tag, 26
  double-integer(DINT)-type tag, 26
  integer(INT)-type tag, 26
  REAL-type tag, 27
  single-integer (SINT)-type tag, 26
tag editor, 28–29, 185
tag members, 71
  function blocks, 260
  steps, 228–230
tag names, 24
Tag Properties screen, 27
tagout, 414
tagout devices, 414, 416–418
tags
  adding values, 266
  addressing, 23–36
  alternative name (alias), 27
  assigning value to, 203
  automatically created, 37–38
  base-type, 25–27
  case insensitivity, 24
  configuring, 27
  consumed-type, 32
  controller scope, 28
  creation, 28–29
  data types, 328
  diagnostic modules, 37–38
  editing, 29
  function blocks, 260
  holding data, 25–27
  I/O module, 37–39
  input module, 38
  memory allocation, 25
  monitoring, 29
  motion function block, 328
  physical address, 23–24
  produced-type, 32
  program scope, 28
  real-world I/O addressing, 36–39
  scope, 28
  structures, 32
  subtracting values, 267
  symbolic names, 24
  User-Defined structure, 33–35
  values, 29
TAN instructions, 165
TAN (Tangent) function, 206
Target Position Entry screen, 337
task-based risk assessment, 345
task-based risk reduction, 345
Task Properties dialog box, 42
tasks, 20–21, 23, 39, 40–42
  risk estimation analysis, 350–354
  technical protective measures, 355–356
temperature, 135
temperature sensors
  RTDs (resistive temperature devices), 138

thermistors, 139
thermocouples, 135–138
temperatures
  automated systems, 389
  industrial environment, 137
test statement, 209–210
thermistors, 139
thermocouple transmitter, 136–137
thermocouples
  colorcoded for polarity and type, 135–136
  resistors, 137
  types, 138
three-phase power, 392
three-wire sensors, 140
thru-beam fiber-optic sensors, 117
thru-beam sensors, 116–117
time-of-flight measurement, 379
time-stamping inputs, 89
timers
  ACC value, 74
  accumulating, 70, 76
  continuously running, 89
  controlling, 69
  delaying actions, 69
  DN (done) bits, 70–71
  EN (enable) bit, 71
  enabling, 272
  function blocks, 272–273
  misusing, 69
  nonretentive, 70, 274
  off-delay, 75–76
  on-delay, 70
  preset value, 69–70
  renaming, 272
  retentive, 70, 274
  RTO (retentive-timer-on) instruction, 76
  ST (structured text), 213–215
  status bits, 71–73
  tagname, 69
  time base, 69–70
  time delay, 70
  TOF (timer-off-delay) instruction, 75–76
  TON (Timer-On-Delay) instruction, 70
  tracking elapsed time, 69
  TT (timer-timing) bit, 72
TLL (transistor-transistor logic) output modules, 93
TOD instruction, 167, 271
TOF (timer-off-delay) instruction, 75–76
TOFR instruction, 274
token-passing bus network, 298–299
tolerable risk, 345
TON (Timer-On-Delay) instruction, 70
TONR instruction, 274
TONR (timer) function block instruction, 274–275
training and lockout/tagout standard, 415
transformers, 393
transistors, 371
transitions, 221, 241–244
transmitted noise, 398
transmitters
  process sensors and, 136
  strain gauges and, 132

trapezoidal profile, 331–332
tree-type structure, 304
trigonometric function blocks, 269
trigonometric instructions, 165–166
TRN (truncate) instruction, 167, 272
troubleshooting, 400
  analog input modules, 402–403
  analog output modules, 404
  CPUs (central processing units), 404
  discrete input modules, 402
  discrete output modules, 403
  example, 405–406
  inputs chart, 406
  logically isolating probable cause, 401–402
  outputs chart, 407
  people skills, 401
  summary, 405
  use of Web, 401
TRUNC (Truncate) function, 206
truncating, 272
trunk line, 298, 303–304
TT (timer-timing) bit, 72
twisted pair wiring, 295
two-hand control and safety relays, 380
two-hand safety switch, 50–52
two-hand switches, 368–369
two-wire sensors, 92, 139

U

UL (Underwriters Laboratories), 361
ultrasonic flow sensors, 134–135
unipolar modules, 99
unlatch instruction, 61
up counters, 77–79
User Defined command, 33
User-Defined structure tags, 33–35

V

values
  averaging, 154–155
  changing sign, 156
  comparing, 161–162, 206–207
  falling between two values, 160
  greater than, 159
  greater than or equal to, 159
  inequality, 161
  less than, 160
  moving, 173
  square root, 156
  testing equality, 158
VDE 0113 standard, 371
Via point, 322

W

warning areas, 372
watchdog time-out fault, 41
watchdog timer, 41–42
WECA (Wireless Ethernet Compatibility Alliance), 299
whole numbers, 26
Wi-Fi Alliance, 299
wireless communication, 299–300
wireless networks, 300
wiring
  AC output module, 95

analog output module, 111
automated systems, 388
CLX DC input modules, 89
color, 394
DC output module, 94
DeviceNet, 303–305
differential, 106–107
function blocks, 262
high-speed-mode, 107
input modules, 89–90
networks, 301
output modules, 94–95
RTDs (resistive temperature devices), 138
sensors, 139–140, 144
single-ended, 105
wiring (connecting) steps, 224
wiring harness, 88
WLANs (wireless local area networks), 299

X
X axis, 315
556x processors, 9

Xero, 310
XIC (examine-if-closed) contacts, 46
XIO contacts, 48
XOR (Logical exclusive OR) operator, 207

Y
Y axis, 315
yellow wire, 394

Z
zero crossing, 181